Insights into Organic Carbon, Iron, Metals and Phosphorus Dynamics in Freshwaters

Insights into Organic Carbon, Iron, Metals and Phosphorus Dynamics in Freshwaters

Editor

Liudmila S. Shirokova

MDPI • Basel • Beijing • Wuhan • Barcelona • Belgrade • Manchester • Tokyo • Cluj • Tianjin

Editor
Liudmila S. Shirokova
Université Paul Sabatier
France

Editorial Office
MDPI
St. Alban-Anlage 66
4052 Basel, Switzerland

This is a reprint of articles from the Special Issue published online in the open access journal *Water* (ISSN 2073-4441) (available at: https://www.mdpi.com/journal/water/special_issues/freshwaters_biogeochemical_cycles).

For citation purposes, cite each article independently as indicated on the article page online and as indicated below:

LastName, A.A.; LastName, B.B.; LastName, C.C. Article Title. *Journal Name* **Year**, *Volume Number*, Page Range.

ISBN 978-3-0365-6744-0 (Hbk)
ISBN 978-3-0365-6745-7 (PDF)

Cover image courtesy of Liudmila S. Shirokova.

Contents

water

MDPI

Editorial

Insights into Organic Carbon, Iron, Metals and Phosphorus Dynamics in Freshwaters

Liudmila S. Shirokova [1,2]

[1] Georesources and Environnement Toulouse GET UMR 5563 CNRS, Université Paul Sabatier, 14 Avenue Edouard Belin, 31400 Toulouse, France; liudmila.shirokova@get.omp.eu

[2] N. Laverov Federal Center for Integrated Arctic Research, UrB RAS, 23 Nab Severnoi Dviny, Arkhangelsk 163000, Russia

Citation: Shirokova, L.S. Insights into Organic Carbon, Iron, Metals and Phosphorus Dynamics in Freshwaters. *Water* **2022**, *14*, 2863. https://doi.org/10.3390/w14182863

Received: 29 August 2022
Accepted: 30 August 2022
Published: 14 September 2022

Organic carbon (OC), iron (Fe), metal, and phosphorus (P) are key aquatic components that largely determine the biotic and abiotic functioning of freshwater systems, including groundwater, soil water, lakes, rivers, and their estuaries. Over the past decade, there has been increasing interest regarding the elevation in organic carbon and iron concentrations in freshwaters in relation to the so-called "browning" effect, caused by climate warming and changes in anthropogenic pressure. As for phosphorus, it is a vital element for all aquatic ecosystems and its aquatic biogeochemical cycle is now undergoing sizable changes linked to eutrophication, invasive species development, and transformations between organic and inorganic forms. This Special Issue combines the articles dedicated to various aspects of the behavior of organic carbon, phosphorus, iron (and other related metals) in a broad range of freshwater environments, from soil solutions and groundwaters to ponds, lakes, rivers, and their riparian zones and estuaries.

Savenko and Savenko [1] presented a review on the geochemistry of phosphorus in continental runoff in the form of both dissolved and solid substances, with a separate consideration for the processes of runoff transformation in river mouth areas. The authors could draw a conclusion about the non-conservative distribution of phosphorus in the estuaries, in most cases associated with biological production and destruction processes. They further argue that conservative behavior of phosphorus was observed only in heavily polluted river mouths with abnormally high concentrations of this element.

Pokrovsky et al. [2] reported a new assessment of riverine fluxes of carbon, nutrients, and metals in the surface waters of permafrost-affected regions; their study is crucially important for constraining adequate models of ecosystem functioning under various climate change scenarios. As a case study, they used the largest permafrost peatland territory on the Earth, the Western Siberian Lowland (WSL). By applying a "substituting space for time" scenario, the WSL south–north gradient was used as a model for future changes due to the permafrost boundary shift and climate warming. The authors demonstrated that, contrary to common expectations, the climate warming and permafrost thaw in the WSL will likely decrease the riverine export of organic C and many other elements; they conclude that modeling of C and element cycles in the Arctic and sub-Arctic should be region-specific, and that neglecting huge areas of permafrost peatlands might produce sizeable bias in our predictions of climate change impact.

In the same territory, Manasypov et al. [3] studied the biogeochemistry of thermokarst lakes and demonstrated that thermokarst lakes and ponds that are formed due to thawing of frozen peat in high-latitude lowlands are very dynamic and environmentally important aquatic systems. They showed that these thermokarst lakes and ponds play a key role in controlling C emission to the atmosphere and organic carbon (OC), nutrient, and metal lateral exports to rivers and streams. Analyses of lake water chemical composition across the permafrost gradient allowed them a first-order empirical prediction of lake hydrochemical changes in the case of climate warming and permafrost thaw, employing a scenario where space is substituted for time. The main conclusion of this study was that the exact magnitude of this response will be strongly seasonally dependent, with the largest effects observable during baseflow seasons.

Extending the knowledge of river water chemical composition in Siberia, Tashiro et al. [4] studied watershed-scale iron dynamics, and coupled the seasonal changes in Fe and dissolved organic carbon (DOC) concentrations in the tributaries of the Amur River basin. They demonstrated that permafrost wetlands in valley areas act as hotspots of dissolved Fe production and greatly contribute to Fe and DOC discharge to rivers, especially during snowmelt and rainy seasons.

Additionally, in Eastern Siberia, Vorobyev et al. [5] described a snapshot study of major and trace element concentration in the Lena River basin during the peak of spring flooding, which revealed a specific group of solutes according to their spatial pattern across the river main stem and tributaries and allowed the establishment of a link to certain landscape parameters. The authors demonstrate that future changes in the river water chemistry linked to climate warming and permafrost thaw at the scale of the whole river basin are likely to stem from changes in the spatial pattern of dominant vegetation as well as the permafrost regime. They further argue that comparable studies of large, permafrost-affected rivers during contrasting seasons, including winter baseflow, should allow efficient prediction of future changes in riverine "inorganic" hydrochemistry induced by permafrost thaw.

Two studies of river water hydrochemical composition in Western Siberia—those of Ivanova et al. [6] and Kolesnichenko et al. [7]—provided new insights on landscape, soil, lithology, climate, and permafrost control of dissolved carbon and other major and trace elements in the Ob River, its tributaries, and other small- and medium-sized rivers of the region. These works demonstrated strong environmental factor control on major and trace element concentrations in Western Siberian rivers and predicted a future increase in the concentration of DIC and labile major and trace elements and a decrease in the transport of DOC and low soluble trace metals in the form of colloids in the main stem of the Ob River. They assert that large-scale, seasonally resolved transect studies of large riverine systems of Western Siberia are needed to assess the hydrochemical response of this environmentally important territory to ongoing climate change. An example of using such an approach for two contrasting rivers of the region—permafrost-affected Taz River and permafrost-free Ket River—is provided by Pokrovsky et al. [8], who showed that climate warming in northern rivers may double or triple the concentration of DIC, Ca, Sr, U, but also increase the concentration of DOC, POC, and nutrients. The study applied a substituting-space-for-time approach for the south–north gradient of the studied river basins.

Further insights on the factors shaping the chemical composition of rivers were provided by Krickov et al. [9] who measured chemical composition of dissolved (<0.22 μm) fractions of snow across a 2800 km south–north gradient in Western Siberia. Based on mass balance calculation, these authors demonstrate that the wintertime atmospheric input represents a sizable contribution to the riverine export fluxes of dissolved (<0.45 μm) Mn, Co, Zn, Cd, Pb, and Sb during springtime and can appreciably shape the hydrochemical composition of the Ob River main stem and tributaries.

The role of river water hydrochemistry in freshwater organism ecology and chemical composition is presented in work of Lyubas et al. [10]. The authors used new data on trace elements accumulation by freshwater mussels in the Severnaya Dvina and the Onega River Basin—the two largest subarctic river basins in the Northeastern Europe. The study revealed that iron and phosphorous accumulation in shells have a strong relationship with their distance from the mouth of the river; additionally, they demonstrated that the accumulation of elements in the shell depends on the environment of the biotope.

Finally, Venelinov and Tsakovski [11] presented a case study of surface waters where they implemented the metal bioavailability concept in the Water Framework Directive (WFD) compliance assessment. Based on comprehensive database including DOC and metals, they used three substitution approaches and demonstrated that BIO-MET can be used as the most appropriate tool for the bioavailability assessment of Cu, Zn, and Pb in Bulgarian surface water bodies.

Funding: This research was funded by RSF, grant number 22-17-00253.

Conflicts of Interest: The author declares no conflict of interest.

References

1. Savenko, V.S.; Savenko, A.V. The Main Features of Phosphorus Transport in World Rivers. *Water* **2022**, *14*, 16. [CrossRef]
2. Pokrovsky, O.S.; Manasypov, R.M.; Kopysov, S.G.; Krickov, I.V.; Shirokova, L.S.; Loiko, S.V.; Lim, A.G.; Kolesnichenko, L.G.; Vorobyev, S.N.; Kirpotin, S.N. Impact of Permafrost Thaw and Climate Warming on Riverine Export Fluxes of Carbon, Nutrients and Metals in Western Siberia. *Water* **2020**, *12*, 1817. [CrossRef]
3. Manasypov, R.M.; Lim, A.G.; Krickov, I.V.; Shirokova, L.S.; Vorobyev, S.N.; Kirpotin, S.N.; Pokrovsky, O.S. Spatial and Seasonal Variations of C, Nutrient, and Metal Concentration in Thermokarst Lakes of Western Siberia Across a Permafrost Gradient. *Water* **2020**, *12*, 1830. [CrossRef]
4. Tashiro, Y.; Yoh, M.; Shiraiwa, T.; Onishi, T.; Shesterkin, V.; Kim, V. Seasonal Variations of Dissolved Iron Concentration in Active Layer and Rivers in Permafrost Areas, Russian Far East. *Water* **2020**, *12*, 2579. [CrossRef]
5. Vorobyev, S.N.; Kolesnichenko, Y.; Korets, M.A.; Pokrovsky, O.S. Testing Landscape, Climate and Lithology Impact on Carbon, Major and Trace Elements of the Lena River and Its Tributaries during a Spring Flood Period. *Water* **2021**, *13*, 2093. [CrossRef]
6. Ivanova, I.; Savichev, O.; Trifonov, N.; Kolubaeva, Y.V.; Volkova, N. Major-Ion Chemistry and Quality of Water in Rivers of Northern West Siberia. *Water* **2021**, *13*, 3107. [CrossRef]
7. Kolesnichenko, I.; Kolesnichenko, L.G.; Vorobyev, S.N.; Shirokova, L.S.; Semiletov, I.P.; Dudarev, O.V.; Vorobev, R.S.; Shavrina, U.; Kirpotin, S.N.; Pokrovsky, O.S. Landscape, Soil, Lithology, Climate and Permafrost Control on Dissolved Carbon, Major and Trace Elements in the Ob River, Western Siberia. *Water* **2021**, *13*, 3189. [CrossRef]
8. Pokrovsky, O.S.; Lim, A.G.; Krickov, I.V.; Korets, M.A.; Shirokova, L.S.; Vorobyev, S.N. Hydrochemistry of Medium-Size Pristine Rivers in Boreal and Subarctic Zone: Disentangling Effect of Landscape Parameters across a Permafrost, Climate, and Vegetation Gradient. *Water* **2022**, *14*, 2250. [CrossRef]
9. Krickov, I.V.; Lim, A.G.; Shevchenko, V.P.; Vorobyev, S.N.; Candaudap, F.; Pokrovsky, O.S. Dissolved Metal (Fe, Mn, Zn, Ni, Cu, Co, Cd, Pb) and Metalloid (As, Sb) in Snow Water across a 2800 km Latitudinal Profile of Western Siberia: Impact of Local Pollution and Global Transfer. *Water* **2022**, *14*, 94. [CrossRef]
10. Lyubas, A.A.; Tomilova, A.A.; Chupakov, A.V.; Vikhrev, I.V.; Travina, O.V.; Orlov, A.S.; Zubrii, N.A.; Kondakov, A.V.; Bolotov, I.N.; Pokrovsky, O.S. Iron, Phosphorus and Trace Elements in Mussels' Shells, Water, and Bottom Sediments from the Severnaya Dvina and the Onega River Basins (Northwestern Russia). *Water* **2021**, *13*, 3227. [CrossRef]
11. Venelinov, T.; Tsakovski, S. How to Implement User-Friendly BLMs in the Absence of DOC Monitoring Data: A Case Study on Bulgarian Surface Waters. *Water* **2022**, *14*, 246. [CrossRef]

water

MDPI

Article

Hydrochemistry of Medium-Size Pristine Rivers in Boreal and Subarctic Zone: Disentangling Effect of Landscape Parameters across a Permafrost, Climate, and Vegetation Gradient

Oleg S. Pokrovsky [1,*], Artem G. Lim [2], Ivan V. Krickov [2], Mikhail A. Korets [3], Liudmila S. Shirokova [1,4] and Sergey N. Vorobyev [2]

1 GET, UMR 5563, CNRS, University of Toulouse (France), 14 Avenue Edouard Belin, 31400 Toulouse, France; liudmila.shirokova@get.omp.eu
2 BIO-GEO-CLIM Laboratory, Tomsk State University, Lenina Av., 36, 634050 Tomsk, Russia; lim_artyom@mail.ru (A.G.L.); krickov_ivan@mail.ru (I.V.K.); soil@green.tsu.ru (S.N.V.)
3 V.N. Sukachev Institute of Forest of the Siberian Branch of Russian Academy of Sciences, KSC SB RAS, 660036 Krasnoyarsk, Russia; mik@ksc.krasn.ru
4 N.P. Laverov Federal Center for Integrated Arctic Research, UrB Russian Academy of Science, 23 Nab Severnoi Dviny, 163000 Arkhangelsk, Russia
* Correspondence: oleg.pokrovsky@get.omp.eu

Abstract: We studied two medium size pristine rivers (Taz and Ket) of boreal and subarctic zone, western Siberia, for a better understanding of the environmental factors controlling major and trace element transport in riverine systems. Our main objective was to test the impact of climate and land cover parameters (permafrost, vegetation, water coverage, soil organic carbon, and lithology) on carbon, major and trace element concentration in the main stem and tributaries of each river separately and when considering them together, across contrasting climate/permafrost zones. In the permafrost-bearing Taz River (main stem and 17 tributaries), sizable control of vegetation on element concentration was revealed. In particular, light coniferous and broadleaf mixed forest controlled DOC, and some nutrients (NO_2, NO_3, Mn, Fe, Mo, Cd, Ba), deciduous needle-leaf forest positively correlated with macronutrients (PO_4, P_{tot}, Si, Mg, P, Ca) and Sr, and dark needle-leaf forest impacted N_{tot}, Al, and Rb. Organic C stock in the upper 30–100 cm soil positively correlated with Be, Mn, Co, Mo, Cd, Sb, and Bi. In the Ket River basin (large right tributary of the Ob River) and its 26 tributaries, we revealed a correlation between the phytomass stock at the watershed and alkaline-earth metals and U concentration in the river water. This control was weakly pronounced during high-water period (spring flood) and mostly occurred during summer low water period. Pairwise correlations between elements in both river systems demonstrated two group of solutes—(1) positively correlated with DIC (Si, alkalis (Li, Na), alkaline-earth metals (Mg, Ca, Sr, Ba), and U), this link originated from groundwater feeding of the river when the labile elements were leached from soluble minerals such as carbonates; and (2) elements positively correlated with DOC (trivalent, tetravalent, and other hydrolysates, Se and Cs). This group reflected mobilization from upper silicate mineral soil profile and plant litter, which was strongly facilitated by element colloidal status, notably for low-mobile geochemical tracers. The observed DOC vs DIC control on riverine transport of low-soluble and highly mobile elements, respectively, is also consistent with former observations in both river and lake waters of the WSL as well as in soil waters and permafrost ice. A principal component analysis demonstrated three main factors potentially controlling the major and TE concentrations. The first factor, responsible for 26% of overall variation, included aluminum and other low mobile trivalent and tetravalent hydrolysates, Be, Cr, Nb, and elements strongly complexed with DOM such as Cu and Se. This factor presumably reflected the presence of organo-mineral colloids, and it was positively affected by the proportion of forest and organic C in soils of the watershed. The second factor (14% variation) likely represented a combined effect of productive litter in larch forest growing on carbonate-rich rocks and groundwater feeding of the rivers and acted on labile Na, Mg, Si, Ca, P, and Fe(II), but also DOC, micronutrients (Zn, Rb, Ba), and phytomass at the watershed. Via applying a substituting space for time approach for south-north gradient of studied river basins, we predict that

Citation: Pokrovsky, O.S.; Lim, A.G.; Krickov, I.V.; Korets, M.A.; Shirokova, L.S.; Vorobyev, S.N. Hydrochemistry of Medium-Size Pristine Rivers in Boreal and Subarctic Zone: Disentangling Effect of Landscape Parameters across a Permafrost, Climate, and Vegetation Gradient. *Water* **2022**, *14*, 2250. https://doi.org/10.3390/w14142250

Academic Editor: Maurizio Barbieri

Received: 29 June 2022
Accepted: 14 July 2022
Published: 18 July 2022

climate warming in northern rivers may double or triple the concentration of DIC, Ca, Sr, U, but also increase the concentration of DOC, POC, and nutrients.

Keywords: metals; carbon; nutrients; trace elements; landscape; permafrost; river; watershed; boreal

1. Introduction

Climate change, which is mostly pronounced in high latitudes, strongly impacts the chemistry of rivers and streams and can bring yet unknown consequences on carbon, nutrient, and metal export from land to ocean thus enhancing the retroactive link to climate change drivers [1–4]. There are numerous case studies of climate change impact on surface and groundwater sustainability and hydrochemistry, such as those performed in temperate/subtropical regions [5,6]. However, huge northern unpopulated territories of the world remain poorly covered by a coupled hydrochemical/hydrological approach. This is especially true for pristine permafrost-bearing regions located at high latitudes and containing sizable amount of carbon in the form of peat, organic litter, and vegetation. An example is Western Siberian Lowland (WSL), located in the gradient of climate and permafrost zones within essentially the same lithological background, minimal variations in river runoff and relief and moderate to negligible anthropogenic activity. The interest of the WSL is that it contains huge peat resources and presents rather shallow, essentially discontinuous to sporadic/isolated permafrost, highly vulnerable to thawing [7–10]. For this relatively large territory (2 million km^2), extensive studies of small [11–19] and large [20] river dissolved, colloidal and particulate load, chemical composition of soil ice and suprapermafrost waters [21–24], and gaseous regime of rivers and lakes [25–27] have been performed. However, the majority of these studies except that of the Ob River [20] dealt with single site sampling of a given river, without addressing the spatial variability of riverine solutes within a watershed. This is a clear shortcomings of current state of knowledge of western Siberian rivers, because without sufficient spatial coverage, one cannot test the impact of various landscape factors on water hydrochemistry within the same river basin.

In the present study, we used a coupled hydrochemical/landscape (land cover) approach that allows revealing the main environmental factors controlling the hydrochemical composition of the river water. Up to now, such an approach has been efficiently used only for single-point sampling of small rivers [13–19] and one large Siberian River [20] but has not been implemented at the scale of the whole watershed of medium-size rivers. Toward better understanding of land cover control on riverine C, nutrient and metal transport, we chose two medium size rivers of western Siberia, Taz ($S_{watershed}$ = 150,000 km^2) and Ket ($S_{watershed}$ = 95,000 km^2) located within permafrost-bearing forest-tundra/tundra and permafrost-free taiga biomes, respectively. Both rivers drain through similar sedimentary deposits overlayed by peatland and forest/tundra; they are virtually pristine (population density < 1 people km^{-2}) and have no industrial or agricultural activity on their watershed. Within the on-going drastic climate change in Siberia, the southern river (Ket) can be considered as an extreme scenario of long-term transformation of the northern River (Taz) in case of complete disappearance of discontinuous permafrost and northward migration of the taiga forest. Therefore, our primary objective in this work was to test the impact of climate and land cover parameters (permafrost, vegetation, water coverage, soil organic carbon, and lithology), on carbon, major and trace element concentration in the main stem and tributaries of each river separately and when considering them together, across contrasting climate/permafrost zones. A flowchart of methodology and approach used in the present study is illustrated in Figure 1. We anticipate that, via employing similar approach for two sufficiently contrasting and yet representative river basin of the WSL, we can provide essential information for coupled land—River C, nutrient and metal fluxes for climate modeling of the region and to foresee future, long-term changes in much large territory of strongly understudied permafrost-affected Eurasian lowlands, which includes North Siberia, Anabar, Kolyma and Yana-Indigirka lowlands, with an overall territory more than 2 million km^2.

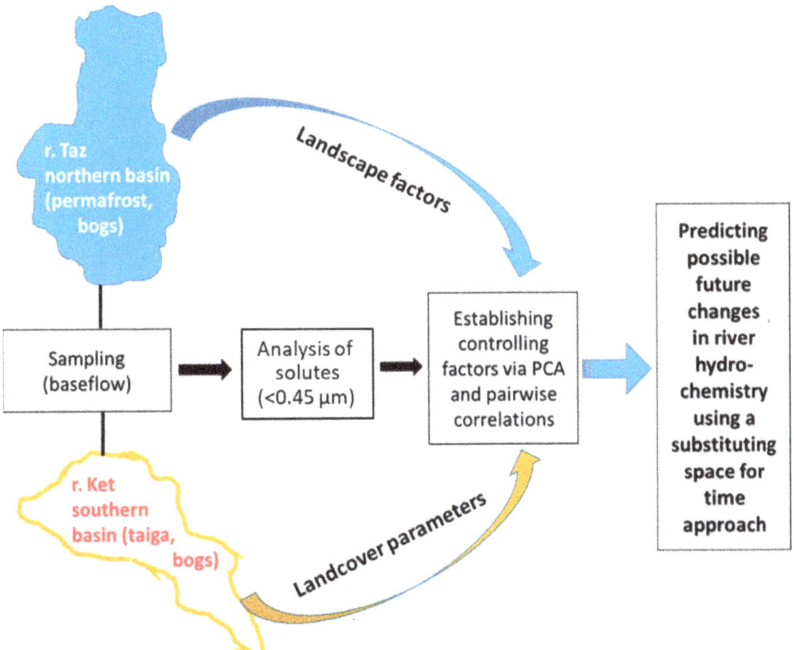

Figure 1. A flow chart of methodological approach used in the present study.

2. Study Site and Methods

2.1. Ket River Basin

We sampled the Ket River main stem and its 26 tributaries during the peak of the spring flood and the end of summer baseflow. The catchment area of sampled watersheds ranged from 94,000 km^2 (Ket's mouth) to 20 km^2 (smallest tributary). The studied watersheds presented rather similar lithology, climate, and vegetation, given its generally west—east orientation, Figure 2; Ref. [28]). The Ket River basin (right tributary of the Ob River) is poorly accessible and can be considered as essentially pristine comprising about 50% forest and 40% of wetlands while having virtually negligible agricultural and forestry activity. The river basin is poorly populated (0.27 person km^{-2}) and, compared to left tributaries of the Ob River, lacks road infrastructure due to absence of hydrocarbon exploration activity. We therefore consider this river as a model one for medium size rivers of the boreal zone of western Siberia Lowland covered by forests and bogs. As such, the results on river water hydrochemistry obtained from this watershed can be extrapolated to much larger territory, comprising several million km^2 of permafrost-free taiga forest and bog biome, extending over 3000 km between the left tributaries of the Yenisei River in the east and Finland in the west. The mean annual air temperatures (MAAT) of the Ket River basin is -0.75 ± 0.15 °C and the mean annual precipitation is 520 ± 20 mm y^{-1}. The lithology is represented by silts and sands with carbonate concretions overlayed by quaternary deposits (loesses, fluvial, glacial, and lacustrine deposits). The dominant soils are podzols in forest areas and histosols in peat bog regions.

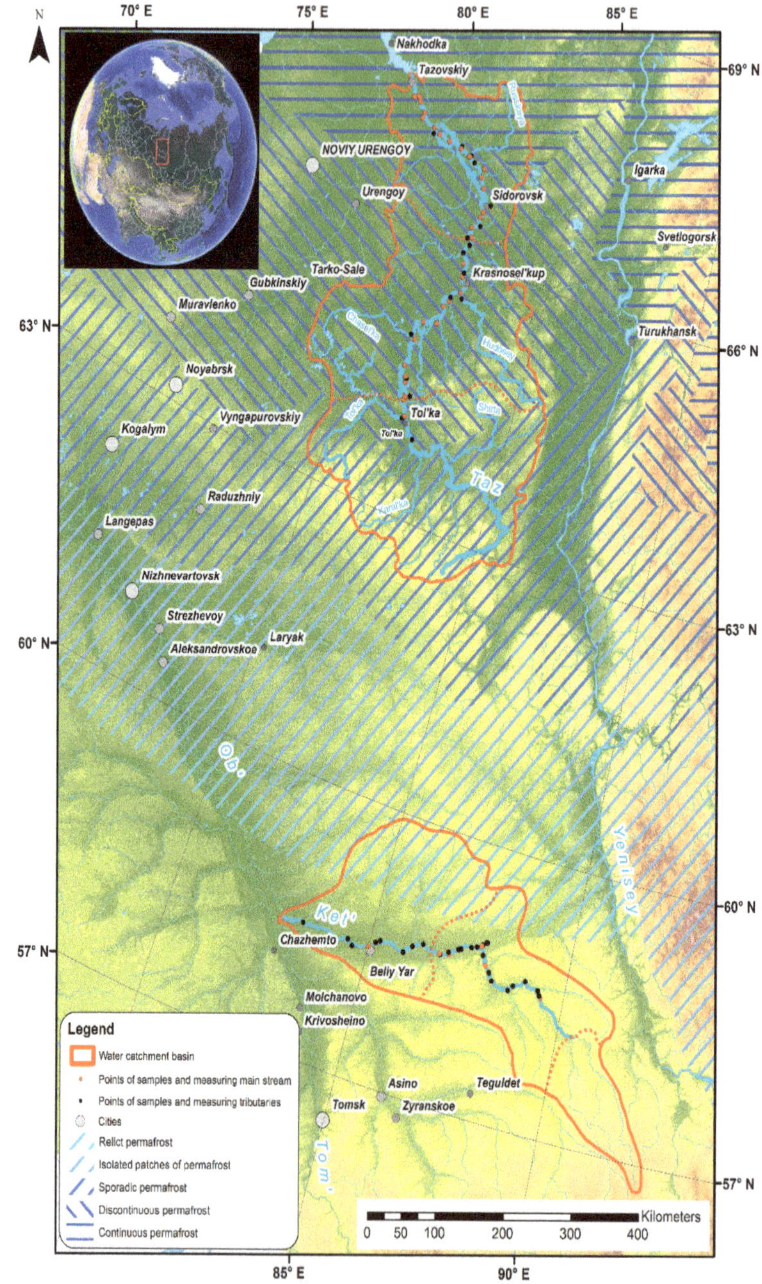

Figure 2. Map of the two studied river basins (Taz in the north and Ket in the south). The sampling points of the main stem and tributaries are shown by red and black circles, respectively.

2.2. Taz River Basin

The Taz River sampled in this work included the main stem and its 17 tributaries whose catchment area ranged from 149,000 km^2 at the Taz's mouth to 25 km^2 of smallest sampled tributary. Alike the Ket River basin, all sampled catchments exhibited similar lithology (clays, silts and sands overlayed by quaternary loesses, fluvial, glacial, and lacustrine deposits), but more contrasting climate and vegetation due to its north to south orientation (Figure 2). The MAAT ranges from −4.6 °C in the headwaters (Tolka village) to −5.4 °C in the low reaches (Tazovskiy town). The mean annual precipitation is 500 mm y^{-1} in the central part of the basin (Krasnoselkup) and 600 mm y^{-1} in the low reaches of the Taz River. Given its landscape and climate features, the Taz River basin includes: (1) the upper (southern) part ("headwaters") which comprises 400–800 km upstream of the river mouth, where the permafrost is sporadic to discontinuous and the dominant vegetation is forest-tundra and taiga, and (2) the low reaches (northern) part, located 0–400 km upstream of the mouth where the permafrost is continuous to discontinuous and the dominant biome is tundra and forest-tundra.

2.3. Sampling

For the Ket River, the peak of annual discharge in 2019 occurred in the end of May whereas in August, the discharge was three to five times lower. From 18 May to 28 May 2019, and from 30 August to 2 September 2019, we started the boat trip in the middle course of the Ket River (Beliy Yar), and moved, first, 475 km upstream the Ket River till its most headwaters, and then moved 834 km downstream till the river mouth. We stopped each 50 km along the Ket River and sampled for major hydrochemical parameters, suspended matter, and bacteria. We also moved several km upstream of selected tributaries to sample for river hydrochemistry.

In the Taz River, the peak of annual discharge in 2019 occurred in the middle of June (5600 m^3 s^{-1}; in August, the discharge was 5 times lower). Therefore, the month of July (average discharge is 2300 m^3 s^{-1}; range 1920–3370 m^3 s^{-1}) can be considered as the end of spring flood period. From 12 July to 16 July 2019 we moved 800 km downstream the Taz River from its most headwaters (Tolka village) to the low reaches (Tazovskiy town). We stopped each ~50 to 100 km of the boat route and sampled for hydrochemical parameters, river suspended matter and total bacterial number of the main stem. For sampling the tributaries, we moved 500–1500 m upstream of the confluence zone.

2.4. Analyses

The dissolved oxygen (CellOx 325; accuracy of ±5%), specific conductivity (TetraCon 325; ± 1.5%), and water temperature (±0.2 °C) were measured in situ at 20 cm depth using a WTW 3320 Multimeter. The pH was measured using portable Hanna instrument via combined Schott glass electrode calibrated with NIST buffer solutions (4.01, 6.86, and 9.18 at 25 °C), with an uncertainty of 0.01 pH units. The river water was sampled in pre-cleaned polypropylene bottle from 20–30 cm depth in the middle of the river and immediately filtered through disposable single-use sterile Sartorius filter units (0.45 µm pore size). The first 20 mL of filtrate was discarded. The DOC and dissolved inorganic carbon (DIC) were determined by a Shimadzu TOC-VSCN Analyzer (Kyoto, Japan) with an uncertainty of 3% and a detection limit of 0.1 mg/L. Blanks of Milli-Q water passed through the filters demonstrated negligible release of DOC from the filter material. Specific ultra-violet absorbance ($SUVA_{254}$) was measured via ultraviolet absorbance at 254 nm using a 10-mm quartz cuvette on a Bruker CARY-50 UV-VIS spectrophotometer and divided by the concentration of DOC (L mg^{-1} m^{-1}). The concentrations of C and N in suspended material (particulate organic carbon and nitrogen (POC and PON, respectively)) were determined via filtration of 1 to 2 L of freshly collected river water (at the river bank or in the boat) with pre-weighted GFF filters (47 mm, 0.8 µm) and Nalgene 250-mL polystyrene filtration units using a Mityvac® manual vacuum pump. Particulate C and N were measured using catalytic combustion with Cu-O at 900 °C with an uncertainty of ≤0.5% using Thermo Flash 2000 CN Analyzer at EcoLab, Toulouse. The samples were analyzed before and after 1:1 HCl treatment to distinguish between total and inorganic C; however, the ratio of $C_{organic}$:$C_{carbonate}$ in the river suspended matter (RSM) was always above 20 and the contribution of carbonate C to total C in the RSM was equal in average 0.3 ± 0.3% (2 s.d., n = 30).

Total microbial cell concentration was measured after sample fixation in glutaraldehyde, by flow cytometry (Guava® EasyCyteTM systems, Merck, Darmstadt, Germany). Cells were stained using 1 μL of a 10 times diluted SYBR GREEN solution (10000×, Merck), added to 250 μL of each sample before analysis. Particles were identified as cells based on green fluorescence and forward scatter [29].

All analytical approaches used in this study for major and trace element analyses followed methods developed for western Siberian organic-rich surface waters [17,30,31]. The samples were preserved via refrigeration 1 month prior to analysis. Major anion (Cl, SO_4^{2-}) concentrations were measured by ion chromatography (HPLC, Dionex ICS 2000) with an uncertainty of 2%. International certified samples ION, PERADE, and RAIN were used for validation of the analyses. Major cations (Ca, Mg, Na, K), Si, and ~40 trace elements were determined with an Agilent iCap Triple Quad (TQ) ICP MS using both argon and helium modes to diminish interferences. About 3 μg L^{-1} of In and Re were added as internal standards along with three various external standards. Detection limits of TE were determined as 3× the blank concentration. The typical uncertainty for elemental concentration measurements ranged from 5–10% at 1–1000 μg/L to 10–20% at 0.001–0.1 μg/L. The Milli-Q field blanks were collected and processed to monitor for any potential contamination introduced by our sampling and handling procedures. The SLRS-6 (Riverine Water Reference Material for Trace Metals certified by the National Research Council of Canada) was used to check the accuracy and reproducibility of analyses [32,33]. Only those elements that exhibited good agreement between replicated measurements of SLRS-6 and the certified values (relative difference < 15%) are reported in this study.

2.5. Landscape Parameters and Water Surface Area of the Ket and Taz River Basin

The physio-geographical characteristics of the Ket and Taz tributaries and several sampling points of the main stem of both rivers were determined by applying available digital elevation model (DEM GMTED2010), soil, vegetation, and lithological maps. The landscape parameters were typified using TerraNorte Database of Land Cover of Russia (Ref. [34]; http://terranorte.iki.rssi.ru, accessed on 1 February 2020). This included various types of forest (evergreen, deciduous, needleleaf/broadleaf), grassland, tundra, wetlands, water bodies, and riparian zones. The climate parameters of the watershed were obtained from CRU grids data (1950-2016) [35] and NCSCD data (Ref [36]; doi:10.5879/ecds/00000001), respectively, whereas the biomass and soil OC content were obtained from BIOMASAR2 [37] and NCSCD databases. The lithology layer was taken from GIS version of Geological map of the Russian Federation (scale 1: 5,000,000, http://www.geolkarta.ru/, accessed on 1 February 2020).

2.6. Data Analysis

Element concentrations for all datasets were tested for normality using the Shapiro–Wilk test. In case of the data were not normally distributed, we used non-parametric statistics. Comparisons of major and trace element concentration in the main stem and the tributaries during two sampling seasons (Ket) and summer baseflow (Taz) were conducted using a non-parametric Mann Whitney test at a significance level of 0.05. For comparison of unpaired data, a non-parametric H-criterion Kruskal–Wallis test was used to reveal the differences between different seasons and between the main stem and tributaries. The Pearson rank order correlation coefficient ($p < 0.05$) was used to determine the relationship between major and trace element concentrations and main landscape parameters of several points of the main stem and all tributaries, as well as other potential drivers of TE concentrations in the river water such as pH, O_2, water temperature, specific conductivity, DOC, SUVA (aromaticity), DIC, Fe, Al, particulate carbon and nitrogen, and total bacterial number.

3. Results and Discussion

3.1. Spatial and Seasonal Variation of Elements in the Ket River and Control of River Hydrochemistry by Landscape Parameters

All the primary data on dissolved (<0.45 µm) element concentration in both rivers together with relevant landscape parameters of the sampling points are available from the Mendeley Repository [28]. The spatial variations of element concentration in the main stem and tributaries of the Ket River were generally lower than the differences between the two seasons. This is illustrated by a plot of several major and trace elements along the full length of the river basin (Figure 3A–E), and further confirmed by a Mann–Whitney U test (Table S1 of the Supplementary Materials) which shows that the maximal differences in element concentrations are observed between seasons in both the main stem and the tributaries. The differences between the Ket River main stem and the tributaries were always lower or not pronounced for the same season. This allowed to calculate the mean concentrations of elements across the full length of the main stem and among all tributaries for each seasons (Figure S1 and Table S2 of the Supplementary Materials). Analyses of the differences in element concentration between the flood and the baseflow of the Ket River allowed distinguishing three group of solutes as illustrated in Figure 4. The first groups comprised elements having a factor of 2 to 10 higher concentrations in both main stem of the Ket River and tributaries during spring flood compared to summer baseflow, and included DOC, low soluble low mobile "lithogenic" hydrolysates (Be, Al, Ti, Cr, Ga, Y, Zr, Nb, REE, Hf, Bi, Th), some divalent heavy metals (Ni, Cu, Cd), oxyanions (Sb), Cs, Tl, and SO_4. The second group typified highly mobile elements exhibiting lower concentrations during spring flood compared to the baseflow and included DIC, POC, alkaline and alkaline earth metals (Li, Na, Mg, Ca, Sr, Ba), labile nutrients (Si, P, Mn), oxyanions (As, Mo), Fe and U(VI). Finally, a group of elements demonstrated rather similar (within 30%) concentrations during both seasons and included Cl, micronutrients (V, Co, Zn, Se, Rb), Ge, W, and Cd.

A pairwise (Pearson) correlation of major and trace element concentration in the main stem and tributaries of the Ket River with main land cover parameters of the watershed was tested for both seasons, but notable correlations were observed only for the summer baseflow (Table S3). These correlations were detected only for alkaline-earth elements (Mg, Ca, Sr) and U, whose concentrations positively correlated ($R_{Pearson} \geq 0.5$; $p < 0.05$) with phytomass stock on the watersheds (Figure 5). However, the observed correlations do not necessarily indicate a direct control but may be the consequence of the fact that carbonate-bearing loesses are most favorable substrates for productive forests compared to clay, sand and peat –rich soils [20]. The carbonate minerals of these substrates are known to act as sizable sources of alkaline-earth metals and uranium [the latter is carried in the form of highly labile uranyl-carbonate complexes] in shallow subsurface and groundwater feeding the river during summer baseflow [16,17]. Other elements did not demonstrate any sizable correlations to the landscape factors. The most likely reason for paucity of such a control is highly homogeneous coverage of the river basins by forest, bogs, and riparian zones with essentially the same climate, runoff, vegetation, soil, and total phytomass stock, represented by similar tree species of the boreal taiga of permafrost-free zone of the WSL.

Overall, the dominance of the season over space in solute control in the river basin let us to consider, for further analysis, only the same season (summer baseflow) for both rivers. This allowed better understanding the main driving factors (landscape parameters) of dissolved major and trace element variation along the full length of the main stem of both rivers as well as within each river tributaries.

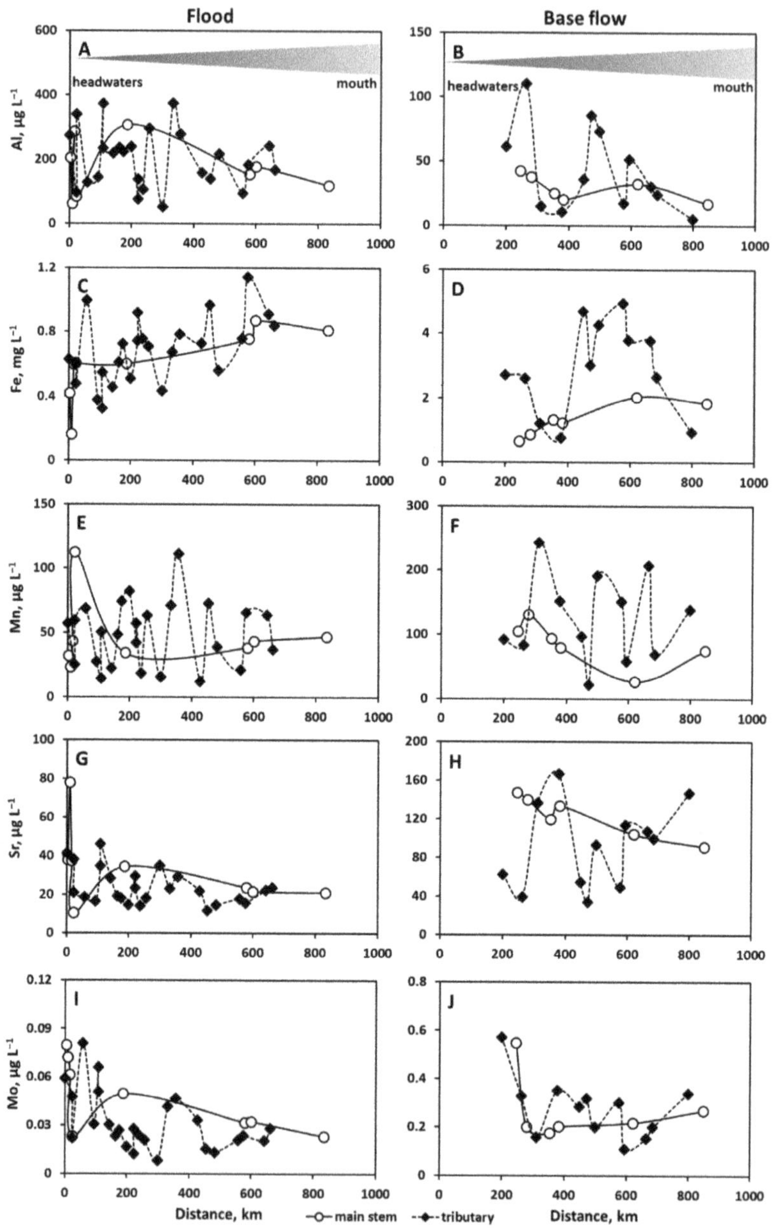

Figure 3. Al (**A**,**B**), Fe (**C**,**D**), Mn (**E**,**F**), Sr (**G**,**H**) and Mo (**I**,**G**) concentration in the Ket River main stem (open circles) and tributaries (solid diamonds) during spring flood (**A**,**C**,**E**,**G**,**I**) and summer baseflow (**B**,**D**,**F**,**H**,**J**).

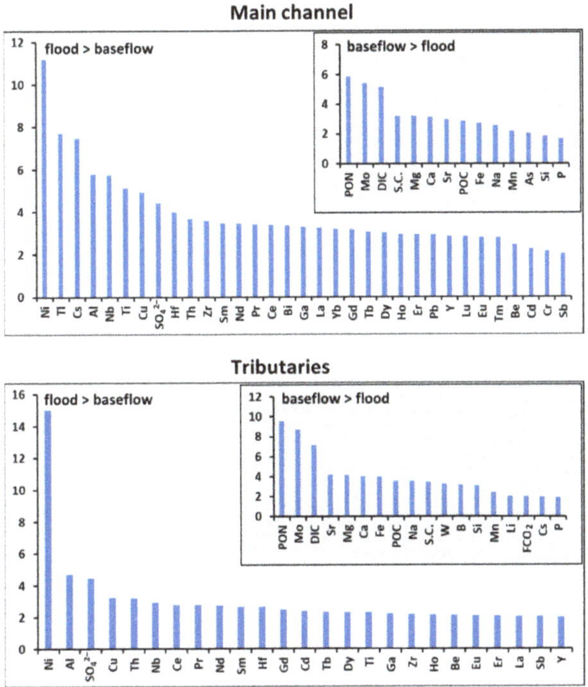

Figure 4. Mean ratio of element concentrations between spring flood period and summer baseflow in the Ket River main stem and tributaries.

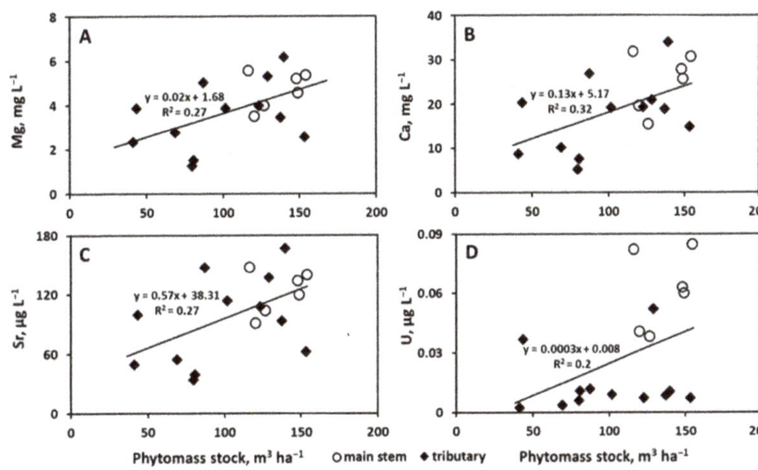

Figure 5. Example of positive correlations between Mg (**A**), Ca (**B**), Sr (**C**), and U (**D**) and landscape factors of the Ket River in summer.

3.2. Major and Trace Element Spatial Variation over the River Main Stem and among Tributaries of the Taz River Basin and Land Cover Control

The variations of major and trace element concentration along the Taz River were also rather low as illustrated in a plot of some major and trace element concentrations over the river distance, from the headwaters to the mouth (Figure 6). The concentration of DOC,

P_{tot}, NO_3, NO_2, Al, Fe, Cd, Ba, and Bi increased southward, with an increase in mean annual temperature. In contrast, in the northward direction, with an increase in tundra and discontinuous permafrost coverage, the concentrations of Cl, SO_4, Li, B, Na, K, V, Ni, Cu, Y, REE, and U increased, which may stem from a combination of sea-salts release from the underlying marine clays and silts and far-range (>300 km) atmospheric transfer from the Norilsk smelters (V, Ni, Cu). The variations among tributaries were generally larger but did not exhibit any systematic evolution between the upper and lower reaches of the Taz River basin (Table S4). Therefore, we can hypothesize that the primary factor controlling solute concentration in the tributaries is land cover (see Section 3.3 below). In the main stem and tributaries, the difference in dissolved element concentration between the upper (southern) and lower (northern) part of the river has not exceeded 20–30% which was often within the standard deviation of the mean values. The only exception is Mn concentrations in the tributaries were two times higher in the south compared to the north.

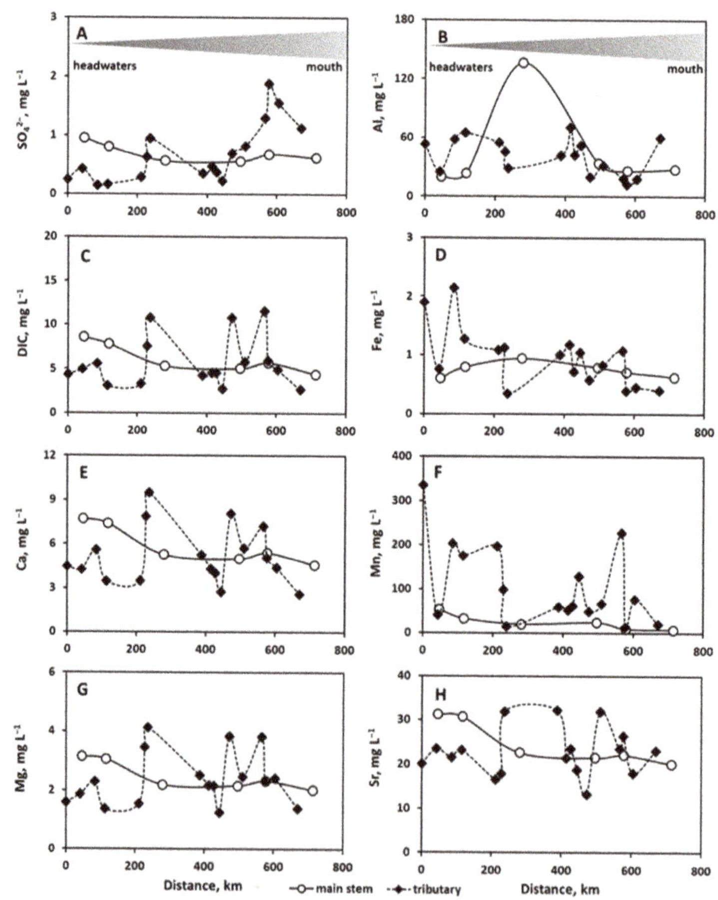

Figure 6. SO_4^{2-} (**A**), Al (**B**), DIC (**C**), Fe (**D**), Ca (**E**) Mn (**F**), Mg (**G**), and Sr (**H**) concentration in the Taz River main stem and tributaries during summer baseflow.

Pairwise correlations of major and trace element concentration in the Taz River basin with main physico-geographical, geocryological, and climatic features of the watershed revealed several potential drivers (Table S5). The tundra coverage of the watershed, corresponding to northward directions and proximity to the sea exhibited strong positive

($R_{Pearson} > 0.60$, $p < 0.01$) correlations with Li, B, Na, K, Cl, SO_4, and U. These elements likely originate from sea salts of the former marine clays dominating the bedrocks of river catchments in the northern part of WSL [17] but also the deposition of marine aerosols in the form of snow [38]. The latter is also known to be enriched in Ni and Cu, reflecting the proximity of the low Taz reaches to the Norilsk Cu-Ni smelters in the northern part of WSL. This can explain positive ($R_{Pearson} > 0.50$, $p < 0.05$) correlations between the tundra coverage and the concentrations of Ni and Cu in the rivers of the Taz basin. Presumably, the atmospheric deposition of metal-rich aerosols can be transferred, via plant litter decay and surface runoff, to the river main stem and tributaries in the northern part of the basin.

The vegetation also exhibited notable impact on elements concentrations in the river water. Thus, presence of larch trees (deciduous needle-leaf forest coverage) positively correlated with concentrations of DOC and macronutrients (Si, Mg, Ca, Sr) and Cs. Light coniferous and broadleaf mixed forest positively impacted concentrations of DOC, macro- (PO_4, NO_3, NH_4) and micro-nutrients (Mn, Fe, Co, Mo, Ba, Cd). Finally, the organic carbon content in upper 0–30 and 0–100 cm of soil positively correlated with some micro-nutrients (Mn, Co, Mo) but also Be, Sc, Cd, Sb, and Bi. Noteworthy is that neither the watershed surface area nor the permafrost coverage, which are the two potential drivers of element geochemistry in surface waters i.e., [39–42], correlated with major and trace elements of the Taz River basin. Furthermore, we were not able to detect any control of other landscape features such as riparian zone, wetlands, and recent burns.

3.3. Common Features of Spatial Distribution of Major and Trace Element in Two River Basins during Summer

Analysis of pairwise correlations described in Sections 3.1 and 3.2 for two river basins individually revealed a number of common features in terms of apparent control of landscape parameters on element concentration in river waters. This allowed correlating the hydrochemical composition of the water column with major environmental factors that are likely to operate on both river basins as illustrated for some elements in Figure 7. The first factor is forest biomass in general and, in particular, the coverage of the watershed by larch trees (deciduous light needle leaf forest), which are most efficient for recycling the nutrients such as Li, B, Si, Zn, Rb, and Ba [43]. The second environmental parameter is bog coverage of the watershed, which was positively correlated with SUVA and low mobile lithogenic elements and divalent transition metals whose transport is facilitated by high DOC concentration (V, Ni, Cu, Y, heavy REE). The impact of bogs on retention of these elements was demonstrated in North European boreal rivers [44,45]. Noteworthy that the bog presence facilitated only the transport of heavy REE and not the light ones. In the WSL surface waters, the former are known to form much stronger complexes with allochthonous DOM from bogs and peat waters and less prone to be carried as organo-ferric colloids [30,46–48].

A small number of elements were associated with total bacterial number (SUVA, P, V, Fe, and As) and presumably marked shallow subsurface discharge of Fe(II)-rich waters from the adjacent peatlands and biotically driven formation of large size Fe hydroxide colloids, stabilized by allochthonous (aromatic) organic carbon. Strong co-precipitation of P, V, and As with Fe hydroxides is known from laboratory experiments with organic-rich (peatland) waters in the presence of soil and aquatic bacteria [49–52]. The area of riparian zones of the river positively correlated with DIC, Sr, Ba, U, and particulate organic nitrogen. While the first four elements could mark an enhanced discharge of deep and subsurface groundwaters in the riparian zones, notably of larger rivers, the correlation with PON could illustrate the generation of N-rich particles in the sediments of highly productive floodplains [13,18]. Among common climate parameters of both river basins, the precipitation did not impact the hydrochemical composition. However, the mean annual air temperature encompassed the contrast between northern and southern rivers and correlated with labile components of the river water (S.C., DIC, Li, B, Si, Ca, Sc, Zn, Se,

Rb, Sb, Cs, Ba, and U). This was fully consistent with previous observations of large (Ob, Ref. [20]; Lena, Ref. [53]) and small [15–17] Siberian rivers.

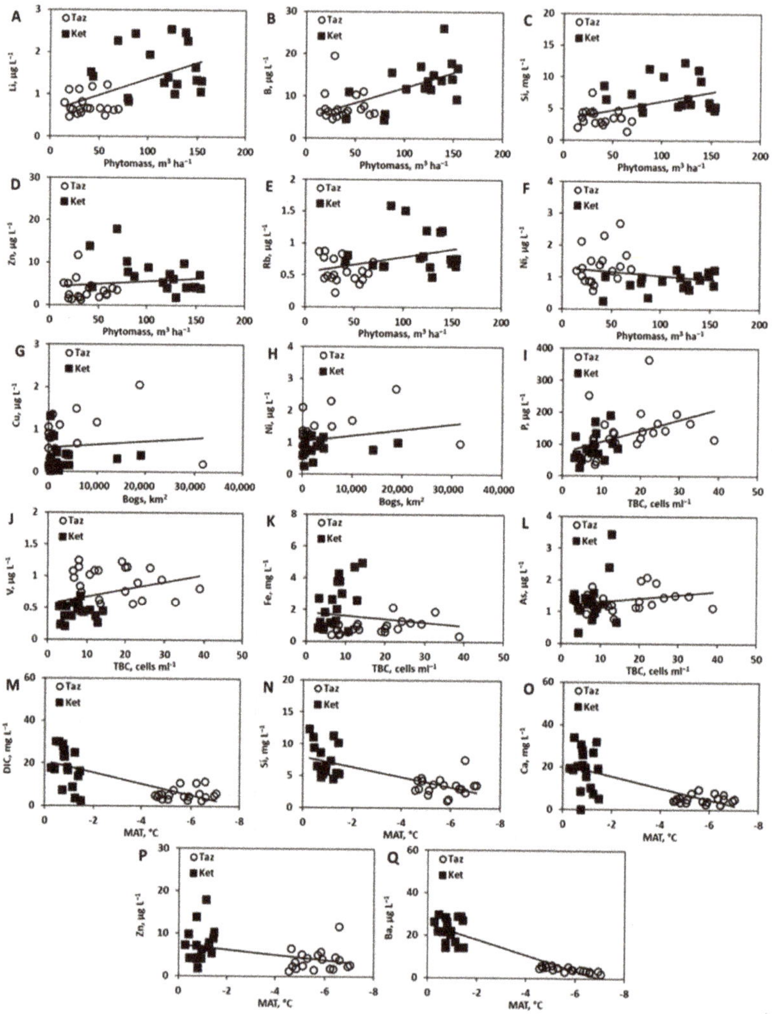

Figure 7. Pairwise correlations of several major and trace elements with two most pronounced landscape parameters—Phytomass stock at the watershed and bog coverage, total bacterial count (TBC) and mean air temperature, for both Taz and Ket river main stem and tributaries.

Given that the season was an important driving factor of element concentrations in the Ket River (see Section 3.1), we attempted a PCA treatment of elementary dataset of both rivers during summer baseflow. This analysis revealed three main factors capable of explaining 26, 14, and 6.6% of total variability, respectively (Figure S2, Table S6). The first factor acted positively on aluminum and other low mobile hydrolysates such as Be, Al, Ti, Cr, Ga, Y, Zr, Nb, Y, REE, Hf, Th and elements which are known to be strongly complexed with DOM such as Cu and Se [54]. Presumably, it reflected mobilization of low soluble elements from lithogenic (silicate) minerals in the form of organo-mineral (essentially organo-aluminum) colloids as it is known from studies in soil porewaters of the WSL regions [24]. The second factor negatively acted on labile Na, Mg, Si, Ca, P, and Fe (probably as Fe(II)). The F2 was also positively linked to DOC, micronutrients (Zn, Rb, Ba), and phytomass at the watershed (primarily, light needle-leaf trees) as well as mean temperature and the proportion of younger (<25 Ma) rocks. In the WSL, these rocks contain carbonate concretions and partially carbonated loesses. The second factor thus likely represented a combined effect of productive litter in larch forest growing on carbonate-rich rocks and groundwater feeding of the rivers by soluble, highly mobile elements. Finally, the third factor, although exhibited limited explanation capacity, was strongly linked to pH, specific conductivity, DIC, Sr, and U and clearly marked a direct impact of bicarbonate-rich, slightly mineralized waters that reside in shallow (Taz) and deeper (Ket) subsurface reservoirs containing carbonate minerals and actively discharge into the river during baseflow.

Noteworthy is the similarity of the two main groups of major and trace elements, DOC- and DIC-related, based on pairwise correlations between element concentrations. The DIC concentration in both river basins significantly ($p < 0.05$) correlated with S.C., Si, alkalis (Li, Na), alkaline-earth metals (Mg, Ca, Sr, Ba), and U. This correlation marks an enhanced mobility of these elements in the form of ionic forms and neutral molecules (uranyl as carbonate/hydroxide complexes), essentially controlled by discharge of DIC and mobile element—rich groundwaters either at the river bank or in the hyporheic zone. These mechanisms of labile element mobilization are fairly well-known across the boreal zone [55]. The DOC correlated with trivalent (Sc, Al, Ga, Y, REE), tetravalent (Ti, Zr, Hf, Th), and other (Be, Cr, Nb) hydrolysates, Se and Cs. These elements are essentially present in the river water in the form of Al-organic colloids or organic complexes [14,54,56–58]. These two main groups of solutes were evidenced not only in large and small rivers and streams [14,16,17,20,59] but in lentic waters of the region, such as lakes and ponds [60,61], supra-permafrost waters [23], peat porewaters [22–24], and peat ice [21].

3.4. Climate Change Consequences on Element Concentration in WSL Rivers (Vegetation and Permafrost/Lithology Control) Using a Substituting Space for Time Approach

The main limitation of our approach is a restricted seasonal coverage in river sampling and lack of daily/weakly discharge measurements, which did not allow calculating elementary export fluxes (yields). As such, predictions of climate change impact on riverine export of this particular pristine basins can be performed solely in terms of river water hydrochemistry. Note however that an approach of element annual yields is well developed for other European Arctic [55] and small western Siberian rivers [13,59] including anthropogenically affected Ob River in its middle course [19]. Therefore, taking the advantage of highly pristine character of Taz and Ket River basins (in contrast to the neighboring Pur River basin and left tributaries of the Ob River, which are strongly impacted by gas and oil industry), one can use these river systems to approximate future changes in river hydrochemistry linked to on-going climate change and permafrost thaw. The validity of this approach is based on rather weak spatial variations within each basin discovered in the present study (notably over the course of the main stem but also among tributaries) in contrast to sizable variations in solute concentrations between the northern and southern river basin.

For this, we can employ a "substituting space for time" approach which postulates, in a broader context, that spatial phenomena which are observed today can be used to describe past and future events [62]. Such an approach has been successfully used in western Siberia since the pioneering work of Frey and Smith [12] to model possible future changes in small rivers [13,59], lakes [61], soil waters [23], and permafrost ice [21]. Indeed, with permafrost and forest boundary shift northward over next decades [9,63–65], the northern part of the Taz River (tundra and forest-tundra of continuous to discontinuous permafrost) can be entirely transformed into southern part-like territory of taiga and forest-tundra biome with discontinuous to sporadic permafrost, whereas the entire Ket River basin can be used as a surrogate for hydrochemistry for the southern part of the Taz River. The mean element concentrations in the main stem and tributaries in the two sub-basins of the Taz and Ket

River are illustrated in Figure 8. It can be seen that only a few elements exhibited sizable contrast in river water concentration between the southern and northern part of the Taz River basin and the Ket river catchment. Therefore, applying a substituting space for time approach, we can anticipate that the concentration of DIC and mobile elements (Li, B, Mg, Ca, Sr, Rb, Sb, W, U) and, to a lesser degree, DOC, Si, and micronutrients (Mn, Fe, Zn, Mo) will increase in the low reaches of the Taz River.

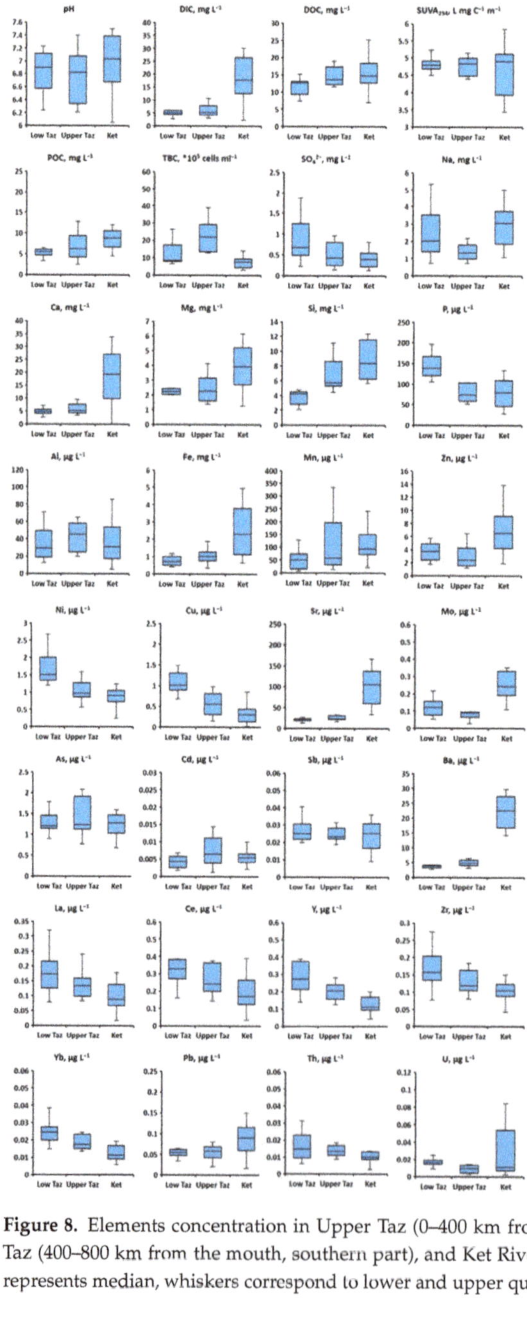

Figure 8. Elements concentration in Upper Taz (0–400 km from the mouth, northern part), Low Taz (400–800 km from the mouth, southern part), and Ket River (main stem and tributaries). Box represents median, whiskers correspond to lower and upper quartile.

Overall, the main consequences of the climate change in western Siberia on middle size river basin hydrochemistry seems to be primarily linked to the changes in dominant vegetation and biomass at the river watershed. In the northern part of the WSL, the change in hydrological pathways due to permafrost thaw might enrich the river waters in soluble elements due to enhanced underground water feeding (DIC, alkaline-earth elements (Ca, Sr), oxyanions (Mo, Sb), and U). The thickening of the active layer and enhanced involvement of thawed peat layer down to underlying mineral horizons may increase the export of trivalent and tetravalent hydrolysates in the form of organo-ferric colloids. However, at the short-term scale, due to two counterbalanced sources and sinks (permafrost thaw and plant uptake), the overall impact of the climate change on inorganic solute export by rivers from the land to the ocean may be smaller than that traditionally viewed for organic carbon.

4. Conclusions

Toward a better understanding of land cover control on dissolved (<0.45 μm) major and trace elements in the river water of high latitude regions, we selected two medium-size pristine rivers of the northern, permafrost-bearing and southern, permafrost-free region of the world's largest peatland, the Western Siberia Lowland (Taz and Ket River, respectively). We sampled the main stem and tributaries while encompassing a large gradient in river basin size, permafrost, and vegetation coverage at essentially similar lithological substrate, runoff, and relief. In the Ket River, the difference in major and trace solute concentration between two seasons was larger than the difference between the main stem and the tributaries. Given that the primary factor controlling the river solute concentration was the season, this allowed straightforward comparison of two river basins during the summer baseflow. Furthermore, sizable variations in solutes among different tributaries of each watershed (depending on land cover) allowed testing the impact of first-order environmental factors on river water hydrochemistry.

The similarity of landscape parameters (mainly, taiga vegetation, partially bogs and permafrost coverage, and, in a lesser degree, the lithology) controlled the major and trace element concentration pattern in two medium-size rivers studied in this work. A number of landscape parameters of the main stem and tributaries correlated with nutrients (B, Si, Zn, Rb, Ba with phytomass), low mobile lithogenic elements, and divalent transition metals whose transport is facilitated by high DOC (V, Ni, Cu, Y, heavy REE with bogs), or specific components reflecting bacterially controlled processing of DOM and groundwater discharge in the water column or within the hyporheic zone (Fe, P, SUVA, V, As). The mean annual air temperature encompassed the contrast between northern and southern rivers and correlated with labile elements of the river water (DIC, Li, B, Si, Ca, Sc, Zn, Se, Rb, Sb, Cs, Ba, and U), which was fully consistent with previous observations of large and small Siberian rivers.

Via quantitative comparison of element concentration along the south–north gradient of studied river basins and applying a substituting space for time approach, we infer that permafrost and vegetation shift northward may sizably (a factor of 2 to 3) enrich the river water in the north in highly mobile elements (DIC, Ca, Sr, U), and, in a lesser degree, in DOC, POC, and nutrients (Si, Mn, Zn, Fe, Mo), whereas the current concentrations of other macro (P) and micro-nutrients (V), and insoluble hydrolysates – geochemical tracers (Y, REE, Zr) in the low reaches of Taz might decrease.

Supplementary Materials: The following are available online at https://www.mdpi.com/article/10.3390/w14142250/s1, Figure S1: Seasonal mean ± SD concentration elements exhibiting higher concentrations during spring flood compared to summer baseflow in tributaries and main channel Ket' in spring flood (blue) and summer (early fall) baseflow (orange). Figure S2: Results of PCA of ~65 hydrochemical abd 16 land cover variables in ~ 60 sampling points of the main stem and tributaries of the Ket and Taz River basin, collected during summer baseflow. Table S1A. Mann-Whitney U Test comparison concentration in different season (flood vs. baseflow) in tributaries and main channel of the Ket River. Red color is for statistically significant differences. Table S1B. Mann-Whitney U

Test comparison concentration tributaries and main channel in different seasons. Table S2. Mean ± SD concentrations of elements and other measured parameters in the Ket River main stem and tributaries. Table S3. Pairwise Pearson correlations of major and trace element concentrations in the Ket River (main stem and tributaries) during summer baseflow and main hydrochemical parameters of the water column and land cover. Significant ($p < 0.05$) correlations are given in red and most significant ($p < 0.01$) are highlighted in pink. Table S4. Hydrochemical parameters of the Taz River basin (stem and tributaries), July 2019. Table S5. Pearson pairwise correlations of major and trace elements of the river water (Taz and tributaries) with landscape parameters, vegetation coverage and climate. Significant ($p < 0.05$) correlations are given in red and most significant ($p < 0.01$) are highlighted in pink. Table S6. Results of PCA of ~65 hydrochemical abd 16 land cover variables in ~ 60 sampling points of the main stem and tributaries of the Ket and Taz River basin, collected during summer baseflow. Significant ($p < 0.05$) correlations are given in red and most significant ($p < 0.01$) are highlighted in pink.

Author Contributions: O.S.P. designed the study and wrote the paper; A.G.L. and I.V.K. performed sampling, analysis, and their interpretation; S.N.V. and O.S.P. were responsible for the choice of sampling objects and statistical treatment. L.S.S. was in charge of DOC and bacterial analyses and their interpretation; M.A.K. provided the GIS data for river watersheds from available databases. All authors have read and agreed to the published version of the manuscript.

Funding: RSF grant 22-17-00253, RFBR grant 20-05-00729, the TSU Development Program "Priority-2030", and grant "Kolmogorov" of MES (Agreement No 075-15-2022-241).

Informed Consent Statement: Not applicable.

Acknowledgments: We acknowledge support from RSF grant 22-17-00253, RFBR grant 20-05-00729, the TSU Development Program "Priority-2030", and grant "Kolmogorov" of MES (Agreement No 075-15-2022-241).

Conflicts of Interest: The authors declare no conflict of interest.

References

1. Frey, K.E.; McClelland, J.W. Impacts of permafrost degradation on arctic river biogeochemistry. *Hydrol. Process.* **2009**, *23*, 169–182. [CrossRef]
2. Vonk, J.E.; Tank, S.E.; Bowden, W.B.; Laurion, I.; Vincent, W.F.; Alekseychik, P.; Amyot, M.; Billet, M.F.; Canário, J.; Cory, R.M.; et al. Reviews and Syntheses: Effects of permafrost thaw on arctic aquatic ecosystems. *Biogeosciences* **2015**, *12*, 7129–7167. [CrossRef]
3. Vonk, J.E.; Tank, S.E.; Walvoord, M.A. Integrating hydrology and biogeochemistry across frozen landscapes. *Nat. Commun.* **2019**, *10*, 5377. [CrossRef] [PubMed]
4. White, D.; Hinzman, L.; Alessa, L.; Cassano, J.; Chambers, M.; Falkner, K.; Francis, J.; Gutowski, W.J., Jr.; Holland, M.; Holmes, R.M.; et al. The arctic freshwater system: Changes and impacts. *J. Geophys. Res.* **2007**, *112*, G04S54. [CrossRef]
5. Barbieri, M.; Barberio, M.D.; Banzato, F.; Billi, A.; Boschetti, T.; Franchini, S.; Gori, F.; Petitta, M. Climate change and its effect on groundwater quality. *Environ. Geochem. Health* **2021**, 1–12. [CrossRef]
6. Eskandari, E.; Mohammadzadeh, H.; Nassery, H.; Vadiati, M.; Zadeh, A.M.; Kisi, O. Delineation of isotopic and hydrochemical evolution of karstic aquifers with different cluster-based (HCA, KM, FCM and GKM) methods. *J. Hydrol.* **2022**, *609*, 127706. [CrossRef]
7. Frey, K.E.; Smith, L.C. How well do we know northern land cover? Comparison of four global vegetation and wetland products with a new ground-truth database for West Siberia. *Glob. Biogeochem. Cycles* **2007**, *21*, GB1016. [CrossRef]
8. Kremenetsky, K.V.; Velichko, A.A.; Borisova, O.K.; MacDonald, G.M.; Smith, L.C.; Frey, K.E.; Orlova, L.A. Peatlands of the West Siberian Lowlands: Current knowledge on zonation, carbon content, and Late Quaternary history. *Quat. Sci. Rev.* **2003**, *22*, 703–723. [CrossRef]
9. Romanovsky, V.E.; Drozdov, D.S.; Oberman, N.G.; Malkova, G.V.; Kholodov, A.L.; Marchenko, S.S.; Moskalenko, N.G.; Sergeev, D.O.; Ukraintseva, N.G.; Abramov, A.A.; et al. Thermal state of permafrost in Russia. *Permafr. Periglac. Processes* **2010**, *21*, 136–155. [CrossRef]
10. Smith, L.C.; Macdonald, G.M.; Velichko, A.A.; Beilman, D.W.; Borisova, O.K.; Frey, K.E.; Kremenetsky, K.V.; Sheng, Y. Siberian peatlands as a net carbon sink and global methane source since the early Holocene. *Science* **2004**, *303*, 353–356. [CrossRef]
11. Frey, K.E.; Siegel, D.I.; Smith, L.C. Geochemistry of west Siberian streams and their potential response to permafrost degradation. *Water Resour. Res.* **2007**, *43*, W03406. [CrossRef]
12. Frey, K.E.; Smith, L.C. Amplified carbon release from vast West Siberian peatlands by 2100. *Geophys. Res. Lett.* **2005**, *32*. [CrossRef]

13. Krickov, I.; Lim, A.; Manasypov, R.M.; Loiko, S.V.; Shirokova, L.S.; Kirpotin, S.N.; Karlsson, J.; Pokrovsky, O.S. Riverine particulate C and N generated at the permafrost thaw front: Case study of western Siberian rivers across a 1700-km latitudinal transect. *Biogeosciences* **2018**, *15*, 6867–6884. [CrossRef]
14. Krickov, I.V.; Pokrovsky, O.S.; Manasypov, R.M.; Lim, A.G.; Shirokova, L.S.; Viers, J. Colloidal transport of carbon and metals by western Siberian rivers during different seasons across a permafrost gradient. *Geochim. Cosmochim. Acta* **2019**, *265*, 221–241. [CrossRef]
15. Krickov, I.V.; Lim, A.G.; Manasypov, R.M.; Loiko, S.V.; Vorobyev, S.N.; Shevchenko, V.P.; Dara, O.M.; Gordeev, V.V.; Pokrovsky, O.S. Major and trace elements in suspended matter of western Siberian rivers: First assessment across permafrost zones and landscape parameters of watersheds. *Geochim. Cosmochim. Acta* **2020**, *269*, 429–450. [CrossRef]
16. Pokrovsky, O.S.; Manasypov, R.M.; Shirokova, L.S.; Loiko, S.V.; Krickov, I.V.; Kopysov, S.; Zemtzov, V.A.; Kulizhsky, S.P.; Vorobyev, S.N.; Kirpotin, S.N. Permafrost coverage, watershed area and season control of dissolved carbon and major elements in western Siberia rivers. *Biogeosciences* **2015**, *12*, 6301–6320. [CrossRef]
17. Pokrovsky, O.S.; Manasypov, R.M.; Loiko, S.; Krickov, I.A.; Kopysov, S.G.; Kolesnichenko, L.G.; Vorobyev, S.N.; Kirpotin, S.N. Trace element transport in western Siberia rivers across a permafrost gradient. *Biogeosciences* **2016**, *13*, 1877–1900. [CrossRef]
18. Vorobyev, S.N.; Pokrovsky, O.S.; Serikova, S.; Manasypov, R.M.; Krickov, I.V.; Shirokova, L.S.; Lim, A.; Kolesnichenko, L.G.; Kirpotin, S.N.; Karlsson, J. Permafrost boundary shift in Western Siberia may not modify dissolved nutrient concentrations in rivers. *Water* **2017**, *9*, 985. [CrossRef]
19. Vorobyev, S.N.; Pokrovsky, O.S.; Kolesnichenko, L.G.; Manasypov, R.M.; Shirokova, L.S.; Karlsson, J.; Kirpotin, S.N. Biogeochemistry of dissolved carbon, major, and trace elements during spring flood periods on the Ob River. *Hydrol. Processes* **2019**, *33*, 1579–1594. [CrossRef]
20. Kolesnichenko, I.; Kolesnichenko, L.G.; Vorobyev, S.N.; Shirokova, L.S.; Semiletov, I.P.; Dudarev, O.V.; Vorobev, R.S.; Shavrina, U.; Kirpotin, S.N.; Pokrovsky, O.S. Landscape, soil, lithology, climate and permafrost control on dissolved carbon, major and trace elements in the Ob River, western Siberia. *Water* **2021**, *13*, 3189. [CrossRef]
21. Lim, A.G.; Loiko, S.V.; Kuzmina, D.; Krickov, I.V.; Shirokova, L.S.; Kulizhsky, S.P.; Vorobyev, S.N.; Pokrovsky, O.S. Dispersed ground ice of permafrost peatlands: A non-accounted for source of C, nutrients and metals. *Chemosphere* **2021**, *266*, 128953. [CrossRef] [PubMed]
22. Raudina, T.V.; Loiko, S.V.; Lim, A.G.; Krickov, I.V.; Shirokova, L.S.; Istigichev, G.I.; Kuzmina, D.M.; Kulizhsky, S.P.; Vorobyev, S.N.; Pokrovsky, O.S. Dissolved organic carbon and major and trace elements in peat porewater of sporadic, discontinuous, and continuous permafrost zones of western Siberia. *Biogeosciences* **2017**, *14*, 3561–3584. [CrossRef]
23. Raudina, T.V.; Loiko, S.V.; Lim, A.; Manasypov, R.M.; Shirokova, L.S.; Istigechev, G.I.; Kuzmina, D.M.; Kulizhsky, S.P.; Vorobyev, S.N.; Pokrovsky, O.S. Permafrost thaw and climate warming may decrease the CO_2, carbon, and metal concentration in peat soil waters of the Western Siberia Lowland. *Sci. Total Environ.* **2018**, *634*, 1004–1023. [CrossRef]
24. Raudina, T.V.; Loiko, S.; Kuzmina, D.M.; Shirokova, L.S.; Kulizhsky, S.P.; Golovatskaya, E.A.; Pokrovsky, O.S. Colloidal organic carbon and trace elements in peat porewaters across a permafrost gradient in Western Siberia. *Geoderma* **2021**, *390*, 114971. [CrossRef]
25. Karlsson, J.; Serikova, S.; Vorobyev, S.N.; Rocher-Ros, G.; Denfeld, B.; Pokrovsky, O.S. Carbon emission from Western Siberian inland waters. *Nat. Commun.* **2021**, *12*, 825. [CrossRef]
26. Serikova, S.; Pokrovsky, O.S.; Ala-Aho, P.; Kazantsev, V.; Kirpotin, S.N.; Kopysov, S.G.; Krickov, I.V.; Laudon, H.; Manasypov, R.M.; Shirokova, L.S.; et al. High riverine CO_2 emissions at the permafrost boundary of Western Siberia. *Nat. Geosci.* **2018**, *11*, 825–829. [CrossRef]
27. Serikova, S.; Pokrovsky, O.S.; Laudon, H.; Krickov, I.V.; Lim, A.G.; Manasypov, R.M.; Karlsson, J. High carbon emissions from thermokarst lakes of Western Siberia. *Nat. Commun.* **2019**, *10*, 1552. [CrossRef]
28. Pokrovsky, O.; Lim, A.; Korets, M.; Krickov, I.; Vorobyev, S. Dissolved (<0.45 μm) major and trace elements and landscape parameters of Ket and Taz Rivers, Western Siberia. *Mendeley Data* **2022**, *V1*. [CrossRef]
29. Marie, D.; Partensky, F.; Vaulot, D.; Brussaard, C. Enumeration of phytoplankton, bacteria, and viruses in marine samples. *Curr. Protoc. Cytom.* **1999**, *10*, 11.11.1–11.11.5. [CrossRef]
30. Pokrovsky, O.S.; Manasypov, R.M.; Loiko, S.V.; Shirokova, L.S. Organic and organo-mineral colloids in discontinuous permafrost zone. *Geochim. Cosmochim. Acta* **2016**, *188*, 1–20. [CrossRef]
31. Shirokova, L.S.; Pokrovsky, O.S.; Kirpotin, S.N.; Desmukh, C.; Pokrovsky, B.G.; Audry, S.; Viers, J. Biogeochemistry of organic carbon, CO_2, CH_4, and trace elements in thermokarst water bodies in discontinuous permafrost zones of Western Siberia. *Biogeochemistry* **2013**, *113*, 573–593. [CrossRef]
32. Heimburger, A.; Tharaud, M.; Monna, F.; Losno, R.; Desboeufs, K.; Nguyen, E.B. SLRS-5 elemental concentrations deduced from SLRS-5/SLRS-4 ratios of thirty-three uncertified elements. *Geostand. Geoanal. Res.* **2013**, *37*, 77–85. [CrossRef]
33. Yeghicheyan, D.; Bossy, C.; Bouhnik Le Coz, M.; Douchet, C.; Granier, G.; Heimburger, A.; Lacan, F.; Lanzanova, A.; Rousseau, T.C.C.; Seidel, J.-L.; et al. A Compilation of Silicon, Rare Earth Element and Twenty-One other Trace Element Concentrations in the Natural River Water Reference Material SLRS-5 (NRC-CNRC). *Geostand. Geoanal. Res.* **2013**, *37*, 449–467. [CrossRef]
34. Bartalev, S.A.; Egorov, V.A.; Ershov, D.V.; Isaev, A.S.; Lupyan, E.A.; Plotnikov, D.E.; Uvarov, I.A. Remote mapping of vegetation land cover of Russia based on data of MODIS spectroradmeter. *Mod. Probl. Earth Remote Sens. Space* **2011**, *8*, 285–302. Available online: http://d33.infospace.ru/d33_conf/2011v8n4/285-302.pdf (accessed on 15 June 2022).

35. Harris, I.; Jones, P.D.; Osborn, T.J.; Lister, D.H. Updated high-resolution grids of monthly climatic observations—The CRU TS3.10 Dataset. *Int. J. Climatol.* **2014**, *34*, 623–642. [CrossRef]
36. Hugelius, G.; Tarnocai, C.; Broll, G.; Canadell, J.G.; Kuhry, P.; Swanson, D.K. The Northern Circumpolar Soil Carbon Database: Spatially distributed datasets of soil coverage and soil carbon storage in the northern permafrost regions. *Earth Syst. Sci. Data* **2013**, *5*, 3–13. [CrossRef]
37. Santoro, M.; Beer, C.; Cartus, O.; Schmullius, C.; Shvidenko, A.; McCallum, I.; Wegmueller, U.; Wiesmann, A. The BIOMASAR algorithm: An approach for retrieval of forest growing stock volume using stacks of multi-temporal SAR data. In Proceedings of the ESA Living Planet Symposium, Bergen, Norway, 28 June–2 July 2010. Available online: https://www.researchgate.net/publication/230662433 (accessed on 15 June 2022).
38. Krickov, I.V.; Lim, A.G.; Shevchenko, V.P.; Vorobyev, S.N.; Candaudap, F.; Pokrovsky, O.S. Dissolved metal (Fe, Mn, Zn, Ni, Cu, Co, Cd, Pb) and metalloid (As, Sb) in snow water across a 2800-km latitudinal profile of western Siberia: Impact of local pollution and global transfer. *Water* **2022**, *14*, 94. [CrossRef]
39. Barker, A.J.; Douglas, T.A.; Jacobson, A.D.; McClelland, J.W.; Ilgen, A.G.; Khosh, M.S.; Lehn, G.O.; Trainor, T.P. Late season mobilization of trace metals in two small Alaskan arctic watersheds as a proxy for landscape scale permafrost active layer dynamics. *Chem. Geol.* **2014**, *381*, 180–193. [CrossRef]
40. Keller, K.; Blum, J.D.; Kling, G.W. Geochemistry of soils and streams on surfaces of varying ages in arctic Alaska. *Arct. Antarct. Alp. Res.* **2007**, *39*, 84–98. [CrossRef]
41. Keller, K.; Blum, J.D.; Kling, G.W. Stream geochemistry as an indicator of increasing permafrost thaw depth in an arctic watershed. *Chem. Geol.* **2010**, *273*, 76–81. [CrossRef]
42. Laudon, H.; Sjoblom, V.; Buffam, I.; Seibert, J.; Morth, M. The role of catchment scale and landscape characteristics for runoff generation of boreal streams. *J. Hydrol.* **2007**, *344*, 198–209. [CrossRef]
43. Viers, J.; Prokushkin, A.S.; Pokrovsky, O.S.; Beaulieu, E.; Oliva, P.; Dupré, B. Seasonal and spatial variability of elemental concentrations in boreal forest larch folliage of Central Siberia on continuous permafrost. *Biogeochemistry* **2013**, *113*, 435–449. [CrossRef]
44. Lidman, F.; Morth, C.M.; Laudon, H. Landscape control of uranium and thorium in boreal streams—Spatiotemporal variability and the role of wetlands. *Biogeosciences* **2012**, *9*, 4773–4785. [CrossRef]
45. Lidman, F.; Kohler, S.J.; Morth, C.-M.; Laudon, H. Metal transport in the boreal landscape—The role of wetlands and the affinity for organic matter. *Environ. Sci. Technol.* **2014**, *48*, 3783–3790. [CrossRef]
46. Ingri, J.; Widerlund, A.; Land, M.; Gustafsson, Ö.; Andersson, P.S.; Öhlander, B. Temporal variations in the fractionation of the rare earth elements in a boreal river, the role of colloidal particles. *Chem. Geol.* **2000**, *166*, 23–45. [CrossRef]
47. Stolpe, B.; Guo, L.; Shiller, A.M.; Aiken, G.R. Abundance, size distribution and trace-element binding of organic and iron-rich nanocolloids in Alaskan rivers, as revealed by field-flow fractionation and ICP-MS. *Geochim. Cosmochim. Acta* **2013**, *105*, 221–239. [CrossRef]
48. Vasyukova, E.V.; Pokrovsky, O.S.; Viers, J.; Oliva, P.; Dupré, B.; Martin, F.; Candaudap, F. Trace elements in organic- and iron-rich surficial fluids of the boreal zone: Assessing colloidal forms via dialysis and ultrafiltration. *Geochim. Cosmochim. Acta* **2010**, *74*, 449–468. [CrossRef]
49. Oleinikova, O.; Drozdova, O.Y.; Lapitskiy, S.A.; Demin, V.V.; Bychkov, A.Y.; Pokrovsky, O.S. Dissolved organic matter degradation by sunlight coagulates organo-mineral colloids and produces low-molecular weight fraction of metals in boreal humic waters. *Geochim. Cosmochim. Acta* **2017**, *211*, 97–114. [CrossRef]
50. Oleinikova, O.; Shirokova, L.S.; Gérard, E.; Drozdova, O.Y.; Lapitsky, S.A.; Bychkov, A.Y.; Pokrovsky, O.S. Transformation of organo-ferric peat colloids by a heterotrophic bacterium. *Geochim. Cosmochim. Acta* **2017**, *205*, 313–330. [CrossRef]
51. Shirokova, L.S.; Bredoire, R.; Rolls, J.-L.; Pokrovsky, O.S. Moss and peat leachate degradability by heterotrophic bacteria: Fate of organic carbon and trace metals. *Geomicrobiol. J.* **2017**, *34*, 641–655. [CrossRef]
52. Shirokova, L.S.; Chupakova, A.A.; Chupakov, A.V.; Pokrovsky, O.S. Transformation of dissolved organic matter and related trace element in the mouth zone of the largest European Arctic river: Experimental modeling. *Inland Waters* **2017**, *7*, 272–282. [CrossRef]
53. Vorobyev, S.N.; Kolesnichenko, Y.; Korets, M.; Pokrovsky, O.S. Testing landscape, climate and lithology impact on carbon, major and trace elements of the Lena River and its tributaries during a spring flood period. *Water* **2021**, *13*, 2093. [CrossRef]
54. Pokrovsky, O.S.; Bueno, M.; Manasypov, R.M.; Shirokova, L.S.; Karlsson, J.; Amouroux, D. Dissolved organic matter controls on seasonal and spatial selenium concentration variability in thaw lakes across a permafrost gradient. *Environ. Sci. Technol.* **2018**, *52*, 10254–10262. [CrossRef] [PubMed]
55. Chupakov, A.V.; Pokrovsky, O.S.; Moreva, O.Y.; Shirokova, L.S.; Neverova, N.V.; Chupakova, A.A.; Kotova, E.I.; Vorobyeva, T.Y. High resolution multi-annual riverine fluxes of organic carbon, nutrient and trace element from the largest European Arctic river, Severnaya Dvina. *Chem. Geol.* **2020**, *538*, 119491. [CrossRef]
56. Bagard, M.L.; Chabaux, F.; Pokrovsky, O.S.; Prokushkin, A.S.; Viers, J.; Dupré, B.; Stille, P.; Rihs, S.; Schmitt, A.-D. Seasonal variability of element fluxes in two Central Siberian rivers draining high latitude permafrost dominated areas. *Geochim. Cosmochim. Acta* **2011**, *75*, 3335–3357. [CrossRef]
57. Dahlqvist, R.; Andersson, K.; Ingri, J.; Larsson, T.; Stolpe, B.; Turner, D. Temporal variations of colloidal carrier phases and associated trace elements in a boreal river. *Geochim. Cosmochim. Acta* **2007**, *71*, 5339–5354. [CrossRef]

58. Lyvén, B.; Hassellöv, M.; Turner, D.R.; Haraldsson, C.; Andersson, K. Competition between iron- and carbon-based colloidal carriers for trace metals in a freshwater assessed using flow field-flow fractionation coupled to ICPMS. *Geochim. Cosmochim. Acta* **2003**, *67*, 3791–3802. [CrossRef]
59. Pokrovsky, O.S.; Manasypov, R.M.; Kopysov, S.; Krickov, I.V.; Shirokova, L.S.; Loiko, S.V.; Lim, A.G.; Kolesnichenko, L.G.; Vorobyev, S.N.; Kirpotin, S.N. Impact of permafrost thaw and climate warming on riverine export fluxes of carbon, nutrients and metals in western Siberia. *Water* **2020**, *12*, 1817. [CrossRef]
60. Manasypov, R.M.; Vorobyev, S.N.; Loiko, S.V.; Kritzkov, I.V.; Shirokova, L.S.; Shevchenko, V.P.; Kirpotin, S.N.; Kulizhsky, S.P.; Kolesnichenko, L.G.; Zemtzov, V.A.; et al. Seasonal dynamics of organic carbon and metals in thermokarst lakes from the discontinuous permafrost zone of western Siberia. *Biogeosciences* **2015**, *12*, 3009–3028. [CrossRef]
61. Manasypov, R.M.; Lim, A.G.; Krickov, I.V.; Shirokova, L.S.; Vorobyev, S.N.; Kirpotin, S.N.; Pokrovsky, O.S. Spatial and seasonal variations of C, nutrient, and metal concentration in thermokarst lakes of Western Siberia across a permafrost gradient. *Water* **2020**, *12*, 1830. [CrossRef]
62. Blois, J.L.; Williams, J.W.; Fitzpatrick, M.C.; Jackson, S.T.; Ferrier, S. Space can substitute for time in predicting climate-change effects on biodiversity. *Proc. Natl. Acad. Sci. USA* **2013**, *110*, 9374–9379. [CrossRef] [PubMed]
63. García Criado, M.; Myers-Smith, I.H.; Bjorkman, A.D.; Lehmann, C.E.R.; Stevens, N. Woody plant encroachment intensifies under climate change across tundra and savanna biomes. *Glob. Ecol. Biogeogr.* **2020**, *29*, 925–943. [CrossRef]
64. Mauclet, E.; Agnan, Y.; Hirst, C.; Monhonval, A.; Pereira, B.; Vandeuren, A.; Villani, M.; Ledman, J.; Taylor, M.; Jasinski, B.L.; et al. Changing sub-Arctic tundra vegetation upon permafrost degradation: Impact on foliar mineral element cycling. *Biogeosciences* **2022**, *19*, 2333–2351. [CrossRef]
65. Tape, K.; Sturm, M.; Racine, C. The evidence for shrub expansion in Northern Alaska and the Pan-Arctic. *Glob. Change Biol.* **2006**, *12*, 686–702. [CrossRef]

water

MDPI

Article

How to Implement User-Friendly BLMs in the Absence of DOC Monitoring Data: A Case Study on Bulgarian Surface Waters

Tony Venelinov [1] and Stefan Tsakovski [2,*]

[1] Chair of Water Supply, Sewerage, Water and Wastewater Treatment, Faculty of Hydraulic Engineering, University of Architecture, Civil Engineering and Geodesy, 1 Hr. Smirnenski Blvd., 1046 Sofia, Bulgaria; TVenelinov_fhe@uacg.bg

[2] Chair of Analytical Chemistry, Faculty of Chemistry and Pharmacy, Sofia University St. Kliment Ohridski, 1 J. Bourchier Blvd., 1164 Sofia, Bulgaria

* Correspondence: STsakovski@chem.uni-sofia.bg

Abstract: The metal bioavailability concept is implemented in the Water Framework Directive (WFD) compliance assessment. The bioavailability assessment is usually performed by the application of user-friendly Biotic Ligand Models (BLMs), which require dissolved metal concentrations to be used with the "matching" data of the supporting physicochemical parameters of dissolved organic carbon (DOC), pH and $Ca_{dissolved}$. Many national surface water monitoring networks do not have sufficient matching data records, especially for DOC. In this study, different approaches for dealing with the missing DOC data are presented: substitution using historical data; the appropriate percentile of DOC concentrations; and combinations of the two. The applicability of the three following proposed substitution approaches is verified by comparison with the available matching data: (i) calculations from available TOC data; (ii) the 25th percentile of the joint Bulgarian monitoring network DOC data (measured and calculated by TOC); and (iii) the 25th percentile of the calculated DOC from the matching TOC data for the investigated surface water body (SWB). The application of user-friendly BLMs (BIO-MET, M-BAT and PNEC Pro) to 13 surface water bodies (3 reservoirs and 10 rivers) in the Bulgarian surface waters monitoring network outlines that the suitability of the substitution approaches decreases in order: DOC calculated by TOC > the use of the 25th percentile of the data for respective SWB > the use of the 25th percentile of the Bulgarian monitoring network data. Additionally, BIO-MET is the most appropriate tool for the bioavailability assessment of Cu, Zn and Pb in Bulgarian surface water bodies.

Keywords: BLM; TOC; DOC; Fe; historical data

Citation: Venelinov, T.; Tsakovski, S. How to Implement User-Friendly BLMs in the Absence of DOC Monitoring Data: A Case Study on Bulgarian Surface Waters. *Water* **2022**, *14*, 246. https://doi.org/10.3390/w14020246

Academic Editor: Liudmila S. Shirokova

Received: 28 November 2021
Accepted: 13 January 2022
Published: 15 January 2022

1. Introduction

The determination of metal species and their bioavailability depends crucially on Dissolved Organic Matter (DOM). The presence of DOM is specified as a Dissolved Organic Carbon (DOC) concentration, contrary to Natural Organic Matter (NOM), which is a collective organic component of water. NOM is a complex mixture of leaf litter, organic acids, proteins and many more complex organic molecules. The Total Organic Matter (TOM) constitutes Particulate Organic Matter (POM) and DOM. The organic carbon results of unfiltered samples are reported as Total Organic Carbon (TOC) and samples filtered through a 0.45 μm filter are reported as DOC. Both results equal the mass of carbon present in the mixture of organic compounds in the raw water/filtrate [1]. Sometimes TOC is reported instead of DOC, but it should be stressed that the DOC concentration will always be less than the concentration of TOC. DOC generally accounts for about 50% of DOM in typical surface water bodies and is in the range between 1 and 15 mg/L. DOC mainly comprises humic substances of natural origin, which is formed as a result of the breakdown of plant and animal tissues by chemical and biological processes or from anthropogenic sources [2].

The bioavailability of Cu, Ni and Zn depends on DOC concentrations [3–5], with higher concentrations of DOC resulting in reduced toxicity. Evidence suggests that the pH also has a significant impact on metal bioavailability and toxicity in aquatic environments [6,7]. Major cations (Ca^{2+}, Na^{+}, K^{+} and Mg^{2+}) can also protect aquatic organisms against free metals by competing on the biotic binding sites, e.g., fish gills [8]. The presence of dissolved Fe and Al can affect the binding of the abovementioned metals to DOC due to their very high affinity for complexation by DOC. In such conditions, the availability of binding sites for the less strongly bound metals is reduced, resulting in an increase in their bioavailability for aquatic organisms. Normally, bodies of water that are high in Fe and Al are associated with a lower water pH (6.5–7.5), under which conditions both metals form insoluble precipitates.

DOC is a composite parameter for a heterogeneous mixture of polyfunctional polymers and contains functional groups (ligands) that bind free metal ions (the most toxic inorganic metal fraction) to reduce the interaction between free metal ions and aquatic organisms [9]. Discharge derived DOC (such as sewage effluents) is likely to be of a different composition to natural DOC and comprises proteins, amino acids, polysaccharides and synthetic chelating agents, such as ethylenediamine tetra acetic acid (EDTA). The current evidence base suggests that this latter type of DOC has a much greater capacity for binding metals compared with similar concentrations of naturally sourced DOC [10]. In bodies of water with low pH, DOM–metal complexes are weaker, resulting in a release of bioavailable ions and, therefore, an increase in the toxicity of the metal to aquatic organisms [11]. Taking DOM binding into account, the effect of considering bioavailability in environmental quality standards ($EQS_{bioavailable}$) [12] has been assessed by the use of user-friendly biotic ligand models (BLMs), which has been proposed as a tool for regulating agencies. In the Netherlands, the assessment of bioavailability is incorporated into policy documents [13]. Feasibility studies on the implementation of a bioavailability-based approach for metals have been undertaken by several member states [14–19]. Guidance on how bioavailability may be incorporated into compliance assessments, classifications and local risk assessments is included in the recent EU Technical Guidance for deriving Environmental Quality Standards (EQS) under the European Water Framework Directive (WFD) [20].

High-quality surface water monitoring data provide tools for the assessment of potential risks to aquatic species. Obtaining such reliable data should follow the principles of surface water chemical monitoring under the WFD [21], including sampling, preservation and analysis [22]. The analytical methods for every analyte of interest should comply with the minimum performance requirements as stated in the QA/QC Directive [23]. A common implementation strategy of the WFD to achieve a good status and prevent any deterioration in the status (ecological and chemical) of European water bodies led to the development of methods for deriving EQSs to be used for the status assessment of surface water bodies (SWBs). Following these requirements, as a general prerequisite for WFD monitoring, EQS compliance assessment and subsequent decision-making are reliably achieved. An EQS is a limit for environmental disturbances, in particular from an ambient concentration of pollutants and wastes, which determines the maximum allowable degradation of environmental media. A water body cannot achieve a good chemical or ecological status if the EQS for any WFD Priority or Specific Substance is exceeded.

The compliance assessment follows a tiered approach, which: (i) compares the annual average concentration, calculated from monitoring data with the $EQS_{bioavailable}$; (ii) uses user-friendly tools (based on Biotic Ligand Models) for the calculation of local metal bioavailability for a comparison between the measured dissolved metal concentration at a site and the EQS; and (iii) considers the local background concentrations of metals as a part of the EQS in the risk assessment [8]. The compliance assessment, which takes into account the bioavailability and uses simplified and user-friendly bioavailability tools (such as BLMs), requires that the concentration of dissolved metals (Cu, Zn, Ni, Pb and Mn) preferably be used along with data for the supporting physicochemical parameters of DOC, pH and dissolved Ca [24]. Due to temporal and spatial variations, data for all of the required input parameters (dissolved metal concentrations, dissolved Ca, DOC and pH) should ideally be "matched" to increase the reliability of the results. Without these, the simplified tools will either not run or not run reliably. The term "matched" means that the required input parameters are all determined in the same sample, obtained from the site of interest, to enable the identification of potential risks. A few of the member states (including Bulgaria) are just starting the implementation of bioavailability-based EQS and some of the required data are not fully available, especially for DOC, which is not routinely measured in Europe. In such cases, two approaches are possible when dealing with the missing DOC data: the use of historical monitoring data (e.g., data for similar and/or neighbouring catchment types) and the use of substitute data. Such approaches should be carefully chosen due to the quality of the available data—the dataset is likely to have been collected with inadequate sampling protocols, insufficient analytical sensitivity and LOQ, for a different purpose than the implementation of an EQS compliance assessment, with different sampling frequency, etc. [25,26]. Different approaches to estimating DOC have been proposed, including: UV absorbance [27,28]; colour measurements [29]; dissolved iron [30]; fluorescence measurements [31]; and a modifying factor that is dependent on the optical properties of DOC [32]. Any of the proposed approaches to estimating DOC should be based on locally derived empirical relationships and proven to be valid for the studied territory. Such data can be used for feasibility and screening assessments, but these need to be assessed on a case-by-case basis.

The present study aims to propose a methodology to the national environmental bodies for the implementation of BLM in compliance assessments when DOC data are missing. The methodology comprises a selection of appropriate substitute approaches, which are applicable for the environmental authorities and their validation, with the matching data using the three most widely used BLMs: BIO-MET; M-BAT; and PNEC Pro. In this manner, the metal bioavailability could be assessed using the most suitable pair of substitute approaches and a user-friendly BLM.

2. Materials and Methods

Data for the concentrations of DOC, $Fe_{dissolved}$ and TOC were obtained by the Bulgarian Environmental Agency surface water monitoring programme. The database returned over 27,000 measurement results for around 950 SWBs for the period 1 January 2010–31 December 2020 (Figure 1).

The methods used by the different laboratories followed standardized procedures. For the determination of pH, EN ISO 10523:2012 was used; for Ca, EN ISO 14911:2002 and ISO 6058:2002 were used; for DOC and TOC, EN 1484:2001 was used, including one result for TOC that was obtained by EN ISO 6468:2006. Cu, Zn, Ni and Pb were measured using EN ISO 17294-2:2016. The basic statistics of the dataset are presented in Table 1. No outlier tests were performed and where the halved limit of quantitation (LOQ) data was presented (for either DOC, $Fe_{dissolved}$ or TOC), the whole dataset was omitted. Where the calculated DOC/TOC ratio exceeded one, the concentrations were not used in the subsequent calculations.

Table 1. The basic statistics of the monitoring data for the Bulgarian surface waters (2010–2020).

	Monitoring Data		
	DOC	**$Fe_{dissolved}$**	**TOC**
Results	1829	13,234	11,943
Water bodies (n = 952)	143	513	545
Average	5.50 mg/L	68.61 mg/L	6.71 mg/L
Standard deviation	6.92 mg/L	602.61 mg/L	11.74 mg/L
Median	3.80 mg/L	22.00 mg/L	4.80 mg/L

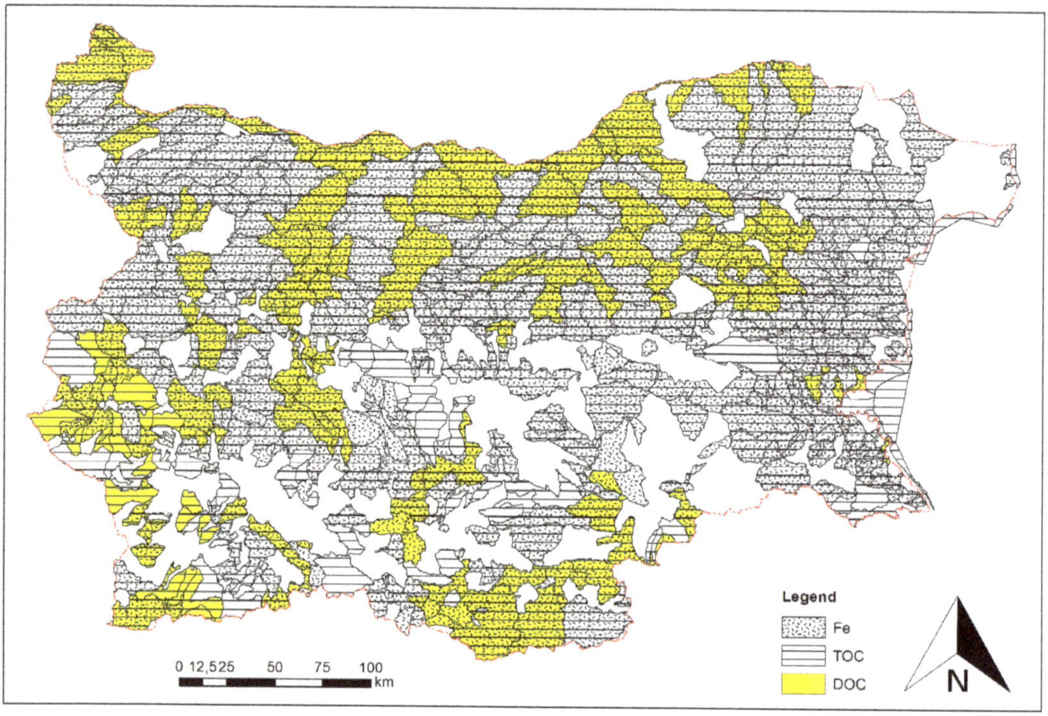

Figure 1. Data availability map.

User-friendly tools (based on the BLMs) were used in the metal bioavailability calculations. For the application of BIO-MET (https://bio-met.net/, last accessed 25 October 2021), M-BAT (https://www.wfduk.org, last accessed 25 October 2021) and PNEC Pro (http://www.pnec-pro.com/, last accessed 25 October 2021), data for the concentrations of pH, DOC, TOC and $Ca_{dissolved}$ were obtained by the open data portal of the Bulgarian Environmental Agency surface water monitoring programme (https://data.egov.bg/data/view/9ee2ca78-051d-4ec8-b8f8-1a1e93ee637e, last accessed 25 October 2021). A total of 13 water bodies—3 reservoirs (Dospat Dam, Pchelina Dam and Iskar Dam) and 10 rivers (Strumeshnitsa, Mesta, Yantra, Dospat, Dragovishtitsa, Negovanka, Eleshnitsa, Struma, Rusenski Lom and Iskar) were used for the comparison during the implementation of the BLMs, and for which the matching datasets for DOC, TOC, pH, Ca, Cu, Zn, Ni and Pb were available for 2020.

3. Results

3.1. Substitutes for Missing DOC

The Bulgarian Environmental Agency surface water monitoring programme revealed a total of 1829 measurement results for DOC for all monitored surface water bodies since 2010 and 13,234 measurement results for dissolved Fe. The "matching" of the data indicated 805 results for DOC and $Fe_{dissolved}$. Unfortunately, nearly half of the results could not be used for correlation between the parameters since most of the $Fe_{dissolved}$ concentrations were reported as LOQ/2. The analysis showed a poor correlation (R^2 = 0.38, n = 445) when "matched" data was used (Figure 2).

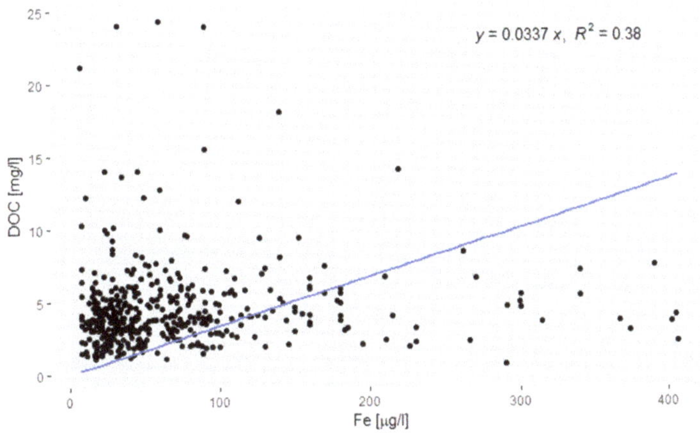

Figure 2. The relationship between the measured concentrations of DOC and dissolved Fe in Bulgarian surface waters.

The Bulgarian Environmental Agency surface water monitoring programme contained 11,945 results for TOC concentrations in the 2010–2020 period. The "matching" data indicated 736 results for DOC and TOC. The analysis showed a significant correlation (R^2 = 0.95, n = 736) when matched data was used (Figure 3) and a conversion factor DOC = 0.81 × TOC could be applied for missing DOC data.

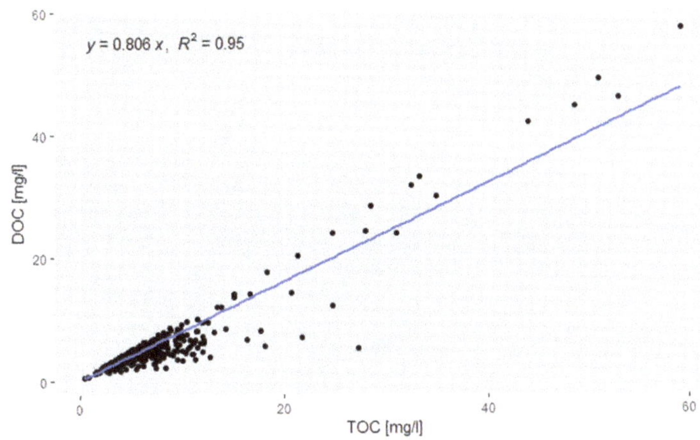

Figure 3. The relationship between the measured concentrations of DOC and TOC in Bulgarian surface waters.

3.2. Substitutes in the Absence of Matching Data

The UK [9] and France [18] have adopted an approach to use the 25th percentile of the DOC concentrations for estimating DOC when the matching data are unavailable. In this paper, the calculated DOC concentrations using the TOC data (DOC = 0.81 × TOC) was used as a substitute. Furthermore, the 25th percentile of the DOC concentrations (2.4 mg/L) from the Bulgarian DOC data obtained in 2020 (n = 1711) was used and compared with the results of the matching data. Additionally, the 25th percentile of the calculated DOC concentrations using the measured TOC concentration for the respective water body was used and compared with the results of the matching data.

3.3. Application of User-Friendly Models (BIO-MET, M-BAT and PNEC Pro) in Compliance Assessment

Following the tiered approach, the user-friendly BLMs were implemented on 13 water bodies (3 reservoirs and 10 rivers) with the available matching datasets for DOC, TOC, pH, Ca, Cu, Zn, Ni and Pb for 2020. The average annual concentrations that were calculated from the monitoring data were compared with the $EQS_{bioavailable}$ (1 µg/L for Cu, 4 µg/L for Ni, 8 µg/L for Zn and 1.2 µg/L for Pb), which was derived from the WFD [21] and national ordinance [33].

The results show 13 exceedances for Cu, 6 exceedances for Zn and 4 exceedances for Pb (Table 2). No exceedances of the average annual concentrations were observed for Ni, compared with the EQS.

Table 2. The exceedances for Cu, Zn, Ni and Pb in the studied Bulgarian surface water bodies.

	Water Body	Cu	Zn	Ni	Pb
1	Dospat Dam	+	+		
2	Pchelina Dam	+	+		+
3	Iskar Dam	+			+
4	Strumeshnitsa	+			
5	Mesta	+			
6	Yantra	+			
7	Dospat	+	+		
8	Dragovishtitsa	+	+		
9	Negovanka	+			
10	Eleshnitsa	+			+
11	Struma	+	+		+
12	Rusenski Lom	+	+		
13	Iskar	+			

When the average annual concentrations exceeded the EQS, the samples moved to the second tier where the Biotic Ligand Model-based tools (BIO-MET, M-BAT and PNEC Pro) were used for the calculation of local metal bioavailability and a comparison between the measured dissolved metal concentration at a site and the EQS was made. The DOC concentration played an important role in determining the bioavailable dissolved metal concentration.

For the calculations, the following data were used: the measured DOC concentrations; the calculated DOC from the matching TOC data (DOC = 0.81 × TOC); the 25th percentile (2.4 mg/L) of the DOC and calculated TOC for 2020 of all surface water bodies (n = 1711); and the 25th percentile of the calculated DOC from the matching TOC data for 2020 for each of the water bodies studied. An indication of whether there were exceedances was expressed as the Risk Characterization Ratio (RCR). If the RCR was lower than one, the site was considered as having "passed" the assessment and no further investigation was necessary. If the value of the RCR was greater than one, this indicated an exceedance of the EQS and led to a progression to Tier 3.

3.3.1. BLM Comparisons for Copper

The user-friendly BLM models were implemented for the 13 water bodies where the annual concentrations of Cu for 2020 exceeded the generic $EQS_{bioavailable}$ (1 µg/L). First, the BLM models were performed with the matching data (n = 63) for dissolved Cu, pH, dissolved Ca and DOC, and after that, the DOC values were substituted with: the calculated DOC from the available TOC data (DOC = 0.81 × TOC); the 25th percentile of the DOC data from the Bulgarian monitoring network results from 2020 (P25_BG); and by the 25th percentile of DOC data that was calculated from the TOC for the respective water body (P25_SWB), as described earlier. The calculated RCRs for all of the approaches are presented in the Supplementary Material (Figures S1–S3).

The comparisons between the Cu bioavailability results obtained by the BLM models are presented in Figures 4–6. The correlations between the BIO-MET calculated RCR using matching and substituted data were very good (Figure 4). For the RCR values using percentiles approaches: P25_BG and P25_SWB were overestimated with slopes of 2.7 and 1.8, respectively. These environmental conservative estimates were more pronounced when the 25th percentile of the data from the Bulgarian monitoring network was used. The results from the M-BAT model followed a similar pattern to those of the BIO-MET (Figure 5), but the R^2 values dropped to 0.54 for P25_BG and 0.69 for P25_SWB. The comparison between the RCR values obtained by the PNEC Pro model revealed lower correlations and slopes that were significantly different from one for all pairs (Figure 6). It should be mentioned that the comparison was restricted by the validated applicability domain of the PNEC Pro model, which decreased the number of calculated RCR values.

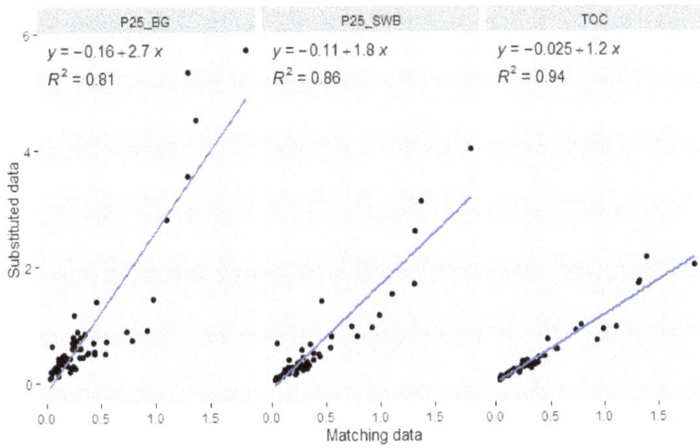

Figure 4. The comparison of the BIO-MET calculated RCR for Cu between the matching data and substituted data (P25_BG—the 25th percentile of the available DOC data and/or the DOC data calculated by the TOC from the Bulgarian monitoring network results during 2020; P25_SWB—the 25th percentile of the DOC data calculated by the TOC for the respective water body; TOC—the matching DOC data calculated by TOC).

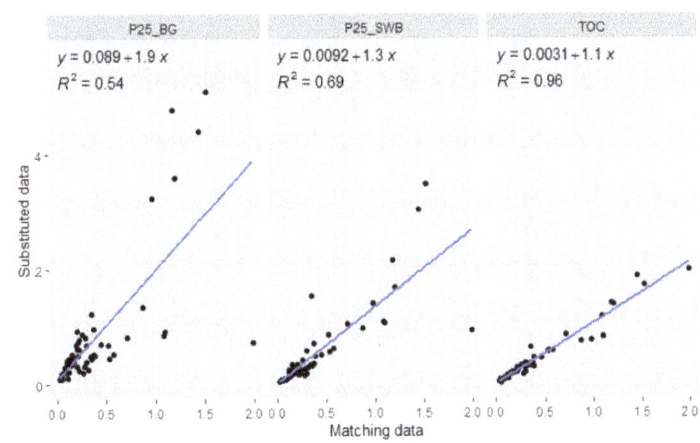

Figure 5. The comparison of the M-BAT calculated RCR for Cu between the matching data and substituted data (P25_BG—the 25th percentile of the available DOC data and/or the DOC data calculated by the TOC from the Bulgarian monitoring network results during 2020; P25_SWB—the 25th percentile of the DOC data calculated by the TOC for the respective water body; TOC—the matching DOC data calculated by TOC).

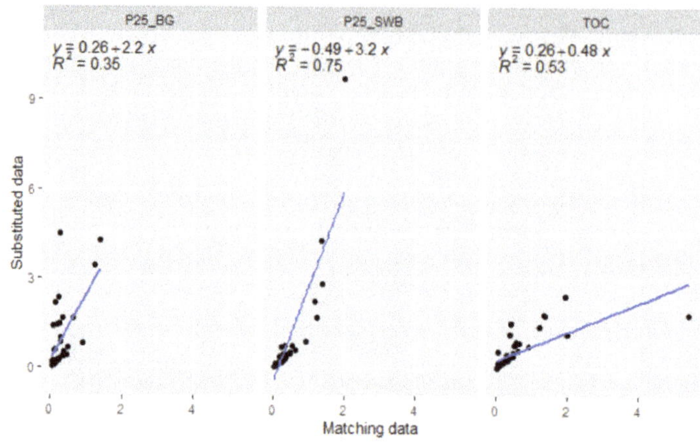

Figure 6. The comparison of the PNEC Pro calculated RCR for Cu between the matching data and substituted data (P25_BG—the 25th percentile of the available DOC data and/or the DOC data calculated by the TOC from the Bulgarian monitoring network results during 2020; P25_SWB—the 25th percentile of the DOC data calculated by the TOC for the respective water body; TOC—the matching DOC data calculated by TOC).

3.3.2. Comparisons for Zn

The user-friendly BLM models were implemented for the six water bodies (two reservoirs and four rivers) where the annual concentrations of Zn for 2020 exceeded the generic $EQS_{bioavailable}$ (8 μg/L) and the matching data (n = 34) for dissolved Zn, pH, dissolved Ca and DOC were available. The calculated RCRs are presented in the Supplementary Material (Figures S4–S6).

The comparison between the Zn bioavailability results obtained by the three BLM models using the matching and substituted DOC data is presented in Figures 7–9. The correlations between the calculated RCRs using the matching and substituted data for all BLMs were very good (R^2 > 0.90). The slopes varied between 1 and 1.5, decreasing in the order: P25_BG > P25_SWB > TOC.

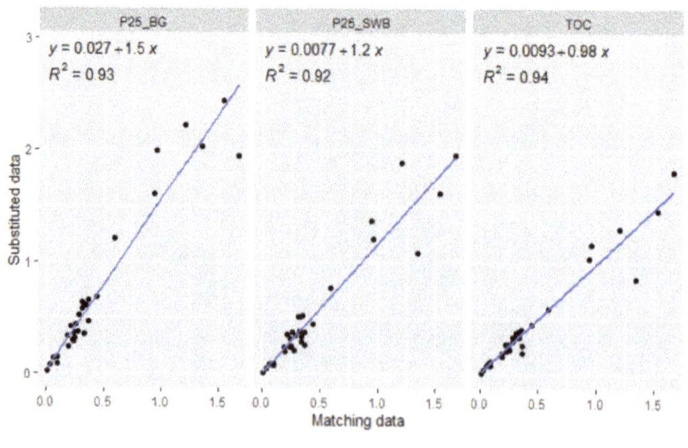

Figure 7. The comparison of the BIO-MET calculated RCR for Zn between the matching data and substituted data (P25_BG—the 25th percentile of the available DOC data and/or the DOC data calculated by the TOC from the Bulgarian monitoring network results during 2020; P25_SWB—the 25th percentile of the DOC data calculated by the TOC for the respective water body; TOC—the matching DOC data calculated by TOC).

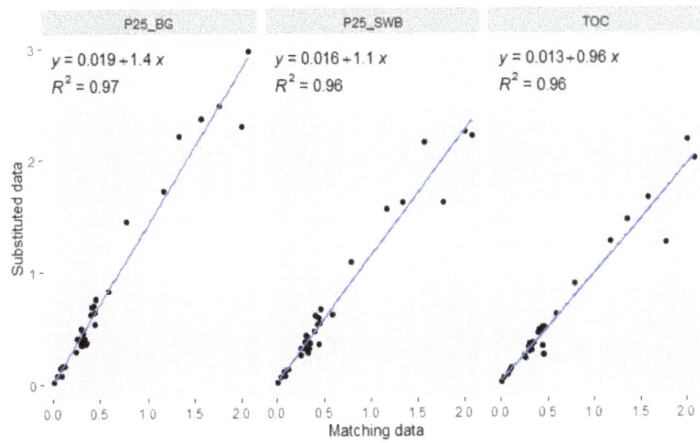

Figure 8. The comparison of the M-BAT calculated RCR for Zn between the matching data and substituted data (P25_BG—the 25th percentile of the available DOC data and/or the DOC data calculated by the TOC from the Bulgarian monitoring network results during 2020; P25_SWB—the 25th percentile of the DOC data calculated by the TOC for the respective water body; TOC—the matching DOC data calculated by TOC).

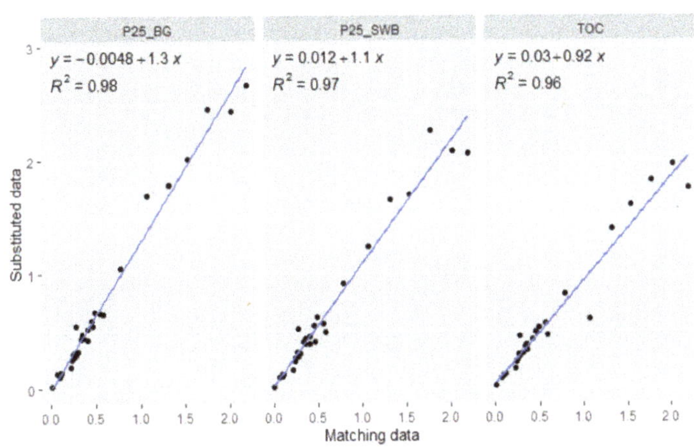

Figure 9. The comparison of the PNEC Pro calculated RCR for Zn between the matching data and substituted data (P25_BG—the 25th percentile of the available DOC data and/or the DOC data calculated by the TOC from the Bulgarian monitoring network results during 2020; P25_SWB—the 25th percentile of the DOC data calculated by the TOC for the respective water body; TOC—the matching DOC data calculated by TOC).

3.3.3. Comparisons for Lead

The user-friendly BLM models were implemented for the four water bodies (two reservoirs and two rivers) where the annual concentrations of Pb for 2020 exceeded the $EQS_{bioavailable}$ (1.2 μg/L) and the matching data (n = 22) for dissolved Pb, pH, dissolved Ca and DOC were available. The calculated RCRs are presented in the Supplementary Material (Figures S7–S9).

The comparison between the Pb bioavailability results obtained by the three BLM models using the matching and substituted DOC data are presented in Figures 10–12. The performed linear regression analysis only showed a good agreement between the RCRs obtained using the matching data and the DOC calculated by TOC. The best result achieved by the percentile approaches was for the BIO-MET calculated RCR using P25_SWB as the substituted data.

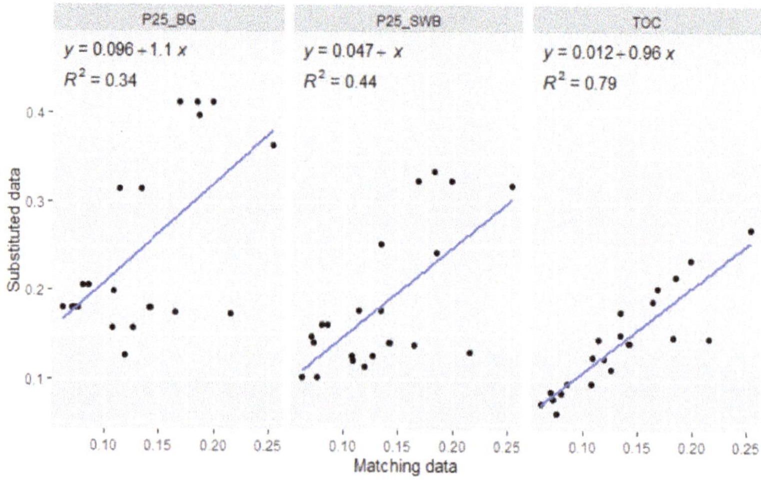

Figure 10. The comparison of the BIO-MET calculated RCR for Pb between the matching data and substituted data (P25_BG—the 25th percentile of the available DOC data and/or the DOC data calculated by the TOC from the Bulgarian monitoring network results during 2020; P25_SWB—the 25th percentile of the DOC data calculated by the TOC for the respective water body; TOC—the matching DOC data calculated by TOC).

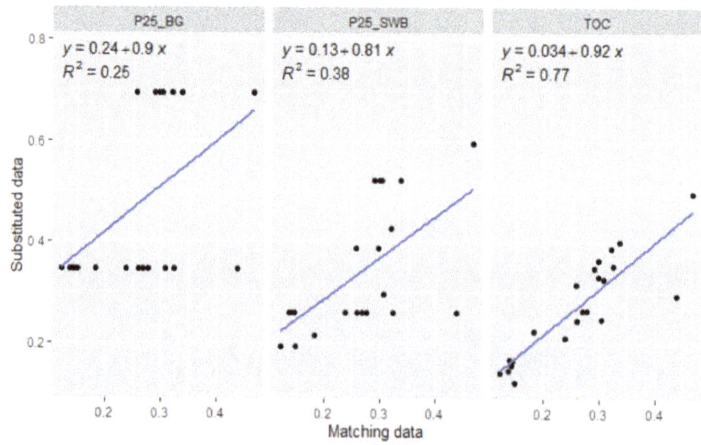

Figure 11. The comparison of the M-BAT calculated RCR for Pb between the matching data and substituted data (P25_BG—the 25th percentile of the available DOC data and/or the DOC data calculated by the TOC from the Bulgarian monitoring network results during 2020; P25_SWB—the 25th percentile of the DOC data calculated by the TOC for the respective water body; TOC—the matching DOC data calculated by TOC).

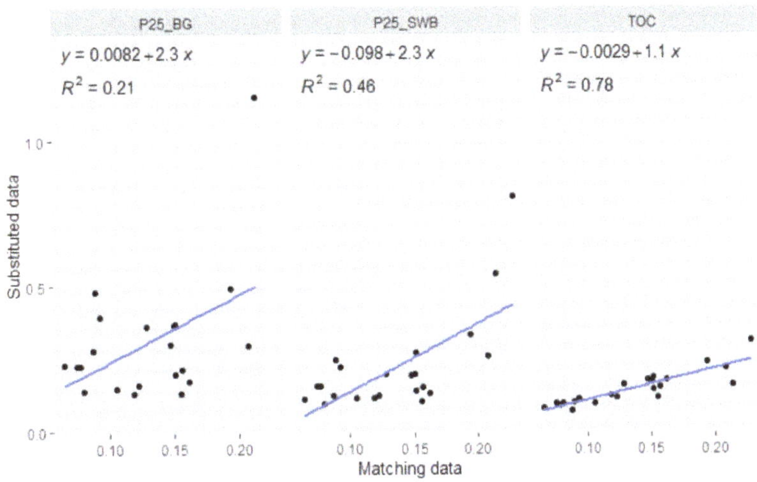

Figure 12. The comparison of the PNEC Pro calculated RCR for Pb between the matching data and substituted data (P25_BG—the 25th percentile of the available DOC data and/or the DOC data calculated by the TOC from the Bulgarian monitoring network results during 2020; P25_SWB—the 25th percentile of the DOC data calculated by the TOC for the respective water body; TOC—the matching DOC data calculated by TOC).

4. Discussion

The first step in the implementation of BLMs in the case of missing DOC monitoring data is the possible substitution of the available historical data from the Bulgarian Environmental Agency surface water monitoring programme. Contrary to the practice used in Great Britain [29], the correlation obtained between the DOC and $Fe_{dissolved}$ concentrations (Figure 2) shows that it cannot be used for the substitution of missing DOC data. The result obtained from the comparison between the available DOC and TOC data reveals that a conversion factor of DOC = 0.81 × TOC can be applied when DOC data are missing (Figure 3). This outcome is in agreement with the approach used in Finnish [34] and Swedish [11] networks, where TOC is traditionally measured and the applied conversion factors are 0.94 and 0.90, respectively. Similar to the Bulgarian monitoring data, a relationship (DOC = 0.83×TOC) is used for BLM purposes in the State of Oregon, USA [35]. Similar conversion factors were reported for rainfall and runoff samples in the USA (DOC = 0.84 × TOC) [36] and in Polish groundwater samples (DOC = 0.93×TOC) [37].

The second step of the proposed scheme for the implementation of BLMs deals with cases where no matching data are available, which is caused by the different monitoring frequencies used for metals and DOC. To overcome this problem, the environmental authorities of England and Wales [7] and France [18] and the EPA of the USA [38] propose the use of the 25th percentile of measured DOC concentrations as the input for the user-friendly BLMs. The proposed approaches in this study for dealing with the absence of matching data includes: (i) the substitution of missing DOC data using available TOC data (DOC = 0.81 × TOC) and (ii) the use of the 25th percentile of the data consisting of measured and substituted DOC values. This allows for the significant broadening of the implementation of BLMs in the Bulgarian monitoring network (Figure 1).

The comparison of the following suggested substitution approaches was performed for 13 water bodies where all data records were available.: (i) DOC calculated by TOC; (ii) 25th percentile of the joint (measured and calculated) Bulgarian monitoring network DOC data; and (iii) 25th percentile of the respective SWB with matching data.

The application of BLMs for assessing the Cu bioavailability showed the closest agreement between the BIO-MET and M-BAT results (Figures 4 and 5). While both methods confirmed the applicability of the proposed substitution approaches, those based on the 25th percentile provided more conservative estimates. It should be mentioned that the correlations were better between the BIO-MET calculated RCRs using matching data and data substituted by the 25th percentile. The PNEC Pro results were unsatisfactory for all BMLs and were an indication that none of the substitution approaches could be used for this user-friendly BLM (Figure 6).

The comparison between the Zn bioavailability results obtained by all BLMs using matching and all types of substituted data confirmed the applicability of all of the proposed approaches (Figures 7–9). Again, the most conservative estimates are provided by the 25th percentile of Bulgarian monitoring network data, followed by the 25th percentile of SWB data and the DOC data calculated by TOC.

The application of BLMs for the assessment of Pb bioavailability revealed a clear separation between the substituted data calculated by TOC on one side and the data substituted by the 25th percentile on the other (Figures 10–12). The comparison of the calculated RCRs by all BLMs between the matching data and the DOC data calculated by TOC showed very good correlations and slopes that were close to one. The comparison between the matching data and the 25th percentile substituted data yielded lower R^2 values and the slopes of the PNEC Pro comparison differed significantly from one. When using the BLM calculations in Tier 2, the results showed a great reduction in the exceedances of the samples (Table 3). For Cu, when BIO-MET was applied, 5 out of 63 samples were found to be exceedances when the measured DOC was used, 4 exceedances were found when TOC value was used (DOC = 0.81 × TOC), 8 exceedances were found when the 25th percentile of the DOC concentration was used and 7 exceedances were found when the 25th percentile of the converted TOC for the respective water body was used, in contrast to the direct comparison of the EQS to the measured copper values (57 out of 63).

Similar results were obtained using M-BAT (7, 6, 7 and 9 exceedances out of 63, respectively) and PNEC Pro (9, 8, 10 and 5 exceedances out of 63, respectively).

The results showed a reduction in the exceeding samples for BIO-MET (2, 5, 7 and 6, respectively), M-BAT (6, 6, 7 and 7, respectively) and PNEC Pro (6, 5, 7 and 6, respectively) compared with the direct comparison of the EQS to the zinc values (14 out of 34 exceeding values).

The results from the application of the BLM tools showed the greatest reduction in the exceeding samples for Pb—only one RCR value was found to be above one—when PNEC Pro was used with the 25th percentile of the Bulgarian DOC data. The direct comparison with the EQS (Tier 1) showed 8 exceedances out of 22 samples.

Generally, the user-friendly BLM tools returned a lower number of RCR exceedances when the measured DOC concentration was used compared with the other approaches, with BIO-MET returning the lowest number of exceedances.

Table 3. The exceedances for Cu, Zn, Ni and Pb in the studies of Bulgarian surface water bodies after the application of the BLMs.

Element	BLM	Data Used	Totals	Exceedances
EQS Cu = 1 μg/L		Measured Cu	63	56
	RCR BIO-MET	Measured DOC	63	5
		DOC = 0.81 × TOC	63	4
		25th percentile DOC (P25_BG)	63	8
		25th percentile TOC (P25_SWB)	63	7
	RCR M-BAT	Measured DOC	63	7
		DOC = 0.81 × TOC	63	6
		25th percentile DOC (P25_BG)	63	7
		25th percentile TOC (P25_SWB)	63	9
	RCR PNEC	Measured DOC	63	9
		DOC = 0.81 × TOC	63	8
		25th percentile DOC (P25_BG)	63	10
		25th percentile TOC (P25_SWB)	63	5
EQS Zn = 8 μg/L		Measured Zn	34	14
	RCR BIO-MET	Measured DOC	34	2
		DOC = 0.81 × TOC	34	5
		25th percentile DOC (P25_BG)	34	7
		25th percentile TOC (P25_SWB)	34	6
	RCR M-BAT	Measured DOC	34	6
		DOC = 0.81 × TOC	34	6
		25th percentile DOC (P25_BG)	34	7
		25th percentile TOC (P25_SWB)	34	7
	RCR PNEC	Measured DOC	34	6
		DOC = 0.81 × TOC	34	5
		25th percentile DOC (P25_BG)	34	7
		25th percentile TOC (P25_SWB)	34	6
EQS Pb = 1.2 μg/L		Measured Pb	22	8
	RCR BIO-MET	Measured DOC	22	0
		DOC = 0.81 × TOC	22	0
		25th percentile DOC (P25_BG)	22	0
		25th percentile TOC (P25_SWB)	22	0
	RCR M-BAT	Measured DOC	22	0
		DOC = 0.81 × TOC	22	0
		25th percentile DOC (P25_BG)	22	0
		25th percentile TOC (P25_SWB)	22	0
	RCR PNEC	Measured DOC	22	0
		DOC = 0.81 × TOC	22	0
		25th percentile DOC (P25_BG)	22	1
		25th percentile TOC (P25_SWB)	22	0

5. Conclusions

The metal bioavailability assessment by user-friendly BLMs requires inputs for the dissolved metal concentrations and the supporting physicochemical parameters, such as DOC, pH and dissolved Ca. The main obstacle in front of the environmental authorities for the implementation of BLMs is the lack of matching data, especially for DOC. In cases where the required data are not fully available, alternative approaches are needed to fill the missing inputs.

In this study, various alternative approaches were tested to support the environmental decision-making bodies. The proposed methodology included three approaches for the substitution of missing DOC data: (i) calculation from available TOC data; (ii) the 25th percentile of the joint (measured and calculated by TOC) Bulgarian monitoring network DOC data; and (iii) the 25th percentile of the joint DOC data for the investigated SWB. For the next step, the proposed substitution approaches were validated by a comparison with the available matching data from the Bulgarian monitoring surface waters network. The comparisons were performed for the three most widely used BLMs: BIO-MET; M-BAT; and PNEC Pro. The described methodology would allow the environmental authorities to estimate the bioavailability of certain metals and then choose the best possible substitution approach by the implementation of the most appropriate BLM, which would cover as many SWBs as possible without the matching data being available.

The application of the abovementioned methodology to the Bulgarian surface waters monitoring data outlined that the suitability of the substitution approaches decreases in the following order: DOC calculated by TOC > use of the 25th percentile of the data for the respective SWB > use of the 25th percentile of the Bulgarian monitoring network data. The results obtained by the implementation of BIO-MET rendered it the most appropriate tool for the bioavailability assessment of Cu, Zn and Pb in Bulgarian surface waters.

Supplementary Materials: The following are available online at https://www.mdpi.com/article/10.3390/w14020246/s1, Figure S1: The calculated RCR for Cu using the BIO-MET and measured DOC, the TOC, the 25th percentile of the Bulgarian DOC data and the 25th percentile of the DOC = 0.81 × TOC for the respective water body; Figure S2: The calculated RCR for Cu using the M-BAT and measured DOC, the TOC, the 25th percentile of the Bulgarian DOC data and the 25th percentile of the DOC = 0.81 × TOC for the respective water body; Figure S3: The calculated RCR for Cu using the PNEC Pro and measured DOC, the TOC, the 25th percentile of the Bulgarian DOC data and the 25th percentile of the DOC = 0.81 × TOC for the respective water body; Figure S4: The calculated RCR for Zn using the BIO-MET and measured DOC, the TOC, the 25th percentile of the Bulgarian DOC data and the 25th percentile of the DOC = 0.81 × TOC for the respective water body; Figure S5: The calculated RCR for Zn using the M-BAT and measured DOC, the TOC, the 25th percentile of the Bulgarian DOC data and the 25th percentile of the DOC = 0.81 × TOC for the respective water body; Figure S6: The calculated RCR for Zn using the PNEC Pro and measured DOC, the TOC, the 25th percentile of the Bulgarian DOC data and the 25th percentile of the DOC = 0.81 × TOC for the respective water body; Figure S7: The calculated RCR for Pb using the BIO-MET and measured DOC, the TOC, the 25th percentile of the Bulgarian DOC data and the 25th percentile of the DOC = 0.81 × TOC for the respective water body; Figure S8: The calculated RCR for Pb using the M-BAT and measured DOC, the TOC, the 25th percentile of the Bulgarian DOC data and the 25th percentile of the DOC = 0.81 × TOC for the respective water body; Figure S9: The calculated RCR for Pb using the PNEC Pro and measured DOC, the TOC, the 25th percentile of the Bulgarian DOC data and the 25th percentile of the DOC = 0.81 × TOC for the respective water body.

Author Contributions: Conceptualization, T.V. and S.T.; methodology, T.V. and S.T.; software, T.V.; formal analysis, T.V. and S.T.; investigation, T.V.; resources, T.V.; data curation, T.V.; writing—original draft preparation, T.V.; writing—review and editing, S.T.; visualization, T.V.; supervision, S.T.; project administration, T.V. and S.T.; funding acquisition, T.V. and S.T. All authors have read and agreed to the published version of the manuscript.

Funding: This research was funded by the Ministry of Education and Science, Bulgarian National Science Fund, Grant DN 19/15 and by the National Science Program "Environmental Protection and Reduction of Risks of Adverse Events and Natural Disasters", approved by the Resolution of the Council of Ministers № 577/17.08.2018 and supported by the Ministry of Education and Science (MES) of Bulgaria (Agreement № Д01-363/17.12.2020).

Institutional Review Board Statement: Not applicable.

Informed Consent Statement: Not applicable.

Data Availability Statement: The data presented in this study are available on request from the corresponding author.

Acknowledgments: The authors gratefully acknowledge the financial support of the Ministry of Education and Science, Bulgarian National Science Fund (Grant DN 19/15) and the National Science Program "Environmental Protection and Reduction of Risks of Adverse Events and Natural Disasters" (Agreement № Д01-363/17.12.2020).

Conflicts of Interest: The authors declare no conflict of interest. The funders had no role in the design of the study; in the collection, analyses or interpretation of data; in the writing of the manuscript; or in the decision to publish the results.

References

1. Smith, K.; Balistrieri, L.; Todd, A. Using biotic ligand models to predict metal toxicity in mineralized systems. *Appl. Geochem.* **2015**, *57*, 55–72. [CrossRef]
2. Thurman, E. *Organic Geochemistry of Natural Waters*; Martinus Nijhoff/Dr W. Junk Publishers: Dordrecht, The Netherlands, 1985; p. 497.
3. De Schamphelaere, K.; Janssen, C. Development and field validation of a biotic ligand model predicting chronic copper toxicity to Daphnia magna. *Environ. Toxicol. Chem.* **2004**, *23*, 1365–1375. [CrossRef]
4. Deleebeeck, N.; De Schamphelaere, K.; Janssen, C. A novel method for predicting chronic nickel bioavailability and toxicity to Daphnia magna in artificial and natural waters. *Environ. Toxicol. Chem.* **2008**, *27*, 2097–2107. [CrossRef] [PubMed]
5. Heijerick, D.; De Schamphelaere, K.; Van Sprang, P.; Janssen, C. Development of a chronic zinc biotic ligand model for Daphnia magna. *Ecotoxicol. Environ. Saf.* **2005**, *62*, 1–10. [CrossRef] [PubMed]
6. Di Toro, D.; Allen, H.; Bergman, H.; Meyer, J.; Paquin, P.; Santore, R. Biotic ligand model of the acute toxicity of metals. 1. Technical basis. *Environ. Toxicol. Chem.* **2001**, *20*, 2383–2396. [CrossRef]
7. Ardestani, M.; van Straalen, N.; van Gestel, C. The relationship between metal toxicity and biotic ligand binding affinities in aquatic and soil organisms: A review. *Environ. Pollut.* **2014**, *195*, 133–147. [CrossRef]
8. Traas, T.; Leeuwen, C. Ecotoxicological Effects. In *Risk Assessment of Chemicals*; Leeuwen, C., Vermeire, T., Eds.; Springer: Dordrecht, The Netherlands, 2007; p. 332.
9. Technical Guidance to Implement Bioavailability-Based Environmental Quality Standards for Metals. Available online: https://bio-met.net/wp-content/uploads/2016/10/FINAL-TECHNICAL-GUIDANCE-TO-IMPLEMENT-BIOAVAILABILITYApril-2015.pdf (accessed on 5 January 2022).
10. Sarathy, V.; Allen, H. Copper complexation by dissolved organic matter from surface water and wastewater effluent. *Ecotoxicol. Environ. Saf.* **2005**, *61*, 337–344. [CrossRef]
11. Hoppe, S.; Gustafsson, J.; Borg, H.; Breitholtz, M. Evaluation of current copper bioavailability tools for soft freshwaters in Sweden. *Ecotoxicol. Environ. Saf.* **2015**, *114*, 143–149. [CrossRef]
12. European Commission. Directive 2013/39/EU of the European Parliament and of the Council of 12 August 2013 Amending Directives 2000/60/EC and 2008/105/EC as Regards Priority Substances in the Field of Water Policy. *Off. J. Eur. Union* **2013**, *226*, 1–17.
13. Workshop on Metal Bioavailability under the Water Framework Directive: Policy. Science and Implementation of Regulatory Tools. Available online: https://bio-met.net/wp-content/uploads/2016/10/110905-Draft-Bioavailability-Meeting-Report-FINAL.pdf (accessed on 5 January 2022).
14. Merrington, G.; Peters, A. The Importance of Dissolved Organic Carbon in the Assessment of Environmental Quality Standard Compliance for Copper and Zinc. Available online: https://wfduk.org/sites/default/files/Media/DOC%20Report%20-%20final.pdf (accessed on 5 January 2022).
15. Cousins, A.; Jönsson, A.; Iverfeldt, A. *Testing the Biotic Ligand Model for Swedish Surface Water Conditions—A Pilot Study to Investigate the Applicability of BLM in Sweden*; Swedish Environmental Research Institute Ltd.: Stockholm, Sweden, 2009.
16. Hoppe, S.; Lithner, G.; Borg, H. *Utvärdering av Användbarheten av BLM i Svenska Vatten*; Stockholms Universitet: Stockholm, Sweden, 2009.
17. Geoffroy, L.; Tack, K.; Andres, S. EQS for a metal: Is it possible to determine a lonely legal value? Interest of the Biotic Ligand Model (BLM)? In Proceedings of the Communication in the SETAC Annual Meeting, Seville, Spain, 23–27 May 2010.
18. Tack, K. *Prise en Compte de la Biodisponibilité des Métaux Selon la DCE: Guide Méthodologique*; Institut National de l'Environnement industriel et des Risques: Verneuil-en-Halatte, France, 2012.
19. Hommen, U.; Rüdel, H. Sensitivity Analysis of Existing Concepts for Application of Biotic Ligand Models (BLM) for the Derivation and Application of Environmental Quality Standards for Metals and Evaluation of the Approaches with Appropriate Monitoring Data Sets from German Waters. Final Report for FKZ 363 01 352, Schmallenberg, Germany. 2012. Available online: https://www.researchgate.net/profile/Heinz-Ruedel/publication/260337187_Sensitivity_analysis_of_existing_concepts_for_application_of_biotic_ligand_models_BLM_for_the_derivation_and_application_of_environmental_quality_standards_for_metals_and_evaluation_of_the_approaches_/links/02e7e530cdbd3c81da000000/Sensitivity-analysis-of-existing-concepts-for-application-of-biotic-ligand-models-BLM-for-the-derivation-and-application-of-environmental-quality-standards-for-metals-and-evaluation-of-the-approaches.pdf (accessed on 5 January 2022).
20. Guidance Document No. 27 Technical Guidance for Deriving Environmental Quality Standards. In Common Implementation Strategy for the Water Framework Directive (2000/60/EC) Technical Report, Update Version. Available online: https://rvs.rivm.nl/sites/default/files/2019-04/Guidance%20No%2027%20-%20Deriving%20Environmental%20Quality%20Standards%20-%20version%202018.pdf (accessed on 5 January 2022).
21. European Parliament, Council of the European Union. Directive 2000/60/EC of the European Parliament and of the Council establishing a framework for Community action in the field of water policy. *Off. J. Eur. Communities* **2000**, *327*, 1–72.
22. Common Implementation Strategy for the Water Framework Directive (2000/60/EC). Guidance Document No. 19 Guidance on Surface Water Chemical Monitoring under the WFD. Available online: https://circabc.europa.eu/sd/a/e54e8583-faf5-478f-9b11-41fda9e9c564/Guidance (accessed on 5 January 2022).

23. European Commission. Directive 2009/90/EC of 31 July 2009 laying down, pursuant to Directive 2000/60/EC of the European Parliament and of the Council, technical specifications for chemical analysis and monitoring of water status. *Off. J. Eur. Union* **2009**, *201*, 36–38.
24. European Commission. Directive 2008/105/EC of the European Parliament and of the Council of 16 December 2008 on environmental quality standards in the field of water policy, amending and subsequently repealing Council Directives 82/176/EEC, 83/513/EEC, 84/156/EEC, 84/491/EEC, 86/280/EEC and amending Directive 2000/60/EC of the European Parliament and of the Council. *Off. J. Eur. Union* **2008**, *348*, 84–97.
25. Nriagu, J.; Lawson, G.; Wong, H.; Azcue, J. A Protocol for Minimizing Contamination in the Analysis of Trace Metals in Great Lakes Waters. *J. Great Lakes Res.* **1993**, *19*, 175–182. [CrossRef]
26. Horowitz, A.; Lum, K.; Garbarino, J.; Hall, G.; Lemieux, C.; Demas, C. Problems Associated with Using Filtration to Define Dissolved Trace Element Concentrations in Natural Water Samples. *Environ. Sci. Technol.* **1996**, *30*, 954–963. [CrossRef]
27. Tipping, E.; Corbishley, H.; Koprivnjak, J.; Lapworth, D.; Miller, M.; Vincent, C.; Hamilton-Taylor, J. Quantification of natural DOM from UV Absorption at Two Wavelengths. *Environ. Chem.* **2009**, *6*, 472–476. [CrossRef]
28. De Schamphelaere, K.; Vasconcelos, F.; Tack, F.; Allen, H.; Janssen, C. The effect of dissolved organic matter source on acute copper toxicity to Daphnia magna. *Environ. Toxicol. Chem.* **2004**, *23*, 1248–1255. [CrossRef]
29. Comber, S.; Georges, K. Tiered Approach to the Assessment of Metal Compliance in Surface Waters; Environment Agency, Bristol, UK. Available online: https://www.wfduk.org/sites/default/files/Media/Environmental%20standards/Tiered%20approach%20metal%20compliance.pdf (accessed on 5 January 2022).
30. Peters, A.; Crane, M.; Adams, W. Effects of iron on benthic macroinvertebrate communities in the field. *Bull. Environ. Contam. Toxicol.* **2011**, *86*, 591–595. [CrossRef] [PubMed]
31. Mueller, K.; Fortin, C.; Campbell, P. Spatial Variation in the Optical Properties of Dissolved Organic Matter (DOM) in Lakes on the Canadian Precambrian Shield and Links to Watershed Characteristics. *Aquatic. Geochem.* **2012**, *18*, 21–44. [CrossRef]
32. Schwartz, M.; Curtis, P.; Playle, R. Influence of natural organic matter source on acute copper, lead, and cadmium toxicity to rainbow trout (*Oncorhynchus mykiss*). *Environ. Toxicol. Chem.* **2004**, *23*, 2889–2899. [CrossRef]
33. DV. Ordinace N 4 for characterisation of surface waters. *DV* **2013**, *22*, 9–46. (In Bulgarian)
34. Mattsson, T.; Kortelainen, P.; Räike, A. Export of DOM from Boreal Catchments: Impacts of Land Use Cover and Climate. *Biogeochemistry* **2005**, *76*, 373–394. [CrossRef]
35. DEQ. *Implementation of the Freshwater Aquatic Life Water Quality Standards for Copper*; Oregon Department of Environmental Quality: Portland, OR, USA, 2016.
36. Kalev, S.; Toor, G. Concentrations and Loads of Dissolved and Particulate Organic Carbon in Urban Stormwater Runoff. *Water* **2020**, *12*, 1031. [CrossRef]
37. Gooddy, D.; Hinsby, K. Organic Quality of Groundwaters. In *Natural Groundwater Quality*; Edmunds, W., Shand, P., Eds.; Blackwell: Oxford, UK, 2008; pp. 59–72.
38. *Draft Technical Support Document: Recommended Estimates for Missing Water Quality Parameters for Application in EPA's Biotic Ligand Model*; US EPA: Washington, DC, USA, 2016. Available online: https://archive.epa.gov/epa/sites/production/files/2016-02/documents/draft-tsd-recommended-blm-parameters.pdf (accessed on 5 January 2022).

water

MDPI

Article

Dissolved Metal (Fe, Mn, Zn, Ni, Cu, Co, Cd, Pb) and Metalloid (As, Sb) in Snow Water across a 2800 km Latitudinal Profile of Western Siberia: Impact of Local Pollution and Global Transfer

Ivan V. Krickov [1,*], Artem G. Lim [1], Vladimir P. Shevchenko [2,*], Sergey N. Vorobyev [1], Frédéric Candaudap [3] and Oleg S. Pokrovsky [3,4,*]

1 BIO-GEO-CLIM Laboratory, Tomsk State University, 36 Lenina, 634050 Tomsk, Russia; lim_artyom@mail.ru (A.G.L.); soil@green.tsu.ru (S.N.V.)
2 Shirshov Institute of Oceanology, Russian Academy of Sciences, 36 Nakhimovsky Pr., 117997 Moscow, Russia
3 Geosciences Environment Toulouse, UMR 5563 CNRS, University of Toulouse, 14 Avenue Edouard Belin, 31400 Toulouse, France; Candaudap@get.omp.eu
4 N Laverov Center for Integrated Arctic Research, Institute of Ecological Problems of the North, 23 Nab Severnoi Dviny, RAS, 163000 Arkhangelsk, Russia
* Correspondence: krickov_ivan@mail.ru (I.V.K.); vshevch@ocean.ru (V.P.S.); oleg.pokrovski@get.omp.eu (O.S.P.)

Citation: Krickov, I.V.; Lim, A.G.; Shevchenko, V.P.; Vorobyev, S.N.; Candaudap, F.; Pokrovsky, O.S. Dissolved Metal (Fe, Mn, Zn, Ni, Cu, Co, Cd, Pb) and Metalloid (As, Sb) in Snow Water across a 2800 km Latitudinal Profile of Western Siberia: Impact of Local Pollution and Global Transfer. *Water* **2022**, *14*, 94. https://doi.org/10.3390/w14010094

Academic Editor: Domenico Cicchella

Received: 1 December 2021
Accepted: 31 December 2021
Published: 4 January 2022

Abstract: Snow cover is known to be an efficient and unique natural archive of atmospheric input and an indicator of ecosystem status. In high latitude regions, thawing of snow provides a sizable contribution of dissolved trace metals to the hydrological network. Towards a better understanding of natural and anthropogenic control on heavy metals and metalloid input from the atmosphere to the inland waters of Siberian arctic and subarctic regions, we measured chemical composition of dissolved (<0.22 μm) fractions of snow across a 2800 km south–north gradient in Western Siberia. Iron, Mn, Co, Ni, and Cd demonstrated sizable (by a factor of 4–7) decrease in concentration northward, which can be explained by a decrease in overall population density and the influence of dry aerosol deposition. Many elements (Mn, Ni, Cu, Cd, Pb, As, and Sb) exhibited a prominent local maximum (a factor of 2–3) in the zone of intensive oil and gas extraction (61–62° N latitudinal belt), which can be linked to gas flaring and fly ash deposition. Overall, the snow water chemical composition reflected both local and global (long-range) atmospheric transfer processes. Based on mass balance calculation, we demonstrate that the winter time atmospheric input represents sizable contribution to the riverine export fluxes of dissolved (<0.45 μm) Mn, Co, Zn, Cd, Pb, and Sb during springtime and can appreciably shape the hydrochemical composition of the Ob River main stem and tributaries.

Keywords: snow; heavy metal; trace element; river flux; gas flaring; pollution; Western Siberia

1. Introduction

Snow cover is known to be an efficient and unique natural archive and indicator of ecosystem status and atmospheric input [1–6]. Snow delivers from the atmosphere to the ground various soluble compounds, including metal pollutants [7,8]. Given its persistence over the frozen period of the year, snow records integral atmospheric input of solutes and delivers invaluable information on both short-range and long-range atmospheric transfer [9–17].

The importance of studying the chemical composition of dissolved (<0.22 μm or 0.45 μm) fraction of the snow water is that the snow input represents a non-negligible source of metal and metalloids in the river water and contributes to trace element delivery from the land to the ocean. Therefore, in order to evaluate the current status and possible future changes in dissolved metal export flux from the land to the ocean, high spatial resolution measurement of snow chemical composition are needed. This is especially true for arctic and subarctic regions, which are subjected to unprecedented changes due to climate warming and susceptible to strong anthropogenic impact due to their vulnerability and low capacity to recover. According to the Arctic Monitoring and Assessment Programme [18], anthropogenic emissions are responsible for more than 50% of the total trace metal content in Arctic soils. For example, the Monchegorsk and Norilsk smelters affect the environment across several thousands of kilometers [19–22]. Among all Arctic regions, Western Siberia offers a unique possibility of performing large-scale spatial sampling due to its well-developed road infrastructure, linked to substantial oil and gas exploration activity. On the one hand, the Western Siberian Lowland (WSL), notably its northern part, is relatively weakly affected by human and heavy industrial activity (large towns, smelters, ore processing plants, forestry, and paper industry), compared with Northern Europe or NW European Russia. On the other hand, hydrocarbon extraction industry, notably accompanied by gas flaring, is capable of delivering various pollutants including metal and metalloids to adjacent territories [23]. These anthropogenic impacts on snow chemical composition can be further complemented by far-range atmospheric transfer of solid aerosols from southern desert and semi-desert regions of Central Asia [24].

In the present study, we used existing road infrastructure to sample a sizable (2800 km from the south to the north) transect of the WSL. We collected 36 integral (from the surface to the ground) snow cores and attempted to distinguish the effect of local (gas flaring, urban settlements) and global (long-range) atmospheric transfer. Via applying a mass balance approach for depth-averaged metal concentration, we estimated possible impact of winter time atmospheric input on metal (Fe, Mn, Ni, Co, Zn, Cd, and Pb) and metalloid (As and Sb) fluxes during the spring flood in the Ob River. We concluded that there is a sizable effect of atmospheric deposition on riverine export fluxes of some trace metals in Western Siberia.

Note that, compared with a previous study of snow chemical composition (upper 0–5 cm) in another part of Western Siberia [24], the originality of the present study consists in (i) sampling of much larger (~2800 km compared with 1700 km in previous work) latitudinal gradient in relatively pristine zones comprising forest, forest tundra, and tundra within the permafrost-free, discontinuous, and continuous permafrost regions; (ii) assessment of dissolved trace elements in integral (from the surface to the ground) snow samples.

2. Materials and Methods

The snow cores were collected along the latitudinal gradient south $\rightarrow$ north, from the vicinity of the Barnaul city (zone of forest steppe) to the Ob estuary (tundra zone) from 8 February 2020 to 19 February 2020. The integral 50–70 cm (from the surface to the ground, except the bottom 2–3 cm layer) snow core was collected at 36 locations, which were evenly distributed along the 2800 km transect (Figure 1). All sampled points were located more than 500 m from the road.

Sampling was performed using a metal-free technique, in a protected environment, using only plastic equipment and vinyl single-use gloves. First, a pit in the snow was dug with a plastic shovel to ensure the representability of the studied location and the evenness of the ground. Then, a two-person team sampled the snow in such a way that no contamination from external surfaces (other than the plastic shovel) was possible. The plastic shovel was inserted 2 cm above the ground in order to avoid touching the ground during sampling. The snow was collected via a 10 cm diameter PVC tube which was inserted into the snow until it made contact with the shovel surface, which was a few cm above the litter, to avoid any contamination from plant debris. Prior to sampling, the tube and the shovel were "rinsed" several times with fresh snow via inserting them into the snow cover 1–2 m from the sampling site.

Approximately 5 L of snow was collected into single-use polyethylene bags, which were double closed with a PVC collar. The polyethylene bags were thoroughly washed with 1 M HCl and abundant MilliQ water in a clean room, class A 10,000. Collected snow samples were transported to the laboratory in the frozen state and processed within 2 weeks after sampling. In the laboratory, the snow was melted at 18–20 °C of ambient temperature, and immediately processed for analyses and filtration. The pH and conductivity were measured on unfiltered samples using Hanna portable instruments. The snow water was filtered through acetate cellulose filters (Millipore, 47 mm diameter) of 0.22 μm pore size. Blanks of MilliQ water from the clean room were also placed in polyethylene bags for the same amount of time as the melted snow (<3 h) and processed via filtration similar to the snow samples. The filtrates were acidified with double distilled HNO_3 acid and stored in pre-cleaned HDPE tubes for ICP MS analysis.

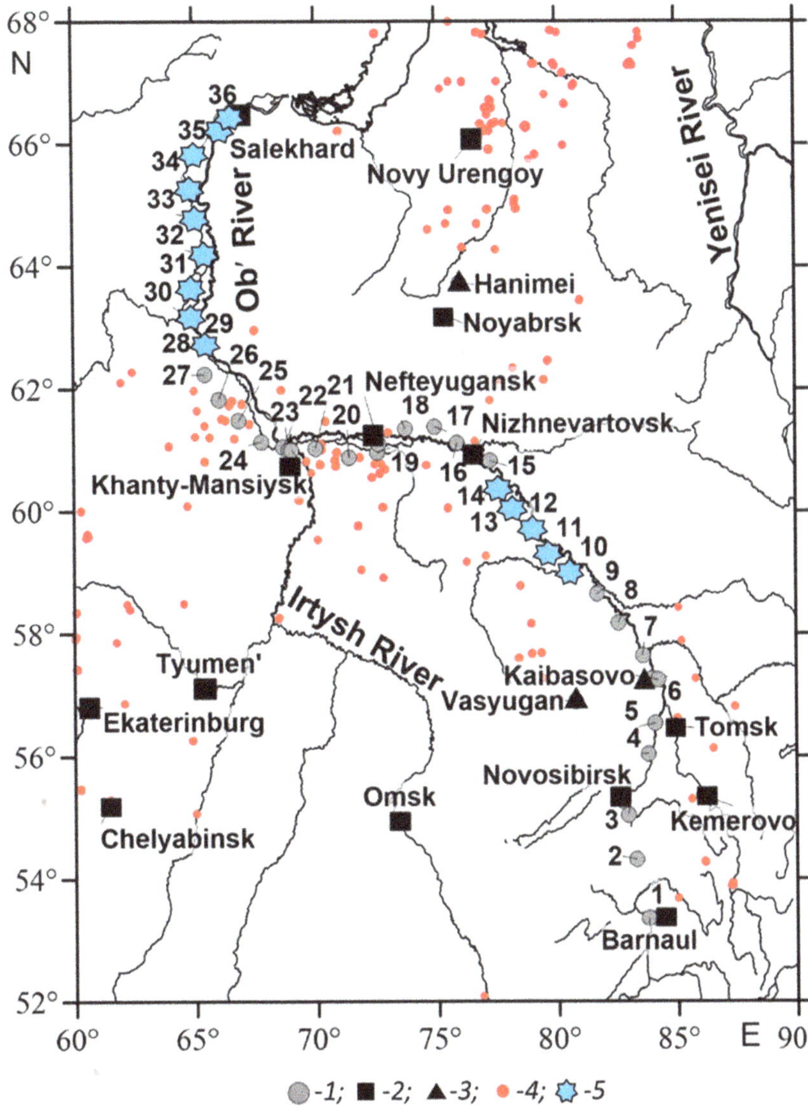

Figure 1. Studied area of the Western Siberian Lowland (WSL): 1—position of the sampling sites; 2—large cities; 3—scientific stations of Tomsk State University; 4—the position of gas flaring from site (https://firms.modaps.eosdis.nasa.gov/download/accessed on 25 July 2020); 5—sampling points taken from the so-called 'zimnik' which has no solid cover and is used only several months per year, with rather low traffic density.

Selected trace elements (TE) in the dissolved fractions were measured without pre-concentration with an ICP-MS Triple Quad using both Ar and He modes to diminish the interference. Indium and rhenium were used as internal standards at concentrations of ~3 µg/L and corrections for oxide and hydroxide ions were made for the REEs and the other trace elements [25]. The typical uncertainty for elemental concentration measurement ranged from 3–5% at 0.1–100 µg/L to 5–10% at 0.001–0.01 µg/L. During the ICP-MS analyses, the international geostandards SLRS-6 (Riverine Water References Material for Trace Metals certified by the National Research Council of Canada) were measured after every 20 samples to assess the validity and reproducibility of the analyses. All certified trace element (Fe, Mn, Ni, Co, Cu, Zn, Cd, Pb, As, and Sb) concentrations of the SLRS-6 standard (e.g., [26]) and the measured concentrations agreed with an uncertainty of 10–20%. For all trace elements except Zn, the concentrations in the blanks were below or comparable with analytical detection limits (~0.1 ng L^{-1}). These values were at least 10 times lower than the average concentration of trace elements in snow samples. Zinc exhibited non-negligible concentrations in the blanks (0.03–1.3 µg L^{-1}); however, these concentrations were several times lower than those in snow water samples and as such Zn concentration corrections did not exceed 10% of the measured values.

Statistical analysis of the average values and the link between element concentration in the dissolved fraction and the latitude was carried out by comparison of different sampling locations using ANOVA and H-criterion of the Kruskal–Wallis and Mann–Whitney U tests, which allowed one to estimate the difference between two independent sets of data based on one given parameter following the approaches developed for lakes and rivers of Western Siberia [27–30]. The normality of data distribution was verified by a Shapiro–Wilk test; because the data were not distributed normally, we used nonparametric statistics. To identify potential drivers of snow water chemical composition, we performed a principal component analysis (PCA) in XLSTAT which is a statistical software that works as an add-on to Excel.

3. Results and Discussion

3.1. Spatial Variation and Possible Sources of Trace Element Concentration in the Snow

The latitude-averaged concentrations of dissolved and particulate fraction of snow samples are listed in Table 1. Concentrations of dissolved element in individual samples are listed in Table S1 of the Supplementary Materials.

Table 1. Physicochemical properties of dissolved (<0.22 µm) fraction of the snow water and mean (±SD) trace element concentrations (n = 36) in Western Siberia latitudinal transect.

pH	SC, µS cm^{-1}	Mn µg/L	Fe µg/L	Co µg/L	Ni µg/L	Cu µg/L	Zn µg/L	As µg/L	Cd µg/L	Sb µg/L	Pb µg/L
4.57 ± 0.15	10.2 ± 2.05	1.65 ± 1.55	2.86 ± 2.39	0.013 ± 0.014	0.057 ± 0.036	0.236 ± 0.167	8.32 ± 13.7	0.324 ± 0.192	0.030 ± 0.012	0.029 ± 0.039	0.670 ± 0.315

The specific conductivity (S.C.) ranged between 10–14 µS cm^{-1} in the southern part of the profile (54–62° N) and decreased to 6–8 µS cm^{-1} in the northern part of the profile. The concentration of solid particles [31] ranged from 1–4 mg/L in the southern part and remained relatively stable (1.27 ± 1.21 mg/L) in the northern part of the Ob River basin. The pH did not exhibit any systematic variation with latitude (4.53 ± 0.17, median ± IQR), Figure S1.

Iron, Mn, Co, Ni, and Cd demonstrated sizable (by a factor of 4–7) decrease in concentration northward (Figure 2a–e). This can be tentatively explained by a combination of both the decrease in overall population density and the degree of influence of dry aerosol deposition. Similar explanation has been put forward during our recent study of the uppermost (0–5 cm) snow cover in another latitudinal profile of Western Siberia [32]. Many elements (Ni, Mn, Cd, Cu, As, Sb, and Pb) exhibited a prominent local maximum (a factor of 2–3, see Figures 2b–d and 3a–d) in the zone of intensive oil and gas extraction (61–62° N latitudinal belt), which can be linked to gas flaring and fly ash deposition.

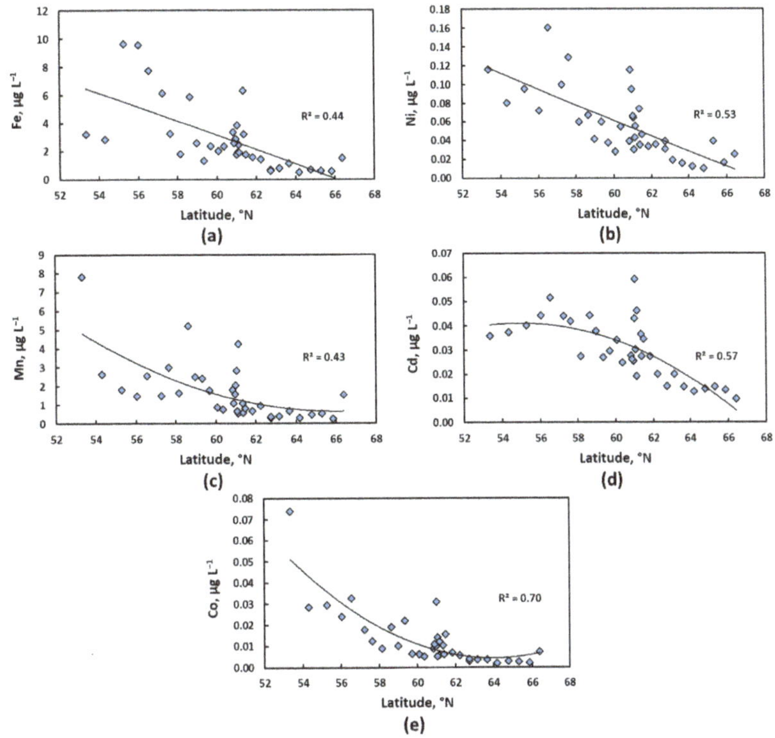

Figure 2. Examples of dissolved (<0.22 μm) metal concentrations in snow water as a function of latitude across the WSL. Fe (**a**); Ni (**b**); Mn (**c**); Cd (**d**); Co (**e**).

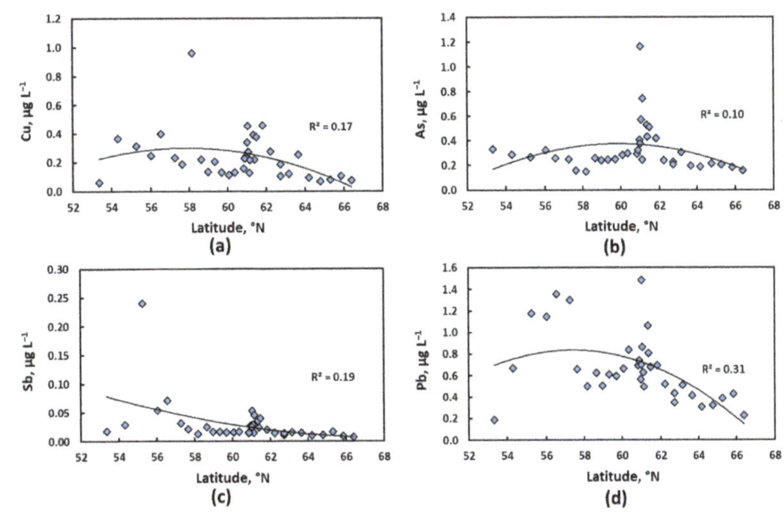

Figure 3. Examples of dissolved (<0.22 μm) metal concentrations in snow water as a function of latitude in the WSL. Cu (**a**); As (**b**); Sb (**c**); Pb (**d**).

Given that the soluble salt content (reflected by S.C.) exhibited much lower decrease northward compared with divalent metals (compare Figure S1e, Figures 2 and 3), the latter are unlikely to originate from marine aerosols or soluble fraction of carbonate minerals. Such carbonate minerals (calcite and dolomite) are known to be present in Western Siberian solid aerosols and likely originate from far-range atmospheric transfer from the Kazakhstan and Mongolia desert regions [32]. Therefore, we hypothesize a local anthropogenic rather than global source of dissolved divalent metals (Mn, Co, Ni, Cd, and Pb). This source is most likely the products of flying ash dissolution or desorption from some solid particles. Note here that a northward decrease in concentration of solid particles was more pronounced than that of S.C. (a factor of 3.1 and 1.4, respectively).

In our previous study [31], we presented the variation of insoluble particles in the snowpack of the Ob River basin and showed that particulate matter consisted of biogenic debris and spores (evenly distributed over the transect) and lithogenic minerals (plagioclase and clay minerals) preferentially enriched in the southern part of the transect. The carbonate minerals (calcite and dolomite) were also reported in the upper layer of the snow pack from Western Siberia [32]. Note that in a previous study of the surface (0–5 cm) snow layer across another transect of the WSL, we found that Ca, Mg, Sr, Mn, and Co increase their concentration with an increase in particle concentration by a factor of maximum 10 [32]. A similar pronounced effect can be observed in the present Table S1, which encompasses the full depth of the snow core and thus represents more integral assessment of leaching potential of these elements from the mineral particles.

In the transect studied in this work, the proportion of anthropogenic particles (fly ash and black carbon) did not exhibit any systematic latitudinal pattern but was the highest near big towns and areas of hydrocarbon production [31]. In accord with these observations, here we hypothesize that a general northward decreasing trend in divalent metal (Mn, Co, Ni, and Cd) reflects possible leaching of these elements from aerosol minerals and desorption from anthropogenic particles linked to cities. In contrast, local maxima in concentration of elements not showing any systematic latitudinal trend (for As, Sb, Zn, Cu, and Pb) or elements exhibiting such a trend (Fe, Ni, and Cd) likely represent the point source of atmospheric pollution such as gas flaring. This is further confirmed by positive correlations ($p < 0.05$) between dissolved elements (Mn, Fe, Co, Ni, Zn, Cd, and Sb) and solid particle concentration in snow (Table S2, Figure S2).

Note that, in another part of Western Siberia, the surface (0–5 cm) layer of the snow pack sampled in 2014 demonstrated a single maximum of Sb, Cd, and Ni concentrations at c.a. 63–65° N, whereas As exhibited two maxima, at 63.5° N and 67.5° N [32]. Similar to the present study which demonstrated maxima of As, Ni, and Cd at 61° N, these patterns most likely reflect prominent ground source of local pollution such as gas flaring. To further examine these impacts, we tested a quantitative link between the distance to the gas flaring site and the element concentration in the snow water (Figure S3). None of the studied elements yielded significant ($p > 0.05$) correlation with the distance to the gas flare position. Note, however, that in this treatment we could not take into account the number of different gas flares around the sampling point and the dominant wind direction. Furthermore, one can expect a strong interference of gas flares' impact with that of large towns, especially in the southern part of Western Siberia.

Because of its exceptionally flat orographic context, extensive vegetation cover, and relative remoteness from the Arctic Coast, the atmospheric precipitates in winter are likely to bear a signature of remote desert and semi-desert regions of Central Asia (chiefly Kazakhstan and partially the Gobi desert). At the same time, the local centers of potential pollution include: (i) permanent (asphalt) roads (only in the southern part of studied transect, south of Khanty-Mansiisk); (ii) large towns with substantial heating centrals (operating on diesel and gas); and point gas flaring stations, especially abundant between Strezhevoy (station 14) and Priobie (station 28).

To better constrain possible local sources of pollution, we defined the distance to the nearest human-related object (gas flare, small settlement (within first 20 km from the sampling point), and large town) taking into account the dominant wind direction during winter (Table S3). In this case, a significant ($p < 0.05$; $R^2 > 0.3$) decrease in five element

(Fe, Co, Ni, Mn, and Cd) concentrations was observed as illustrated in Figure 4. The concentration of other trace metals (Cu, Zn, and Pb) and metalloids (As and Sb) did not show any link to the distance to local pollution. We thus suggest that these elements are essentially controlled by far-range atmospheric transfer and/or by multiple sources of local (<20 km) and near-local (20–100 km) pollution.

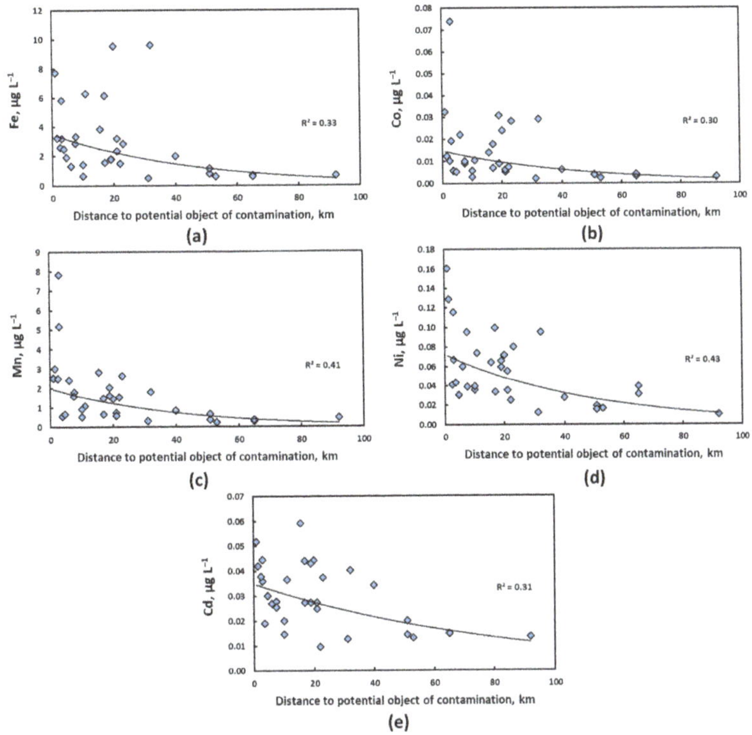

Figure 4. Examples of dissolved (<0.22 μm) metal concentrations in snow water as a function of distance to potential object of contamination in the WSL. Fe (**a**); Co (**b**); Mn (**c**); Ni (**d**); Cd (**e**).

A PCA treatment demonstrated two groups of parameters according to their spatial distribution in the snow water, although of low explicatory capacity (Figure S4, Table S4). The first group comprised Mn and Co solid particle concentration and latitude. The second factor acted on Fe, Cd, and Pb, and specific conductivity.

3.2. Assessment of Possible Snow Thaw Impact on Trace Element Concentration and Export by WSL Rivers

Obtained concentrations of dissolved fraction in the snow water can be compared with typical concentrations of trace elements in river water of the Ob River and tributaries, sampled during the end of the spring flood period [33]. This comparison (Figure 5, Table S5) revealed three groups of elements in the snow water depending on their potential to affect the riverine concentration. Iron, Co, and Ni in snow exhibited concentrations which were 10–15% lower than those in the Ob River water. Therefore, snowpack deposition of these metals could not appreciably affect their riverine export. In the northern sections of the transect, Mn, Co, Cu, As, and Sb from the snow could contribute between 10–30% to the river water concentrations. Finally, Zn, Cd, and Pb across the entire transect, as well as Mn and Co in the southern section of the transect, exhibited comparable or exceeded the concentration of these elements in the river water. This comparison illustrates the important possible contribution of snow water to the riverine transport of trace metals.

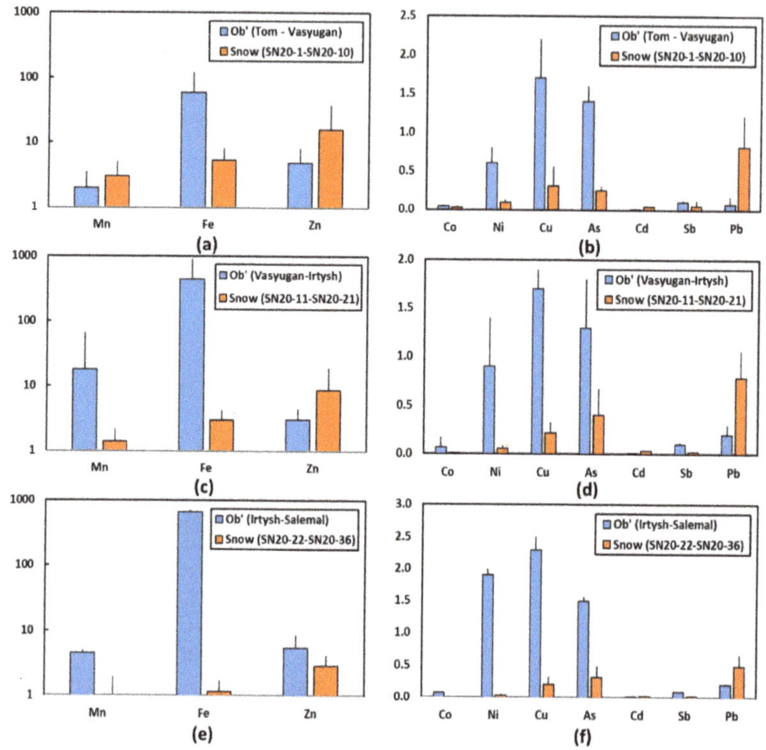

Figure 5. Histogram of metal concentrations in snow and river water (mean ± sd) of Western Siberia: (**a**) concentration of Mn, Fe, and Zn (SN20-1–SN20-10); (**b**) concentration of Co, Ni, Cu, As, Cd, Sb, and Pb (SN20-1–SN20-10); (**c**) concentration of Mn, Fe, and Zn (SN20-11–SN20-21); (**d**) concentration of Co, Ni, Cu, As, Cd, Sb, and Pb (SN20-11–SN20-21); (**e**) concentration of Mn, Fe, and Zn (SN20-22–SN20-36); (**f**) concentration of Co, Ni, Cu, As, Cd, Sb, and Pb (SN20-22–SN20-36).

Overall, the snow water chemical composition reflects both local and global (long-range) atmospheric transfer processes. A northward decrease in Fe and divalent transition metals (Mn, Co, and Ni) and Cd in the snow water contrasts the latitudinal pattern of these elements in the Ob River and tributaries. The latter demonstrates a sizable increase in concentrations of Fe and divalent transition metals from the south to the north [33]. To quantify the snow water input to the springtime dissolved flux of the Ob River water, we accounted for the river runoff in spring and the snow pool of Western Siberia. The water stock in snow (in mm snow water accumulated during winter) is fairly well known for Western Siberia [32,34] and ranges from 110–120 mm in the zone 56–60° N to 140–150 mm in the northern part of the Ob River basin (60–68° N). The springtime river runoff (in mm during May and June) was approximated to be the same as for medium and small rivers of Western Siberia, as calculated in [35]. Here, we considered the average concentration of trace elements in the snow water in the southern (<60° N) and the northern (>60° N) parts of the Ob River basin.

The ratios of river fluxes in May–June to the amount of elements accumulated in the snow stock on the same territory are presented in the form of histograms (Figure 6). In the southern, permafrost-free zone, Mn, Co, Zn, Cd, Pb, and Sb fluxes in rivers can be provided essentially by snowmelt. The riverine flux of Zn, Cd, and Pb can be entirely controlled by snowmelt in the northern, discontinuous and continuous permafrost zones (north of 60–62° N). These comparisons demonstrate a paramount impact of wintertime atmospheric depositions on freshet riverine fluxes of dissolved metals in the WSL. Because the flood period accounts for a sizable fraction of total annual lateral export of solutes in general and trace metals in particular [35], when foreseeing the consequences of climate change on river fluxes of metals, one has to account for collateral changes in snow quantity and chemical composition. This illustrates the intrinsic complexity of the response of elementary fluxes from the WSL to the Arctic Ocean due to ongoing environmental changes.

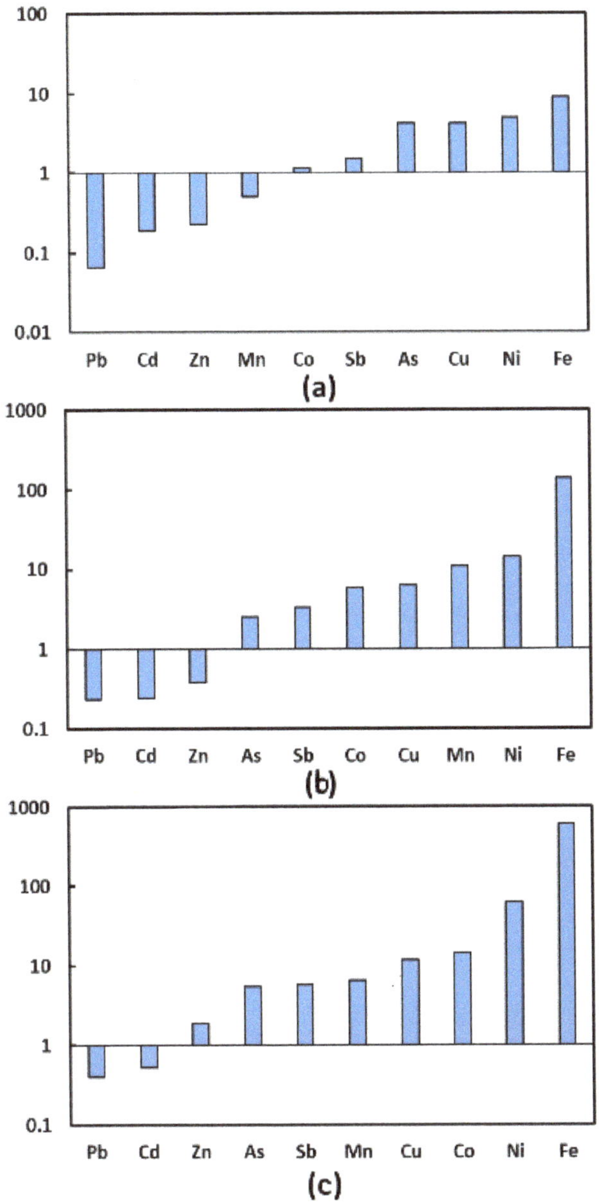

Figure 6. The ratio of mean dissolved flux of rivers in three latitudinal zones (53–60 (**a**), 60–62 (**b**), and 62–66° N (**c**)) of the WSL to the stock of dissolved fraction of snow. For this calculation, the snow volume (in millimeters of water) accumulated over full winter and mean river runoff during May and June were used.

4. Conclusions

The chemical composition of the full depth of the snow pack was studied across a 2800 km latitudinal gradient of the Ob River basin. There was a decreasing trend of Mn, Fe, Co, Ni, and Cd concentration in the snow water (<0.45 µm) northward. Some divalent metals (Cu, Ni, Pb, and Cd) and metalloids (As, Sb) demonstrated a local enrichment at 61–62° N, presumably originated from anthropogenic sources of pollution. However, we could not establish a quantitative link between the distance to gas flaring sites and the element concentration in the snow water. This may be explained by strong interferences with large towns, especially in the southern part of Western Siberia.

A northward decrease in Fe and divalent transition metals (Mn, Co, and Ni) and Cd in the snow water contrasted the latitudinal pattern of these elements in the Ob River and tributaries (according to available literature data). In the southern, permafrost-free zone, Mn, Co, Zn, Cd, Pb, and Sb fluxes in the Ob River during the May–June period can be supplied essentially by dissolved fraction of the snowmelt. The impact of snowmelt on river export fluxes in spring strongly increases northward for Zn, Cd, and Pb. In the permafrost zone, entire riverine fluxes of these elements during spring flood can be provided by the snowmelt.

Altogether, the obtained results suggest large complexity of the riverine export and atmospheric deposition of trace metals under climate warming scenario and require comprehensive analysis of dissolved metals in rain waters during different seasons across the region.

Supplementary Materials: The following are available online at https://www.mdpi.com/article/10.3390/w14010094/s1, Figure S1: Examples of the concentration of SC, pH in snow water as a function of latitude, Figure S2: Examples of Mn, Fe, Co, and Ni in snow water as a function of particulate fraction, Figure S3: Examples of the concentration of SC, Mn, Ni, Fe, Cu, Cd, Co, Zn, As, Pb, and Sb in snow water as a function of distance to gas flaring, Figure S4: PCA factorial map F1 × F2 of elements of a reconstructed table for the dissolved fraction, Table S1: Physicochemical properties of dissolved (<0.22 µm) fraction of the snow water trace element concentrations (µg/L) in Western Siberia latitudinal transect, Table S2: Spearmen correlations $p < 0.05$, Table S3: The distance between the sampling point and some nearest possible sources of pollution, Table S4: Results of PCA treatment of all data, Table S5: Mean (±SD) concentration (µg L^{-1}) of metals in river water and snow water in the three distinct parts of the Ob River main stem upstream and downstream of its confluence with Vasyugan and Irtysh.

Author Contributions: Conceptualization, V.P.S. and O.S.P.; field sampling, S.N.V., I.V.K., A.G.L. and O.S.P.; analysis of collected samples, I.V.K., A.G.L. and F.C.; visualization, I.V.K.; writing and editing O.S.P., I.V.K. and V.P.S. All authors have read and agreed to the published version of the manuscript.

Funding: Expeditionary work was supported from the RFBR grant No 19-05-50096. S.N.V. acknowledge support from RSCF (RNF) grant No 18-17-00237-P. Partial support from the Tomsk State University Development Programme («Priority-2030») is also acknowledged. I.V.K. and A.G.L. acknowledge support from the state assignment of the Ministry of Science and Higher Education of the Russian Federation (project No. 0721-2020-0019).

Institutional Review Board Statement: Not applicable.

Informed Consent Statement: Not applicable.

Data Availability Statement: All the obtained data are presented in the manuscript and its Supplementary Materials.

Acknowledgments: The authors are grateful to A.V. Sorochinsky for his invaluable help in the expedition. The authors are grateful to the comments of the anonymous reviewers whose effort and suggestions helped to greatly improve the manuscript.

Conflicts of Interest: The authors declare no conflict of interest.

References

1. Vasilenko, V.N.; Nazarov, I.M.; Fridman, S.D. *Snow Pollution Monitoring*; Hydrometeoizdat: Leningrad, Russia, 1985; p. 181. (In Russian)
2. Boyarkina, A.P.; Baikovsky, V.V.; Vasiliev, N.V.; Glukhov, G.G.; Medvedev, M.A.; Pisareva, L.F.; Rezchikov, V.I.; Sheludko, S.I. *Aerosols in Natural Tablets of Siberia*; TSU Publishing House: Tomsk, Russia, 1993; p. 157. (In Russian)
3. Raputa, V.F.; Kokovkin, V.V. Methods for interpreting snow cover pollution monitoring data. *Chem. Sustain. Dev.* **2002**, *10*, 669–682. (In Russian)
4. Lisitzin, A.P. *Sea-Ice and Iceberg Sedimentation in the Ocean: Recent and Past*; Springer Science & Business Media: Berlin/Heidelberg, Germany, 2002; ISBN 978-3-540-67965-3.
5. Callaghan, T.V.; Johansson, M.; Brown, R.D.; Groisman, P.Y.; Labba, N.; Radionov, V.; Bradley, R.S.; Blangy, S.; Bulygina, O.N.; Christensen, T.R.; et al. Multiple Effects of Changes in Arctic Snow Cover. *Ambio* **2011**, *40*, 32–45. [CrossRef]
6. Singh, V.P.; Singh, P.; Haritashya, U.K. *Encyclopedia of Snow, Ice and Glaciers*; Encyclopedia of Earth Sciences Series; Springer: Berlin/Heidelberg, Germany, 2011; p. 1253. [CrossRef]
7. Telmer, K.; Bonham-Karter, G.F.; Kliza, D.A.; Hall, G.E.M. The atmospheric transport and deposition of smelter emissions: Evidence from the multi-element geochemistry of snow, Quebec, Canada. *Geochim. Cosmochim. Acta* **2004**, *68*, 2961–2980. [CrossRef]
8. Barrie, L.A. Arctic air pollution: An overview of current knowledge. *Atmos. Environ.* **1986**, *20*, 643–663. [CrossRef]
9. Shevchenko, V.P.; Lisitsyn, A.P.; Vinogradova, A.A.; Smirnov, V.V.; Serova, V.V.; Stein, R. Arctic aerosols. Results of ten-year investigations. *Opt. Atmos. Okeana* **2000**, *13*, 551–576. (In Russian)
10. Shevchenko, V. The influence of aerosols on the oceanic sedimentation and environmental conditions in the Arctic. *Ber. Polar-Und Meeresforsch.* **2003**, *464*, 149.
11. Shevchenko, V.P. *Effect of Aerosols on the Environment and Marine Sedimentation in the Arctic*; Nauka: Moskow, Russia, 2006; p. 226. (In Russian)
12. Ermolov, Y.V.; Makhatkov, I.D.; Khudyaev, S.A. Background concentrations of chemical elements in the snow cover of the central sector of Western Siberia. *Opt. Atmos. Okeana* **2014**, *27*, 790–800. (In Russian)
13. Guéguen, C.; Cuss, C.W.; Cho, S. Snowpack deposition of trace elements in the Athabasca oil sands region, Canada. *Chemosphere* **2016**, *153*, 447–454. [CrossRef]
14. Talovskaya, A.V.; Yazikov, E.G.; Filimonenko, E.A.; Lata, J.-C.; Kim, J.; Shakhova, T.S. Characterization of solid airborne particles deposited in snow in the vicinity of urban fossil fuel thermal power plant (Western Siberia). *Environ. Technol. (UK)* **2018**, *39*, 2288–2303. [CrossRef]
15. Zakharchenko, A.V.; Tigeev, A.A.; Pasko, O.A.; Kolesnichenko, L.G.; Moskovchenko, D.V. Transboundary, regional and local geochemical transfer of chemicals in snow cover. *Geoecol. Eng. Geol. Hydrogeol. Geocryol.* **2020**, *6*, 41–53. (In Russian)
16. Moskovchenko, D.V.; Pozhitkov, R.Y.; Soromotin, A.V. Geochemical characteristics of snow cover in Tobolsk. *Bull. Tomsk. Polytech. Univ. Geo Assets Eng.* **2021**, *332*, 156–169. [CrossRef]
17. Moskovchenko, D.; Pozhitkov, R.; Zakharchenko, A.; Tigeev, A. Concentrations of Major and Trace Elements within the Snowpack of Tyumen, Russia. *Minerals* **2021**, *11*, 709. [CrossRef]
18. AMAP. AMAP Assessment 2002: Heavy Metals in the Arctic. *Arctic Monitoring and Assessment Programme (AMAP)*. 2005. Available online: https://www.amap.no/documents/doc/amap-assessment-2002-heavy-metals-in-the-arctic/97 (accessed on 24 November 2021).
19. Boyd, R.; Barnes, S.J.; De Caritat, P.; Chekushin, V.A.; Melezhik, V.A.; Reimann, C.; Zientek, M.L. Emissions from the copper–nickel industry on the Kola Peninsula and at Noril'sk, Russia. *Atmos. Environ.* **2009**, *43*, 1474–1480. [CrossRef]
20. Jaffe, D.; Cerundolo, B.; Rickers, J.; Stolzberg, R.; Baklanov, A. Deposition of sulfate and heavy metals on the Kola Peninsula. *Sci. Total Environ.* **1995**, *160–161*, 127–134. [CrossRef]
21. Niskavaara, H.; Reimann, C.; Chekushin, V.; Kashulina, G. Seasonal variability of total and easily leachable element contents in topsoils (0–5 cm) from eight catchments in the European Arctic (Finland, Norway, and Russia). *Environ. Pollut.* **1997**, *96*, 261–274. [CrossRef]
22. Zhulidov, A.V.; Robarts, R.D.; Pavlov, D.F.; Kamari, J.; Gurtovaya, T.Y.; Merilainen, J.J.; Pospelov, I.N. Long-term changes of heavy metal and Sulphur concentrations in ecosystems of the Taymyr Peninsula (Russian Federation) North of the Norilsk Industrial Complex. *Environ. Monit. Assess.* **2011**, *181*, 539–553. [CrossRef]
23. Lagutin, A.A.; Mordvin, E.Y.; Volkov, N.V.; Tuchina, N.V. Estimation of natural gas flaring volume at the Western Siberia flares using satellite night-time data in the visible and near-infrared range. *CEUR—WP* **2019**, *2534*, 22–26.
24. Kotova, E.I.; Shevchenko, V.P. Influence of long-range atmospheric transport on formation of ionic composition of atmospheric precipitation and snow cover in coastal zone of western Russian Arctic. *Fundam. Res.* **2014**, *12*, 2378–2382. (In Russian)
25. Ariés, S.; Valladon, M.; Polvé, M.; Dupré, B. A routine method for oxide and hydroxide interference corrections in ICP-MS chemical analysis of environmental and geological samples. *Geostandard. Newslett.* **2000**, *24*, 19–31. [CrossRef]
26. Heimburger, A.; Tharaud, M.; Monna, F.; Losno, R.; Desboeufs, K.; Nguyen, E. B. SLRS-5 elemental concentrations of thirty-three uncertified elements deduced from SLRS-5/SLRS-4 ratios. *Geostand. Geoanal. Res.* **2013**, *37*, 77–85. [CrossRef]
27. Manasypov, R.M.; Pokrovsky, O.S.; Kirpotin, S.N.; Shirokova, L.S. Thermokarst lake waters across the permafrost zones of Western Siberia. *Cryosphere* **2014**, *8*, 1177–1193. [CrossRef]

28. Manasypov, R.M.; Vorobyev, S.N.; Loiko, S.V.; Kritzkov, I.V.; Shirokova, L.S.; Shevchenko, V.P.; Kirpotin, S.N.; Kulizhsky, S.P.; Kolesnichenko, L.G.; Zemtzov, V.A.; et al. Seasonal dynamics of organic carbon and metals in thermokarst lakes from the discontinuous permafrost zone of Western Siberia. *Biogeosciences* **2015**, *12*, 3009–3028. [CrossRef]
29. Pokrovsky, O.S.; Manasypov, R.M.; Shirokova, L.S.; Loiko, S.; Krickov, I.; Kopysov, S.; Kolesnichenko, L.G.; Zemtsov, V.A.; Kulizhsky, S.P.; Vorobyev, S.N.; et al. Permafrost coverage, watershed area and season control of dissolved carbon and major elements in Western Siberia rivers. *Biogeosciences* **2015**, *12*, 6301–6320. [CrossRef]
30. Pokrovsky, O.S.; Manasypov, R.M.; Loiko, S.; Krickov, I.A.; Kopysov, S.G.; Kolesnichenko, L.G.; Vorobyev, S.N.; Kirpotin, S.N. Trace elements transport in western Siberia rivers across a permafrost gradient. *Biogeosciences* **2016**, *13*, 1877–1900. [CrossRef]
31. Shevchenko, V.P.; Vorobyev, S.N.; Krickov, I.V.; Boev, A.G.; Lim, A.G.; Novigatsky, A.N.; Starodymova, D.P.; Pokrovsky, O.S. Insoluble particles in the snowpack of the Ob river basin (Western Siberia) a 2800 km submeridional profile. *Atmosphere* **2020**, *11*, 1184. [CrossRef]
32. Shevchenko, V.P.; Pokrovsky, O.S.; Vorobyev, S.N.; Krickov, I.V.; Manasypov, R.M.; Politova, N.V.; Kopysov, S.G.; Dara, O.M.; Auda, Y.; Shirokova, L.S.; et al. Impact of snow deposition on major and trace element concentrations and elementary fluxes in surface waters of the Western Siberian Lowland across a 1700′km latitudinal gradient. *Hydrol. Earth Syst. Sci.* **2017**, *21*, 5725–5746. [CrossRef]
33. Kolesnichenko, Y.; Kolesnichenko, L.G.; Vorobyev, S.N.; Shirokova, L.S.; Semiletov, I.P.; Dudarev, O.Y.; Vorobyev, R.S.; Shavrina, U.; Kirpotin, S.N.; Pokrovsky, O.S. Landscape, soil, lithology, climate and permafrost control on dissolved carbon, major and trace elements in the Ob River, Western Siberia. *Water* **2021**, *13*, 3189. [CrossRef]
34. Karnatsevich, I.V.; Khrushchev, S.A. *Computer System for Mass Calculations of Current Water Balances of River Catchments of Unexplored Land Areas*; OmGPU: Omsk, Russia, 2014; p. 174. (In Russian)
35. Pokrovsky, O.S.; Manasypov, R.M.; Kopysov, S.G.; Krickov, I.V.; Shirokova, L.S.; Loiko, S.V.; Lim, A.G.; Kolesnichenko, L.G.; Vorobyev, S.N.; Kirpotin, S.N. Impact of permafrost thaw and climate warming on riverine export fluxes of carbon, nutrients and metals in Western Siberia. *Water* **2020**, *12*, 1817. [CrossRef]

water

MDPI

Article

Iron, Phosphorus and Trace Elements in Mussels' Shells, Water, and Bottom Sediments from the Severnaya Dvina and the Onega River Basins (Northwestern Russia)

Artem A. Lyubas [1,*], Alena A. Tomilova [1], Artem V. Chupakov [1], Ilya V. Vikhrev [1], Oksana V. Travina [1], Alexander S. Orlov [1], Natalia A. Zubrii [1], Alexander V. Kondakov [1], Ivan N. Bolotov [1] and Oleg S. Pokrovsky [1,2,3]

[1] N. Laverov Federal Center for Integrated Arctic Research of the Ural Branch of the Russian Academy of Sciences, Northern Dvina Emb. 23, 163000 Arkhangelsk, Russia; tomilova_alyona@mail.ru (A.A.T.); artem.chupakov@gmail.com (A.V.C.); vikhrevilja@gmail.com (I.V.V.); travina.oksana136@yandex.ru (O.V.T.); alseror@yandex.ru (A.S.O.); 9052930111@mail.ru (N.A.Z.); akondakv@yandex.ru (A.V.K.); inepras@yandex.ru (I.N.B.); oleg.pokrovsky@get.omp.eu (O.S.P.)

[2] Geosciences and Environment Toulouse, UMR 5563 CNRS, 31400 Toulouse, France

[3] BIO-GEO-CLIM Laboratory, Tomsk State University, 634050 Tomsk, Russia

* Correspondence: lyubas@ro.ru; Tel.: +7-906-283-9810

Abstract: Trace elements in freshwater bivalve shells are widely used for reconstructing long-term changes in the riverine environments. However, Northern Eurasian regions, notably the European Russian North, susceptible to strong environmental impact via both local pollution and climate warming, are poorly studied. This work reports new data on trace elements accumulation by widespread species of freshwater mussels *Unio* spp. and *Anodonta anatina* in the Severnaya Dvina and the Onega River Basin, the two largest subarctic river basins in the Northeastern Europe. We revealed that iron and phosphorous accumulation in *Unio* spp. and *Anodonta anatina* shells have a strong relationship with a distance from the mouth of the studied river (the Severnaya Dvina). Based on multiparametric statistics comprising chemical composition of shells, water, and sediments, we demonstrated that the accumulation of elements in the shell depends on the environment of the biotope. Differences in the elemental composition of shells between different taxa are associated with ecological preferences of certain species to the substrate. The results set new constraints for the use of freshwater mussels' shells for monitoring riverine environments and performing paleo-reconstructions.

Keywords: freshwater mussels; trace elements; biominerals; bioindicators; Northeastern Europe; boreal; subarctic; river

Citation: Lyubas, A.A.; Tomilova, A.A.; Chupakov, A.V.; Vikhrev, I.V.; Travina, O.V.; Orlov, A.S.; Zubrii, N.A.; Kondakov, A.V.; Bolotov, I.N.; Pokrovsky, O.S. Iron, Phosphorus and Trace Elements in Mussels' Shells, Water, and Bottom Sediments from the Severnaya Dvina and the Onega River Basins (Northwestern Russia). *Water* **2021**, *13*, 3227. https://doi.org/10.3390/w13223227

Academic Editor: Michael Twiss

Received: 28 September 2021
Accepted: 9 November 2021
Published: 14 November 2021

1. Introduction

Trace elements (TE) in carbonate shells of bivalve mollusks have been widely used for environmental studies [1–9]. Measuring TE concentration in shells of mollusks helps to track changes of different parameters in watercourses during long-time periods and to reveal local and global environmental impacts on riverine habitats [6]. For example, Zhao et al. [10] found that patterns of the Mn/Ca index in the shells of *Hyriopsis cumingii* (Lea, 1852) are associated with the occurrence of reduction conditions at the bottom sediments water interface. Therefore, the use of the Mn/Ca ratio can potentially serve as a high-resolution marker of Mn mobility in the anoxic layer. Studying this indicator in the shells of bivalve mollusks can help to retrospectively track environmental changes in aquatic ecosystems caused by eutrophication. Bolotov et al. [6] determined that the accumulation of trace elements in the shells of freshwater pearl mussels depends on the environmental conditions of the biotope and has no taxonomic control. The ratio of manganese and iron concentrations in relation to the environmental conditions of a freshwater body was analyzed by Naeher et al. [11]. These authors demonstrated applicability of this indicator for the reconstruction of redox conditions using values from the shells of modern bivalve mollusks and showed that sedimentation factors can reduce the applicability of the Mn/Fe ratio for reconstructing O_2 concentrations in the bottom water of lakes. Ravera et al. [12] evaluated the possibility of assessing the state of abiotic and biotic components of a freshwater ecosystem using concentrations of chemical elements in the shells and tissues of the bivalve mollusks *Anodonta anatina* (Linnaeus, 1758) and *Unio* sp., as well as in water, bottom sediments, and aquatic plants. These authors demonstrated that these indicators can be a useful tool for long-term studies of environmental pollution.

Over recent years, the use of new methods allowed analyzing changes in the chemical composition of shells with high spatial resolution. A number of indicatory elements were used to assess seasonal trends and hydrological conditions of aquatic ecosystems. For example, van Plantinga and Grossman [8] used shells of two species (*Amblema plicata* (Say, 1817) and *Cyrtonaias tampicoensis* (Lea, 1838)) of naiad as natural archives. These authors established a relationship between the Mn/Ca ratio and the river discharge. Watanabe et al. [13] compared the values of geochemical indicators of hydrological conditions with the values of annual increments in the freshwater pearl mussel *Margaritifera laevis* (Haas, 1910). A correlation was found between the Ba/Ca ratio and the shell growth rate, which depended on the volume of winter snow and spring meltwater.

The present work is aimed to characterize the accumulation of elements in the shells of widespread species of freshwater bivalve mollusks, together with the components of their habitats (bottom sediments and water) using a case study of two large river basins in Northwestern Russia. Subarctic rivers of European Russian North are subjected to strong impact of both local (pollution sources, forestry, industry) and global (climate change, vegetation shifts) factors but they remain poorly studied using complex biogeochemical ecosystem-based approaches that can include river water, sediments, and aquatic organisms. Here, we intended to test the impact of environmental conditions on TE distribution coefficients and characterize the difference between taxa in order to better define the scope of the use of bivalve shells as archives of riverine aquatic environments.

2. Materials and Methods

2.1. Sampling Locations

Sampling sites are located in the basins of the Severnaya Dvina and the Onega rivers (Figure 1, Table 1). The names of sampling localities are given relative to nearby settlements. Mollusks' samples were taken manually from a depth of 0.5 to 1.5 m.

River water samples were collected at sites of shells' sampling from 0.5 m depth in pre-cleaned polypropylene bottles. The water was immediately filtered through a single-use sterile acetate cellulose filter (Sartorius, 0.45 µm), into pre-cleaned polypropylene Nalgene bottles. Samples for TE analyses were acidified with ultrapure double-distilled HNO_3 and stored in the refrigerator pending analyses.

The sediment samples were taken from the water-sediment interface to a depth of 3–4 cm (which corresponds to the depth of burying of *Anodonta anatina* and *Unio* spp. mollusks in the ground), placed in a double zip polyethylene bag, preserved in cold dark environment, and transported within several days to the laboratory where they were dried at 90 °C in the oven.

Shells from 10 sampling sites located in different parts of the Severnaya Dvina and the Onega River basins were used to analyze trace elements' composition (Figure 2). Bottom sediment and water samples were taken at 9 localities (Table 1).

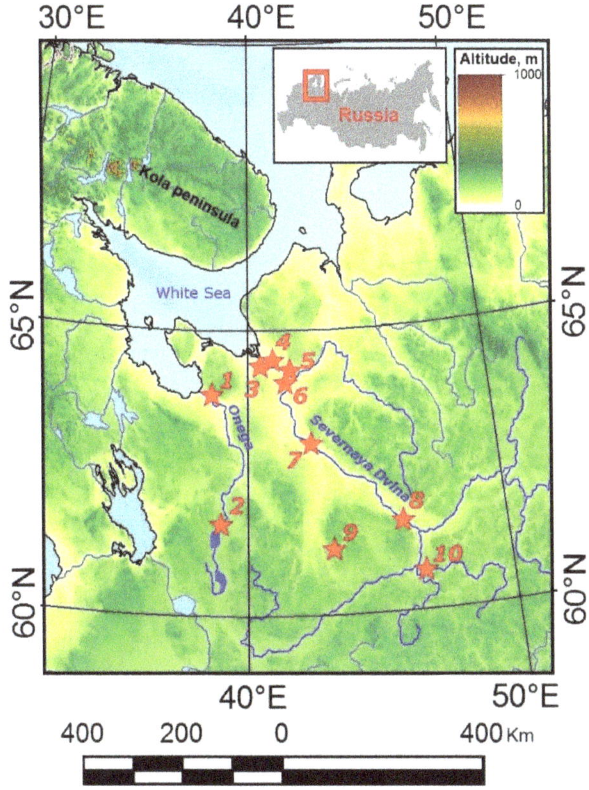

Figure 1. Location map of the field study areas: 1—Porog (POR), 2—Kuftyrikha (Kar), 3—Isakogorka (Z), 4—Krasnoflotsky (Kras), 5—Kholmogory (Hol), 6—Marilovo (Mar), 7—Bereznik (Br), 8—Krasnoborsk (KBOR), 9—Shangaly (ShB), 10—Ust'-Alekseevo (UG).

Table 1. List of studied localities.

№	Locality	River Basin	Coordinates	Bottom Sediments and Short Site Description	Number of Samples
1	Porog village, right bank before the rifts	Onega River	63°49′44.5″ N 38°28′32.3″ E	Silted area of the bottom before the river bend, bottom sediments—sandy loam	Shells—4 Water—1 Bottom Sediment—1
2	Kuftyrikha boundary near Kargopol town	Onega River	61°32′46.9″ N 38°59′54.4″ E	Coarse sand, depth 1–1.5 m, slight current, bottom sediments—sand with stones.	Shells—4 Water—1 Bottom Sediment—1
3	Perkhachevsky settlement, Isakogorka anabranch	Severnaya Dvina River	64°27′55.2″ N 40°29′22.2″ E	Silty bottom, smell of hydrogen sulfide	Shells—4 Water—1 Bottom Sediment—1
4	Krasnoflotsky island, beach in the southwest of the island	Severnaya Dvina River	64°30′07″ N 40°37′02″ E	Willow beach, bottom sediments—medium-grained sand	Shells—4 Water—1 Bottom Sediment—1

Table 1. *Cont.*

№	Locality	River Basin	Coordinates	Bottom Sediments and Short Site Description	Number of Samples
5	Kholmogory village	Severnaya Dvina River	64°14′39.3″ N 41°36′29.9″ E	The bottom is silted in places. The depth increases rapidly near the coast, bottom sediments—sand and silt	Shells—4 Water—1 Bottom Sediment—1
6	Marilovo village	Severnaya Dvina River	64°08′29.0″ N 41°45′47.0″ E	Sandy coast along the village, bottom sediments—rocky-sandy bottom with clay	Shells—4 Water—0 Bottom Sediment—0
7	Bereznik village, western side of the beach	Severnaya Dvina River	62°50′38.0″ N 42°48′01.1″ E	Beach, bottom sediments—fine-grained sand	Shells—3 Water—1 Bottom Sediment - 1
8	Krasnoborsk village, west of Zeleny Island	Severnaya Dvina River	61°34′58.0″ N 45°50′42.8″ E	Beach northwest of the village, bottom sediments—medium-grained sand	Shells—4 Water—1 Bottom Sediment—1
9	Shangaly village Ust'ya river, beach at the suspension bridge	Severnaya Dvina River	61°07′52.3″ N 43°20′46.0″ E	Depth 1.2–1.5 m, straight section of the river, beach, bottom sediments—medium-grained sand pebbles stones.	Shells—4 Water—1 Bottom Sediment—1
10	Yug river, beach in the village Ust-Alekseevo,	Severnaya Dvina River	60°27′54.1″ N 46°30′21.9″ E	The flow rate is low, bottom sediments—bottom sediments—medium-grained sand.	Shells—4 Water—1 Bottom Sediment—1

Figure 2. Shells of freshwater bivalve mollusks *Anodonta anatina* (Linnaeus, 1758), *Unio pictorum* (Linnaeus, 1758) and *Unio tumidus* Retzius, 1788: **A**—*A. anatina,* the Severnaya Dvina River near Kholmogory village; **B**—*U. pictorum,* the Severnaya Dvina near Kholmogory village; **C**—*A. anatina,* the Onega River near Kargopol town (the Kuftyrikha boundary); **D**—*U. tumidus,* the Onega River near Kargopol town (the Kuftyrikha boundary). Scale bar = 10 mm.

2.2. Determination of the Taxonomic Position of Mollusks' Individuals and Preparation Samples for Trace Elements' Analysis

The primary species identification of the collected samples of bivalve mollusks was carried out based on standard and special keys [14–16]. The comparative analysis of the shell morphology was carried out taking into account the structure of the pseudocardinal and lateral teeth, muscle attachment scars, shell shape, and umbo position [17,18]. Details of the shell structure were studied using an Axio Lab.A1 light microscope (Carl Zeiss, Jena, Germany) and a Leica M165C stereomicroscope (Leica Microsystems, Wetzlar, Germany). Shells were sawn from the umbo to the ventral margin perpendicularly to the winter lines and along the axis of minimum growth with a diamond saw, and the obtained fragment to 10 mm width was taken for trace element analysis.

2.3. Analysis of the Trace Element Composition of Shells

For the analysis of major and trace elements, parts of the shell were washed using MilliQ water, dried and ground in an agate mortar. Acidic dissolution of crushed shells was carried out by processing them in ultrapure water Merck H_2O_2, bidistilled ultrapure water Aldrich HNO_3, HNO_3 + HCl and, finally, in HNO_3 at 80 °C in Savillex Teflon containers housed in separate evaporation chambers (class A 100) located inside the cleanroom (class ISO A 10,000). This made it possible to dissolve only the carbonate and organic parts of the shells without leaching silicate mineral impurities. Together with samples, we also processed two certified carbonate reference materials (Coral JCp-1, Giant Clam JCt-1). The leachate products were evaporated to dryness, re-dissolved in 10% HNO_3 and diluted 5000 times for the analysis of major and trace elements using an Agilent 7500ce (Agilent Technologies, Inc., Santa Clara, United States)inductively coupled plasma mass-spectrometer [6,19].

2.4. Analysis of the Elemental Composition of Water Samples

The elemental composition of the river water was measured by mass spectrometric and atomic emission analyzes with inductively coupled plasma (ICP-MS and ICP-AES). The detection limits of the method were 0.1–10 ng L^{-1} for Y, rare earth elements (REE), U, Th, Zr, Li, Mo, Be, Rb, Cd, W, Nb, Hf, Sb, Cs, Tl, and 10–100 ng L^{-1} for (Co, V, Sr, Ba, Mn, Pb, As) and 100–10,000 ng L^{-1} for other elements (S, Si, P, Br, K, Ca, Fe, Na, Mg, Al, B, Ti, Cr, Zn, Cu, Se, Ni) [20].

2.5. Analysis of the Elemental Composition of Bottom Sediments Samples

The mineralization of sediment samples was carried out by acid digestion. Along with the analyzed samples, the digestion of international certified sediment materials (LKSD) and one in-house standard sample was carried out. One hundred mg of sample were placed in Teflon beakers (volume 50 mL), with 0.1 mL of a solution containing 8 μg dm^{-3} of ^{145}Nd, ^{161}Dy, and ^{174}Yb which is used as a control of the chemical yield during the sample decomposition procedure), and moistened with several drops of deionized water. After that, 0.5 mL of $HClO_4$ (Perchloric acid fuming 70% Supratur, Merck, Darmstadt, Germany), 3 mL (HF Hydrofluoric acid 40% GR, ISO, Merck, Darmstadt, Germany), 0.5 mL of HNO_3 (Nitric acid 65%, GR, ISO, Merck, Darmstadt, Germany) were added and evaporated until intense white fumes appeared. The vials were cooled, their walls were washed with water, and the solution was again evaporated to wet salts. Then, 2 mL of HCl (hydrochloric acid fuming 37% GR, ISO, Merck, Darmstadt, Germany) and 0.2 mL of 0.1 M H_3BO_3 solution (analytical grade) were added and evaporated to a volume of 0.5–0.7 mL. The resulting solutions were transferred into polyethylene bottles, amended with 0.1 mL of a solution containing 10 mg L^{-1} In (internal standard), diluted with deionized water to 20 mL, and analyzed by ICP-MS. As blank controls in Teflon beakers, the procedures described above were carried out without samples, and the resulting solutions were used as control [19].

2.6. Statistical Treatment of the Data

Element concentrations in samples were compared using the nonparametric Mann–Whitney U and Kruskal–Wallace H tests because the data were not distributed normally (Shapiro–Wilk test). To assess the relationship between the concentrations of elements, the Spearman's correlation coefficient was used, and linear, logarithmic, and exponential regression models were employed.

A principal component analysis (PCA) was used to determine the factors controlling the pattern of element accumulation in shells, water, and sediments. The selection of factors was carried out according to the screen test and on the basis of eigenvalues (Kaiser criterion). The suitability of the data for factorization was assessed based on the Kaiser–Meier–Olkin measure of sample adequacy (minimum fitness value was higher than 0.6) and using the Bartlett's test of sphericity. Metal pollution index was calculated according with Usero et al. [21] and Sedeño-Díaz et al. [22].

3. Results

3.1. Elemental Composition of Shells

The collected shells of bivalve mollusks belonged to two genera (*Anodonta* and *Unio*). All analyzed individuals of *Anodonta* belonged to the species *A. anatina*. The individuals belonging to the genus *Unio*, according to the teeth structure and the outline of the shell contours, were identified as *U. pictorum* (Linnaeus, 1758) and *U. tumidus* Retzius, 1788 [23]. The average ontogenetic ages of individuals were 4.17 ± 0.20 years for specimens of *Anodonta* and 4.67 ± 0.20 years for specimens of *Unio*.

The average chemical composition of mollusk shells is presented in Table 2. After Ca, the most abundant elements were Na, Fe, Mn, Sr. Average concentrations of Na, Fe, Mn, Sr in *Anodonta anatina* (n = 18) and *Unio* spp. (n = 21) shells were 2340 ± 31 mg/kg and 2300 ± 27 mg/kg, 1580 ± 653 mg/kg and 1030 mg/kg, 680 ± 79 mg/kg and 424 ± 38 mg/kg, 635 ± 48 mg/kg, and 713 ± 68 mg/kg, respectively.

Table 2. Elemental composition of *Anodonta anatina* and *Unio* spp. shells from the Severnaya Dvina and the Onega river basins: the cross into a cell of the table shows the absence of statistically significant differences ($p > 0.05$), the check mark in a cell of the table shows the presence of statistically significant differences ($p < 0.05$).

Element	*Anodonta*, ppm Mean ± SE (n = 18)	*Unio*, ppm Mean ± SE (n = 21)	Statistically Significant Differences		
			by Genera	by Basins	by Localities
Li	0.28 ± 0.04	0.23 ± 0.02	×	✓	✓
Na	2340 ± 30	2300 ± 27	×	✓	✓
Mg	138 ± 18	87.6 ± 9.03	✓	×	✓
Al	304 ± 47	174 ± 30	✓	×	✓
P	246 ± 89	171 ± 62	×	✓	✓
K	129 ± 22	76.8 ± 12.1	×	×	✓
Ti	44 ± 6.9	31.1 ± 3.62	×	×	✓
Cr	1.89 ± 0.3	1.15 ± 0.19	×	×	×
Mn	680 ± 79	423 ± 38	✓	×	✓
Fe	1580 ± 653	1030 ± 493	×	×	✓
Cu	3.32 ± 0.28	2.93 ± 0.23	×	×	×
Zn	3.47 ± 0.60	3.26 ± 0.53	×	×	✓
As	2.49 ± 0.81	2.92 ± 1.36	×	×	×
Rb	0.46 ± 0.10	0.22 ± 0.05	×	×	✓
Sr	635 ± 48	713 ± 68	×	✓	✓
Zr	1.86 ± 0.41	0.99 ± 0.19	×	×	✓
Y	0.35 ± 0.08	0.22 ± 0.06	×	✓	✓
Nb	0.095 ± 0.019	0.063 ± 0.012	×	×	✓
S	314 ± 36	301 ± 25	×	×	✓
Sb	0.12 ± 0.02	0.14 ± 0.02	×	×	✓
Cs	0.022 ± 0.005	0.012 ± 0.002	×	×	✓

Table 2. *Cont.*

Element	*Anodonta*, ppm Mean ± SE (*n* = 18)	*Unio*, ppm Mean ± SE (*n* = 21)	Statistically Significant Differences		
			by Genera	by Basins	by Localities
Ba	64.6 ± 10.1	43.4 ± 5.3	✓	✓	✓
La	0.65 ± 0.14	0.45 ± 0.11	×	✓	✓
Ce	1.2 ± 0.23	0.76 ± 0.18	×	✓	✓
Pr	0.14 ± 0.03	0.04 ± 0.02	×	✓	✓
Nd	0.48 ± 0.1	0.31 ± 0.07	×	✓	✓
Sm	0.094 ± 0.019	0.058 ± 0.013	×	✓	✓
Eu	0.018 ± 0.004	0.012 ± 0.003	×	✓	✓
Gd	0.077 ± 0.018	0.047 ± 0.013	×	✓	✓
Dy	0.069 ± 0.015	0.044 ± 0.009	×	✓	✓
Er	0.032 ± 0.007	0.019 ± 0.005	×	✓	✓
Yb	0.03 ± 0.007	0.016 ± 0.004	×	✓	✓
Pb	0.49 ± 0.03	0.46 ± 0.04	×	×	×
Th	0.062 ± 0.011	0.039 ± 0.011	×	×	✓
U	0.030 ± 0.005	0.019 ± 0.004	✓	×	✓

It was found that the concentrations of only 5 elements (Mg, Al, Mn, Ba, and U) differ between samples of mollusks belonging to different genera. Considering both genera together, the concentrations of Li, Na, P Al, Sr, Y, Ba, REEs differed between the samples from the two river basins. Further, the concentrations of all elements, except for Cr, Cu, As, and Pb, differ statistically significantly among the studied localities when considering two genera together.

3.2. Elemental Composition of Water and Sediments

The concentrations of chemical elements in water and bottom sediments are presented in Tables 3 and 4.

Table 3. Trace element composition of water from the studied biotopes.

Element	C, ppm Mean ± SE (*n* = 9)	Element	C, ppm Mean ± SE (*n* = 9)
B	42.0 ± 11.7	Zr	0.226 ± 0.036
Na	12000 ± 4340	Nb	0.005 ± 0.001
Mg	11300 ± 1590	Mo	0.53 ± 0.12
Al	70.9 ± 19.1	Cd	0.018 ± 0.008
Si	2180 ± 321	Sb	0.07 ± 0.02
S	10200 ± 2590	Cs	0.005 ± 0.001
K	1320 ± 277	La	0.208 ± 0.064
Ca	42000 ± 5410	Ce	0.437 ± 0.15
Ti	2.12 ± 0.54	Pr	0.057 ± 0.017
V	0.904 ± 0.063	Nd	0.234 ± 0.068
Mn	190 ± 84.9	Sm	0.047 ± 0.014
Fe	449 ± 104	Eu	0.0089 ± 0.0025
Co	0.15 ± 0.04	Gd	0.046 ± 0.012
Ni	0.503 ± 0.118	Tb	0.0064 ± 0.0018
Cu	0.933 ± 0.121	Dy	0.0346 ± 0.0090
Zn	10.3 ± 3.4	Ho	0.0069 ± 0.0016
As	1.3 ± 0.2	Er	0.0173 ± 0.0043
Br	34 ± 11	Tm	0.0022 ± 0.0006
Sr	448 ± 111	Yb	0.0142 ± 0.0033
Ba	59 ± 11	Lu	0.0021 ± 0.0005
Pb	0.192 ± 0.053	Hf	0.0055 ± 0.0010
Li	4.8 ± 0.9	W	0.0068 ± 0.0018
Be	0.009 ± 0.002	Tl	0.0029 ± 0.0003
Rb	1.21 ± 0.15	Th	0.0215 ± 0.0043
Y	0.176 ± 0.047	U	0.426 ± 0.122

Table 4. Trace element composition of bottom sediments from the studied biotopes.

Element	C, ppm Mean ± SE (n = 9)	Element	C, ppm Mean ± SE (n = 9)
Na	9130 ± 1520	Sn	0.349 ± 0.104
Al	30200 ± 4260	Cs	0.702 ± 0.214
P	231 ± 51	Ba	337 ± 34
S	1620 ± 1450	La	9.79 ± 1.88
K	8720 ± 910	Ce	21.3 ± 3.98
Ca	16200 ± 665	Pr	2.31 ± 0.43
Ti	1120 ± 257	Nd	8.84 ± 1.63
Mn	362 ± 87	Sm	1.64 ± 0.32
Fe	12100 ± 2980	Eu	0.392 ± 0.064
Li	13.1 ± 6.03	Gd	1.45 ± 0.28
Be	0.786 ± 0.182	Tb	0.198 ± 0.041
Sc	4.00 ± 0.94	Dy	1.25 ± 0.25
V	25.1 ± 6.3	Ho	0.234 ± 0.047
Cr	28.9 ± 5.3	Er	0.733 ± 0.160
Co	4.98 ± 1.12	Tm	0.105 ± 0.022
Ni	16.0 ± 4.3	Yb	0.759 ± 0.148
Cu	9.42 ± 5.22	Lu	0.106 ± 0.021
Zn	34.4 ± 9.1	Hf	1.23 ± 0.22
Ga	5.84 ± 0.83	Ta	0.334 ± 0.094
As	2.71 ± 0.63	W	0.298 ± 0.13
Rb	27.8 ± 2.89	Tl	0.135 ± 0.012
Sr	388 ± 267	Pb	8.03 ± 1.59
Y	6.76 ± 1.52	Bi	0.03 ± 0.01
Zr	49.8 ± 8.8	Th	1.99 ± 0.53
Nb	4.6 ± 1.2	U	0.79 ± 0.28

Correlations between element concentrations in the shells revealed two groups of elements. The largest number of strong correlations was found for aluminum (Mg, K, Rb, U, Th, etc.) and iron (P, K, Rb, Cs). Generally, for *Anodonta anatina*, the correlations between elements were stronger than those for *Unio* species (Figure 3 and Figures S1–S4, Supplementary Materials).

Correlations of REE with other elements were analyzed for *Anodonta* mussels' shells. Strong positive correlations were established for La, Ce, Pr, Nd, Sm, Eu, Gd, Dy, Er, Yb with iron ($R_{Spearman}$ ranged from 0.80 to 0.96, $p < 0.01$) and for Eu, Gd, Er, Yb with aluminium ($R_{Spearman}$ ranged from 0.70 to 0.80, $p < 0.01$). Medium positive correlations were observed between aluminium and Ce, Pr, Nd, Sm, Dy ($R_{Spearman}$ ranged from 0.51 to 0.68, $p < 0.05$).

Correlations of REE with other elements were also examined for *Unio* mussels' shells. Strong positive correlations were identified for Nd, Sm, Ce, Pr, Eu, Gd, Dy, Er, Yb with iron ($R_{Spearman}$ ranged from 0.78 to 0.88, $p < 0.01$) and for Nd, Sm, Gd, Dy, Er with aluminium ($R_{Spearman}$ ranged from 0.71 to 0.75, $p < 0.01$). Medium positive correlations were observed between Al and Ce, Pr, Eu, Yb ($R_{Spearman}$ ranged from 0.64 to 0.69, $p < 0.01$). Medium positive correlations were observed between Fe and La ($R_{Spearman} = 0.69$, $p < 0.01$).

Iron and phosphorus exhibited strong correlations with Mg, Y, Cs ($R_{Spearman} > 0.7$, $p < 0.01$), K, Rb ($R_{Spearman} > 0.7$, $p < 0.05$), and REE ($R_{Spearman} > 0.8$, $p < 0.05$). There were no significant correlations for Na, Al, S, Mn, Sr, and U with the concentrations of iron and phosphorus in the shells of bivalve mollusks ($p > 0.05$).

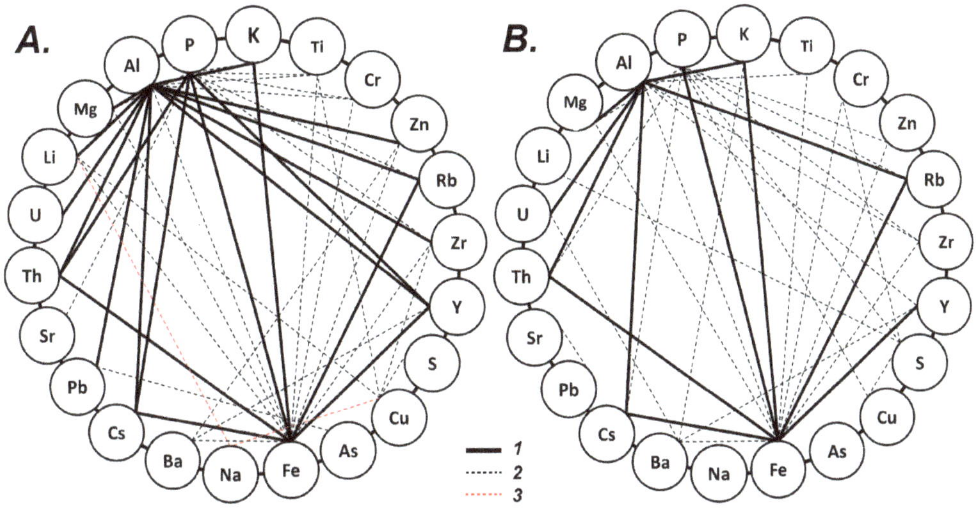

Figure 3. Correlation between trace elements' concentrations in shells of freshwater mussels *Anodonta anatina* (**A**) and *Unio* spp. (**B**), collected from the Onega and the Severnaya Dvina River basins: 1—strong positive correlation ($R_{Spearman} > 0.7$, $p < 0.05$); 2—medium positive correlation ($0.7 \geq R_{Spearman} \geq 0.5$, $p < 0.05$); 3—medium negative correlation ($-0.7 \leq R_{Spearman} \leq -0.5$, $p < 0.05$).

3.3. Multiparametric Statistics

Preliminary assessment of the suitability of concentrations data on Li, Na, Mg, Al, P, S, K, Ti, Cr, Mn, Fe, Cu, Zn, As, Rb, Sr, Y, Zr, Nb, Sb, Cs, Ba, REEs, Pb, Th, U for factorization revealed an acceptable value of the Kaiser–Mayer–Olkin sample adequacy measure (0.719) and a statistically significant indicator of Bartlett's sphericity criterion ($\chi^2 = 2592$; $p < 0.001$). As a result, three factors were identified; in terms of the values of factor loadings, they were interpreted as follows. None of the factors was different between individuals of different genera. Factor 1 had a regression relationship with the sum of the concentrations of elements per sample, normalized to calcium (Figure 4). Factor 2 was associated with the accumulation of iron, barium, and phosphorus during shell growth. Factor 3 described the conditions of a particular biotope associated with the composition of the bottom sediments in which the mollusks lived, including the accumulation of heavy metals in the shell (Sb, Pb, Sr), and differed among localities ($p > 0.001$).

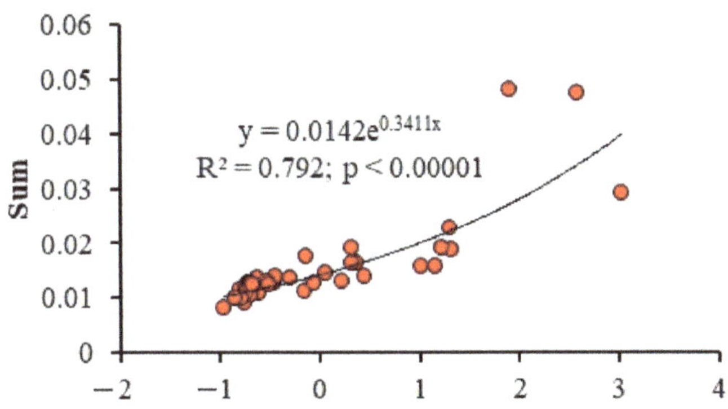

Figure 4. The power relationships between the first principal component and the sum of Ca-normalized elements in each sample.

3.4. Element Distribution Coefficients between the Shell and the Environment

Apparent distribution coefficients of major and trace elements between the river water and the aragonite shells of *Unio* spp. and *A. anatina* were calculated based on several replicates from the same site. Calcium-normalized distribution coefficients K_d (solid/liquid) = ([TE]/[Ca])shell/([TE]/[Ca]) solution varied from 10 to 0.001 (Figure 5). In addition, distribution coefficients of trace elements between shells and bottom sediment samples were determined to range from 6.71×10^{-6} to 0.52 (Figure 6).

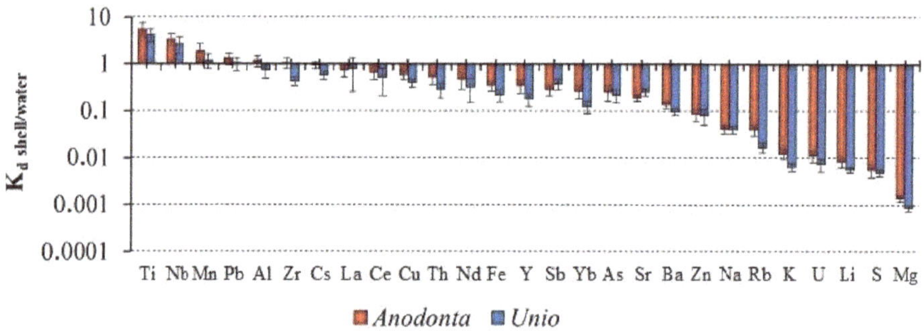

Figure 5. Calcium-normalized distribution coefficients (K_d) of elements between mollusks' shells (*Unio* spp. and *Anodonta anatina*) and river water.

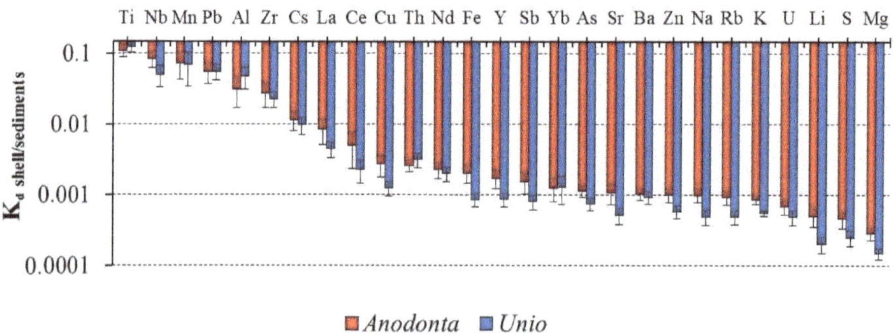

Figure 6. Calcium-normalized distribution coefficients (K_d) of elements between mollusks' shells (*Unio* spp. and *Anodonta anatina*) and bottom sediments.

The metal pollution index was calculated according to Usero et al. (1996) and Sedeño-Díaz et al. (2020) for *U. pictorum, U. tumidus and A. anatina* shells from the Severnaya Dvina River Basin. It varied from 0.59 to 4.75 for *A. anatina* and from 0.67 to 4.22 for *Unio* spp. For the shells from the Onega River Basin this index varied from 0.74 to 2.80 for *A. anatina* and from 0.45 to 0.81 for *Unio* spp.

4. Discussion

4.1. Element Inter-Correlations and Element Ratios as Indicators of Possible Mechanisms and Environmental Conditions

The correlations between element concentrations identified in this study may reflect several main mechanisms of TE accumulation in the shells such as: (*i*) metabolic reactions leading to carbonate mineral generation within the mollusk's body and co-precipitation with aragonite in the "extrapallial" fluid; (*ii*) complexation with organic matter (periostracum, winter lines, nacreous) of the shell structure; (*iii*) co-deposition with iron (III) or manganese (IV) hydroxide on the surface of growing shells and (*iv*) nonspecific passive capture (occlusion) of river water or sediment porewater particles (silicates, sulfides, organic debris) during filtration of the surrounding water (Bolotov et al., 2015).

There was a sizable (a factor of 7 to 11) increase in the Fe concentration in the *U. pictorum* and *U. tumidus* shells and in water from biotopes in the direction from the headwaters of the basin to the mouth of the Severnaya Dvina River (R^2 = 0.928, p < 0.01). In addition, such a regression relationship was also revealed for phosphorus in the shells of mollusks of this genus (R^2 = 0.934, p < 0.01), Figure 7. No such dependence was found for bottom sediments from the corresponding sampling points. In our opinion, this is due to variability of the bottom sediment composition at the sampling sites (from loam to medium-grained sand). In contrast, a strong positive correlation between Fe and P in shells of mollusks may indicate either P co-precipitation with Fe hydroxide or formation of Fe phosphate at the shell surface or impregnated between growing layers. Strong removal of P with Fe hydroxide from aqueous solutions is fairly well known [24–26]. The importance of this process for concomitant Fe and P accumulation in mollusks shells is consistent with the fact that the $K_{d\ Shell/Water}$ for Fe and P did not exhibit statistically significant regression relationships with the distance from the mouth of the Severnaya Dvina River (p > 0.05), despite sizable variations of these element concentrations in the river water.

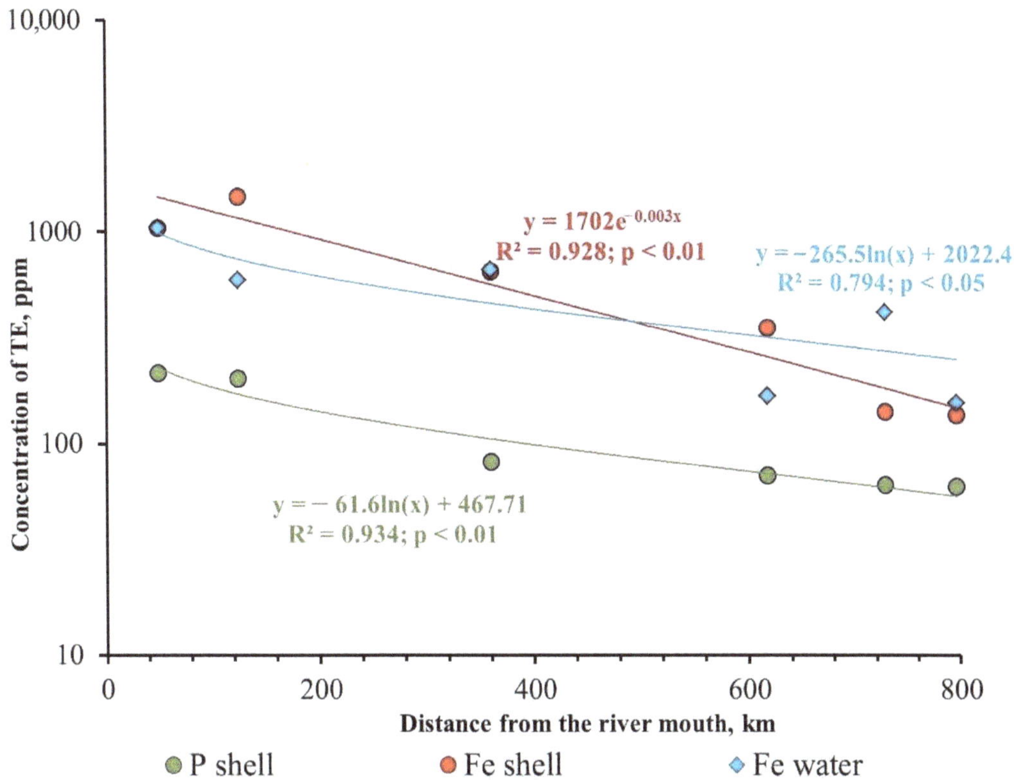

Figure 7. Changes in the concentration of iron and phosphorus in *U. pictorum* and *U. tumidus* shells and iron in the water from biotopes in the Severnaya Dvina Basin. The power relationships between element concentration (shells) (y) and distance from mouth (Severnaya Dvina) (x) are significant at p < 0.01 for shells' samples and at p < 0.05 for samples of water.

For the Severnaya Dvina River, Dzhamalov et al. [27] reported a downstream trend of an increase in Fe dissolved concentration and export fluxes. This trend can be linked to a progressive increase of the bog coverage and a shift in the watershed rock lithology, from carbonates and evaporates in the headwaters and small tributaries to silicate moraine and Archean metamorphic rocks in the low reaches of the basin [28,29]. The data on Fe concentration in shells and ecosystem compartments obtained in this study are consistent with these observations. A northward increasing trend in Fe concentration is evidenced from 6 sampling sites located in the section of the basin from the confluence of the Yug River to the mouth of the Severnaya Dvina River (Figure 7). Therefore, the shells of freshwater mussels with an average lifespan of about five years represent an archive that can be complementary to a long-term river water chemistry monitoring via hydrochemical analyses, i.e., [27,30].

Additional information about the environmental conditions of biotopes can be obtained via the concentration ratios of an element to Ca or Fe (i.e., Mn/Fe, Mn/Ca). A linear regression ($p = 0.0016$) was revealed between the Mn/Fe ratio in the *U. pictorum* and *U. tumidus* shells and in sediments from the corresponding biotopes in the Severnaya Dvina River Basin (Figure 8). Given that Mn/Fe ratio can serve as an indicator of the redox conditions at the sediment-water interface, i.e., [10], the accumulation of Mn relative to Ca in skeletons of mollusks reflects the corresponding redox conditions, which are operating under anoxic/suboxic conditions at the sediment-water interface.

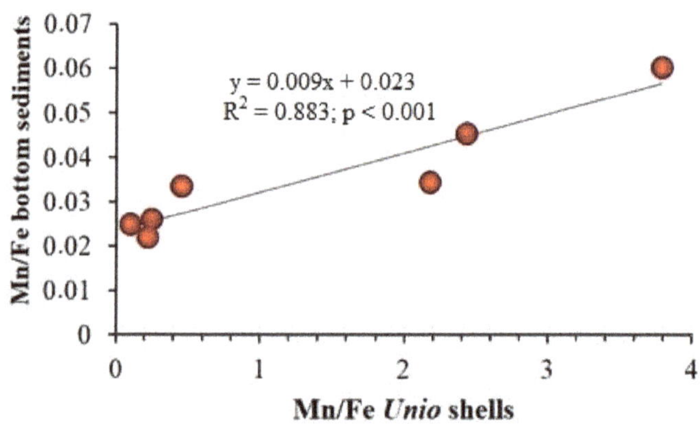

Figure 8. Regression relationships between element Mn/Fe-ratio in *U. pictorum* and *U. tumidus* shells (x) and bottom sediments (y) from Severnaya Dvina basin.

On the one hand, among various mechanisms, Mn, and Fe immobilization in the form of MnS and FeS both in the sediments and shells can be involved. This is consistent with elevated sulfide concentration in upper layer of riverine sediments in the region [31]. On the other hand, preferential uptake of Mn vs. Fe by shells can be explained by higher affinity of Mn^{2+} to the carbonate mineral structure compared to Fe^{3+} in the environmental conditions, which are not sufficiently reduced to convert Fe^{3+} into Fe^{2+}.

Zhao et al. [10] suggested the ranking of different species of freshwater mollusks with respect to the Mn/Ca ratio in watercourses of different trophic status. In terms of their average Mn/Ca (mmol/mol) values in shells, the *U. tumidus*, *U. pictorum* and *A. anatina* samples from the studied basins (the Severnaya Dvina and the Onega rivers) correspond to mesotrophic watercourses (Figure 9). The values of the Mn/Ca ratio in the shells of freshwater bivalve mollusks can reflect the occurrence of reducing conditions at the water-sediment interface; such conditions may occur due to eutrophication [10]. In the studied shells, the Mn/Ca ratios are higher in the *A. anatina* shells ($p < 0.01$), because this species lives under conditions of lower oxygen content, burrowing deeper into softer clay sediments in comparison with *Unio* specimens from the same biotope.

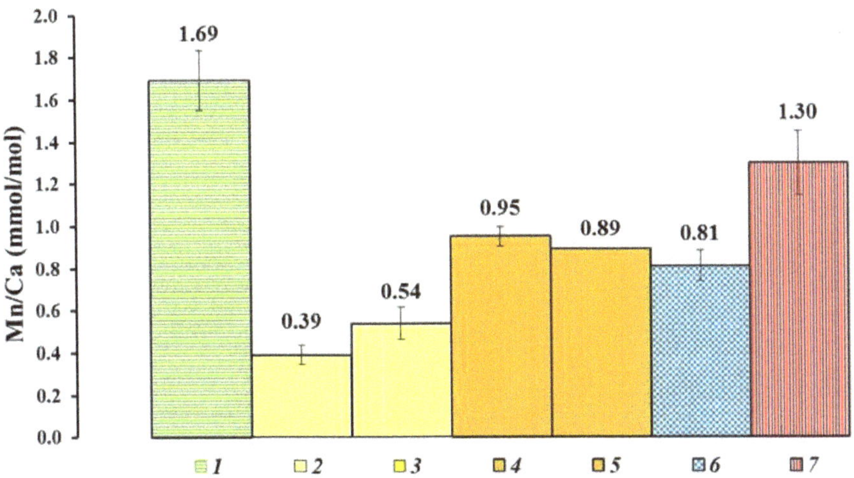

Figure 9. Comparison of Mn/Ca ratio (mmol/mol) in shells of freshwater bivalve mollusks: 1—*Margaritifera margaritifera* (oligotrophic rivers), 2—*M. laevis* (oligotrophic rivers), 3—*M. dahurica* (oligotrophic rivers), 4—*M. laevis* (mesotrophic rivers), 5—*M. dahurica* (mesotrophic rivers), 6—*U. tumidus and U. pictorum* (mesotrophic rivers), 7—*Anodonta anatina* (mesotrophic rivers) (References: 1–5—Bolotov et al., 2015, 6–7—this study).

4.2. Relationship between the Content of Chemical Elements in Shells, Water and Sediments

The distribution coefficients (K_d) of chemical elements between different compartments of the ecosystem were calculated to assess the efficiency of element uptake by mollusks shells from the surrounding milieu. The lowest values of the distribution coefficients ($K_{d\ Shell/Water}$, $K_{d\ Shell/Sediments}$) of highly mobile elements such as Na, S, Sr, Sb, and U were established in the localities of the Severnaya Dvina Basin. These elements are most abundant in the bottom water layer. Lithophile elements (Al, Ti, Mn, Cu, Zr, Pb) exhibited the highest values of the distribution coefficient between the shell and the river water or the sediment.

According to the graphs presented in Figure 10, Rb, Sb, and Pb exhibited linear regression coefficients of Log $K_{d\ Shell/Water}$ (X) and Log $K_{d\ Shell/Sediment}$ (Y). A positive regression was found for Rb ($R^2 = 0.58$; $p < 0.05$) and Sb ($R^2 = 0.54$; $p < 0.05$), whereas a negative regression was observed for Pb ($R^2 = 0.74$; $p < 0.005$).

Overall, an increase in $K_{d\ Shell/Water}$ with $K_{d\ Sediments/Water}$ may reflect (1) a concomitant immobilization of an element from the river water to the shells and the sediments; (2) efficient uptake of an element by mollusks from the surrounding sediments.

Another important observation is that sodium, zinc, barium, and lanthanum demonstrated significant ($p < 0.005$) linear dependence of Log $K_{d\ Shell/Water}$ and Log $K_{d\ Sediment/Water}$ values (Figure 11). These ratios reflect the dependence of the element contents in the shells and sediment concentration in water. Some elements can accumulate in greater amounts in the sediments, with the highest K_d values of Zn and La corresponding to localities with a clay substrate. Sodium has lower values of the distribution coefficients between sediments and water, as well as between shells and water, which is associated with its relatively high concentration and mobility in the liquid phase. The correlations observed between the distribution coefficients of elements between shells/water and sediments by water show that Na, Zn, Ba, and La (and generally all light REE) are more evenly distributed between liquid/solid phases and shells compared to other studied elements, which is consistent with recent results of Sedeño-Díaz et al. [22].

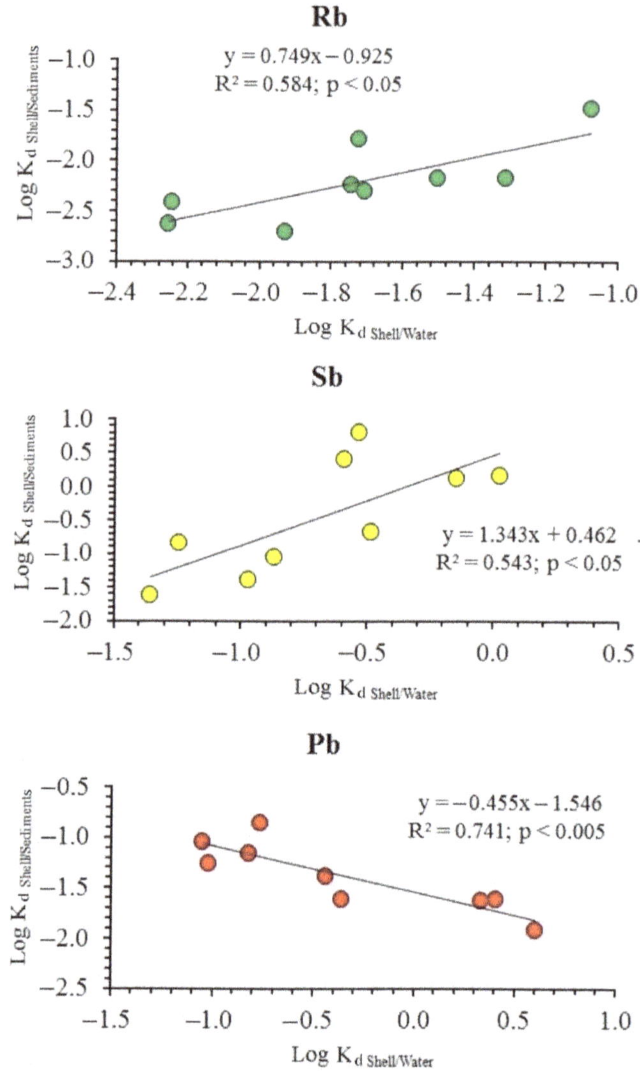

Figure 10. Regression relationships between Log $K_{d\ Shell/Water}$ (X) and Log $K_{d\ Shell/Sediments}$ (Y) for Rb, Sb, and Pb.

A dependence of the metal pollution index for *Anodonta* and *Unio* shells as a function of the distance to the river mouth was strongly pronounced for the Severnaya Dvina River Basin. This dependence can be approximated by a power function ($R^2 = 0.824$, $p < 0.05$, Figure 12). The metal pollution index (MPI) includes the sum of the concentrations of all metal elements (without Ca) accumulated in shells following Usero et al. [21,32] and Stankovic et al. [33]. The MPI value increased in the direction from the headwaters to the mouth of the Severnaya Dvina River. Such a downstream accumulative pattern can be explained by the location of large agglomeration (Arkhangelsk city, population of 354,100 inhabitants) in the low reaches of the river. Statistically significant differences between locations were found for the metal pollution index ($p = 0.003$). The highest index values were found for biotopes located in the mouth of the Severnaya Dvina and the Onega rivers. This likely reflects an increasing local anthropogenic pressure (urban sewage waters, some industry) from the town of Onega and Arkhangelsk, located in the mouth zone of both rivers. At the same time, there was no statistically significant regression relationship for concentrations of Zn, Cd, Pb, As, Sb, Cu, Ni, Co, Cr in shells and for $K_{d\ Shell/Water}$ with the distance from the mouth of the Severnaya Dvina River ($p > 0.05$). The contrast in Fe and metal/metalloid spatial pattern is noteworthy as it suggests that sole urban pollution cannot be responsible for elevated Fe concentration in the low reaches of the Severnaya Dvina River. Rather, enhanced proportion of bogs and wetlands in the mouth zone of Severnaya Dvina may provide high amount of Fe, essentially in the form of organo-ferric colloids i.e., [28,34] to the river main stem.

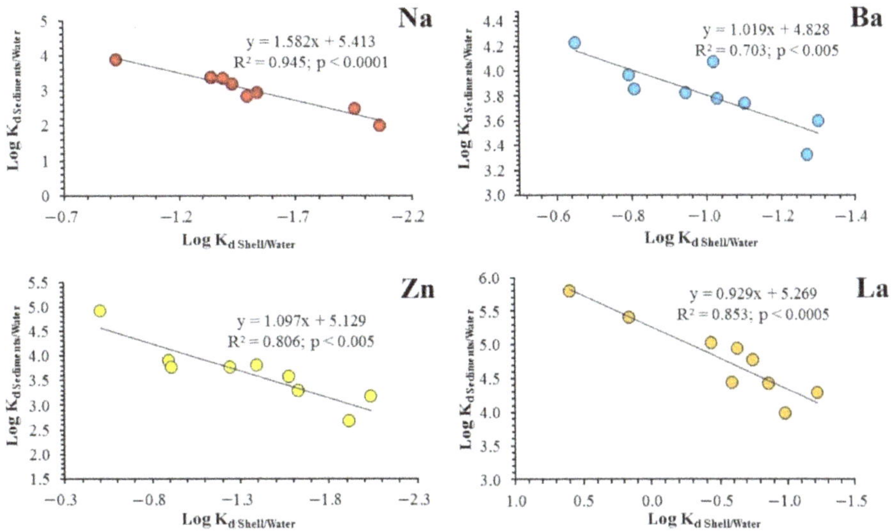

Figure 11. Regression relationships between Log $K_{d\ Shell/Water}$ and Log $K_{d\ Sediments/Water}$ for Na, Zn, Ba and La.

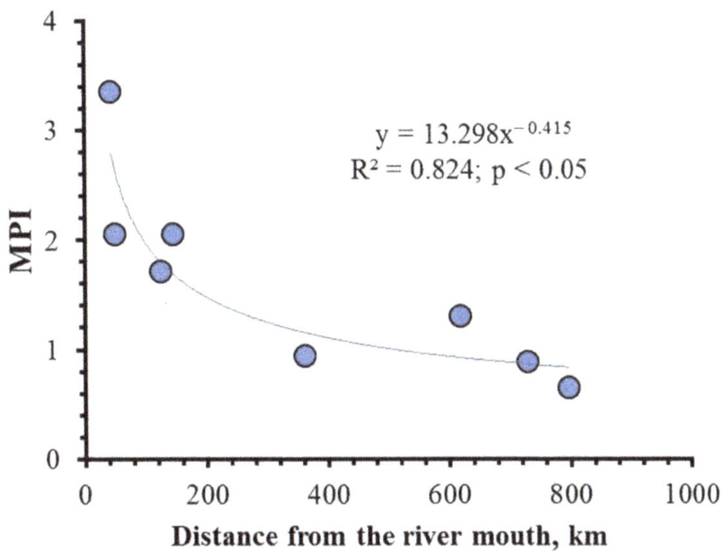

Figure 12. Regression relationships between values of metal pollution index (Y axis) and the distance from the mouth of the Severnaya Dvina River (X axis) are significant at $p < 0.05$.

5. Conclusions

Analyses of the chemical composition of mollusks shells, water, and sediments of the two largest European Arctic Rivers—Severnaya Dvina and Onega—demonstrated a sizable accumulation of a small number of elements (Mg, Al, Mn, Ba, and U) in the shells. These elements exhibited statistically significant differences between studied genera and allowed to trace the ecological preferences of certain species in terms of the substrate. The differences in element concentrations in shells and ecosystem compartments (river water and sediments) between river basins partially stem from the lithological composition of the catchment area and the presence of bogs at the watershed.

The specificity of lithological and mineralogical composition of bottom sediments controls the conditions of chemical element distribution between the components of the riverine ecosystem in each locality. These results suggest that element accumulation in the shell depends on specific conditions of the biotope.

The concentrations of iron in the water and shells and phosphorus in the shells of bivalve mollusks (*Unio* spp.) increased in the direction from the upper parts of the basin to the low reaches of the Severnaya Dvina River. This suggested strong influx of iron from the bogs of the catchment to the river and Fe accumulation by the freshwater mussels.

The principal component analysis revealed three main factors; the first was interpreted as the concentration pattern (1), the second one was responsible for accumulation of iron and phosphorus during shell growth (2), and the third is reflecting the conditions of biotopes associated with the composition of the bottom sediments (3).

The metal pollution indices demonstrated progressive enrichment in the confluence of the Severnaya Dvina River and the Pinega River and also in the vicinity of the river mouth, reflecting possible point source contamination originated from industrial and pollution activity.

Supplementary Materials: The following are available online at https://www.mdpi.com/article/10.3390/w13223227/s1, Figure S1: the power relationships between Fe and P concentrations in *A. anatina* (1) and *Unio* spp. (2) shells; Figure S2: the power relationships between Al and Mg concentrations in *A. anatina* (1) and *Unio* spp. (2) shells; Figure S3: the power relationships between Al and K concentrations in *A. anatina* (1) and *Unio* spp. (2) shells; Figure S4: the power relationships between Fe and Y concentrations in *A. anatina* (1) and *Unio* spp. (2) shells.

Author Contributions: O.S.P., A.A.L. and I.N.B. developed the concept of this study. A.A.T., I.V.V., O.V.T., A.S.O. and I.N.B. collected samples. A.A.L. and A.A.T. carried out preparation of samples for chemical analyses. A.A.L. and N.A.Z. performed statistical processing of the geochemical data. I.V.V. and A.A.T. determined taxonomic position of individuals of studied mollusks' samples. A.A.L. and O.S.P. wrote the paper, with input from A.V.C., A.A.T., A.V.K. and I.N.B. All authors have read and agreed to the published version of the manuscript.

Funding: This research was funded by the Russian Science Foundation, Grant Number 21-17-00126.

Institutional Review Board Statement: Not applicable.

Informed Consent Statement: Not applicable.

Data Availability Statement: The data that support the findings of this study are available from the corresponding author upon reasonable request.

Acknowledgments: We are grateful to those who assisted us in sample collection for chemical analyses: Svetlana E. Sokolova, Yulia V. Bespalaya, Galina D. Popova, Elena D. Lyubas, and Maxim V. Popov. We also wish to express our gratitude to Vasilii K. Karandashev (IMT RAS, Chernogolovka, Moscow Region) for assisting in chemical analyses.

Conflicts of Interest: The authors declare no conflict of interest.

References

1. Schöne, B.R.; Dunca, E.; Mutvei, H.; Norlund, U. A 217-year record of summer air temperature reconstructed from freshwater pearl mussels (*M. margaritifera*, Sweden). *Quat. Sci. Rev.* **2004**, *23*, 1803–1816. [CrossRef]
2. Schöne, B.R.; Page, N.A.; Rodland, D.L.; Fiebig, J.; Baier, S.; Helama, S.O.; Oschmann, W. ENSO-coupled precipitation records (1959–2004) based on shells of freshwater bivalve mollusks (*Margaritifera falcata*) from British Columbia. *Int. J. Earth Sci.* **2007**, *96*, 525–540. [CrossRef]
3. Van der Putten, E.; Dehairs, F.; Keppens, E.; Baeyens, W. High resolution distribution of trace elements in the calcite shell layer of modern *Mytilus edulis*: Environmental and biological controls. *Geochim. Cosmochim. Acta.* **2000**, *64*, 997–1011. [CrossRef]
4. Roditi, H.; Fisher, N.; Sañudo-Wilhelmy, S. Uptake of dissolved organic carbon and trace elements by zebra mussels. *Nature* **2000**, *407*, 78–80. [CrossRef]
5. Dunca, E.; Mutvei, H.; Schöne, B.R. Freshwater bivalves tell of past climates: But how clearly do shells from polluted rivers speak? *Palaeogeogr. Palaeoclimatol. Palaeoecol.* **2005**, *228*, 43–57. [CrossRef]
6. Bolotov, I.N.; Pokrovsky, O.S.; Auda, Y.; Bespalaya, J.V.; Vikhrev, I.V.; Gofarov, M.Y.; Lyubas, A.A.; Viers, J.; Zouiten, C. Trace element composition of freshwater pearl mussels *Margaritifera* spp. across Eurasia: Testing the effect of species and geographic location. *Chem. Geol.* **2015**, *402*, 125–139. [CrossRef]
7. Lyubas, A.A.; Kabakov, M.B.; Kriauciunas, V.V.; Obada, T.F.; Nicoara, I.N.; Tomilova, A.A. Freshwater mollusks from Neogene-Quaternary Dniester and Prut riverine deposits as indicator paleoenvironments: Chemical composition of shells and its palaeoecological interpretation. *Arct. Environ. Res.* **2019**, *19*, 35–42. [CrossRef]
8. Van Plantinga, A.A.; Grossman, E.L. Trace elements in mussel shells from the Brazos River, Texas: Environmental and biological control. *Biogeosci. Discuss.* **2019**, 1–27. [CrossRef]
9. Aljahdali, M.O.; Alhassan, A.B. Spatial variation of metallic contamination and its ecological risk in sediment and freshwater mollusk: *Melanoides tuberculata* (Müller, 1774) (Gastropoda: Thiaridae). *Water* **2020**, *12*, 206. [CrossRef]
10. Zhao, L.; Walliser, E.O.; Mertz-Kraus, R.; Schöne, B.R. Unionid shells (*Hyriopsis cumingii*) record manganese cycling at the sediment–water interface in a shallow eutrophic lake in China (Lake Taihu). *Palaeogeogr. Palaeoclimatol. Palaeoecol.* **2017**, *484*, 97–108. [CrossRef]
11. Naeher, S.; Gilli, A.; North, R.P.; Hamann, Y.; Schubert, C.J. Tracing bottom water oxygenation with sedimentary Mn/Fe ratios in Lake Zurich, Switzerland. *Chem. Geol.* **2013**, *352*, 125–133. [CrossRef]
12. Ravera, O.; Cenci, R.; Beone, G.M.; Dantas, M.; Lodigiani, P. Trace element concentrations in freshwater mussels and macrophytes as related to those in their environment. *J. Limnol.* **2003**, *62*, 61–70. [CrossRef]
13. Watanabe, T.; Suzuki, M.; Komoto, Y.; Shirai, K.; Yamazaki, A. Daily and annual shell growth in a long-lived freshwater bivalve as a proxy for winter snowpack. *Palaeogeogr. Palaeoclimatol. Palaeoecol.* **2021**, *569*, 110346. [CrossRef]
14. Zhadin, V.I. Mollusks of fresh and brackish waters of the USSR. *Keys Fauna USSR Publ. Zool. Inst. USSR Acad. Sci.* **1952**, *46*, 1–376. (In Russian)
15. Klishko, O.K.; Lopes-Lima, M.; Bogan, A.E.; Matafonov, D.V.; Froufe, E. Morphological and molecular analyses of Anodontinae species (Bivalvia, Unionidae) of Lake Baikal and Transbaikalia. *PLoS ONE* **2018**, *13*, e0194944. [CrossRef]
16. Bolotov, I.N.; Kondakov, A.V.; Konopleva, E.S.; Vikhrev, I.V.; Aksenova, O.V.; Aksenov, A.S.; Bespalaya, Y.V.; Borovskoy, A.V.; Danilov, P.P.; Dvoryankin, G.A.; et al. Integrative taxonomy, biogeography and conservation of freshwater mussels (Unionidae) in Russia. *Sci. Rep.* **2020**, *10*, 1–20. [CrossRef]
17. Konopleva, E.S.; Bolotov, I.N.; Vikhrev, I.V.; Gofarov, M.Y.; Kondakov, A.V. An integrative approach underscores the taxonomic status of *Lamellidens exolescens*, a freshwater mussel from the Oriental tropics (Bivalvia: Unionidae). *System. Biodivers.* **2017**, *15*, 204–217. [CrossRef]
18. Konopleva, E.S.; Pfeiffer, J.M.; Vikhrev, I.V.; Kondakov, A.V.; Gofarov, M.Y.; Aksenova, O.V.; Lunn, Z.; Chan, N.; Bolotov, I.N. A new genus and two new species of freshwater mussels (Unionidae) from western Indochina. *Sci. Rep.* **2019**, *9*, 1–14. [CrossRef]
19. Karandashev, V.K.; Turanov, A.N.; Orlova, T.A.; Lezhnev, A.E.; Nosenko, S.V.; Zolotareva, N.I.; Moskvina, I.R. Application of mass-spectrometry with inductively coupled plasma to elemental analysis of environmental objects. *Ind. Lab. Diagn. Mater.* **2007**, *73*, 12–22.
20. Karandashev, V.K.; Leikin, A.Y.; Khvostikov, V.A.; Kutseva, N.K.; Pirogova, S.V. Water analysis by inductively coupled plasma mass spectrometry. *Inorg. Mater.* **2016**, *52*, 1391–1404. [CrossRef]
21. Usero, J.; González, E.; Regalado, L.; Gracia, I. Trace metals in the bivalve mollusc *Chamelea gallina* from the Atlantic coast of Southern Spain. *Baseline* **1996**, *32*, 305–310. [CrossRef]
22. Sedeño-Díaz, J.E.; López-López, E.; Mendoza-Martínez, E.; Rodríguez-Romero, A.J.; Morales-García, S.S. Distribution coefficient and metal pollution index in water and sediments: Proposal of a new index for ecological risk assessment of metals. *Water* **2020**, *12*, 29. [CrossRef]
23. Gural'-Sverdlova, N.V.; Gural', R.I. Mollusks of the Unionidae Family in the Funds of the State Museum of Natural History of the National Academy of Sciences of Ukraine, Their Conchological Variability and Diagnostic Features. 2015. Available online: http://www.pip-mollusca.org/page/epubl/unionidae.php (accessed on 25 March 2021).
24. Takacs, I.; Murthy, S.; Smith, S.; McGrath, M. Chemical phosphorus removal to extremely low levels: Experience of two plants in the Washington, DC area. *Water Sci. Technol.* **2006**, *53*, 21–28. [CrossRef] [PubMed]

25. Liu, Y.-T.; Hesterberg, D. Phosphate Bonding on Noncrystalline Al/Fe-Hydroxide Coprecipitates. *Environ. Sci. Technol.* **2011**, *45*, 6283–6289. [CrossRef] [PubMed]
26. Chen, K.Y.; Hsu, L.C.; Chan, Y.T.; Cho, Y.L.; Tsao, F.Y.; Tzou, Y.M.; Hsieh, Y.C.; Liu, Y.T. Phosphate removal in relation to structural development of humic acid-iron coprecipitates. *Sci. Rep.* **2018**, *8*, 10363. [CrossRef]
27. Dzhamalov, R.G.; Mironenko, A.A.; Myagkova, K.G.; Reshetnyak, O.S.; Safronova, T.I. Space–time analysis of hydrochemical composition and pollution of water in the Severnaya Dvina Basin. *Water Resour.* **2019**, *46*, 188–198. [CrossRef]
28. Pokrovsky, O.S.; Viers, J.; Shirokova, L.S.; Shevchenko, V.P.; Fillipov, A.S.; Dupre, B. Dissolved, suspended, and colloidal fluxes of organic carbon, major and trace elements in Severnaya Dvina River and its tributary. *Chem. Geol.* **2010**, *273*, 136–149. [CrossRef]
29. Malov, A.I. The use of the geological benchmarks to assess the residence time of groundwater in the aquifer using uranium isotopes on the example of the Northern Dvina basin. *Lithol. Mineral Resour.* **2013**, *48*, 254–265. [CrossRef]
30. Chupakov, A.V.; Pokrovsky, O.S.; Moreva, O.Y.; Shirokova, L.S.; Neverova, N.V.; Chupakova, A.A.; Kotova, E.I.; Vorobyeva, T.Y. High resolution multi-annual riverine fluxes of organic carbon, nutrient and trace element from the largest European Arctic river, Severnaya Dvina. *Chem. Geol.* **2020**, *538*, 119491. [CrossRef]
31. Kokryatskaya, N.M.; Volkov, I.I.; Demidova, T.P.; Murzina, T.S. Sulfur compounds in bottom sediments of fresh-water basins (mouth of the Severnaya Dvina River and Rybinsk Water Reservoir). *Lithol. Miner. Resour.* **2003**, *38*, 552–563. [CrossRef]
32. Usero, J.; Gonzalez-Regalado, E.; Gracia, I. Trace metals in the bivalve molluscs *Ruditapes decussatus* and *Ruditapes philippinarum* from the Atlantic coast of southern Spain. *Environ. Int.* **1997**, *23*, 291–298. [CrossRef]
33. Stankovic, S.; Jovic, M.; Mihajlovic, M.L.; Joksimović, D.; Tanaskovski, B. Metal pollution determined by pollution indices for sea grass *P. oceanica* and surface sediments. *Arch. Biol. Sci.* **2015**, *67*, 91–101. [CrossRef]
34. Pokrovsky, O.S.; Shirokova, L.S.; Viers, J.; Gordeev, V.V.; Shevchenko, V.P.; Chupakov, A.V.; Vorobieva, T.Y.; Candaudap, F.; Causserand, C.; Lanzanova, A.; et al. Fate of colloids during estuarine mixing in the Arctic. *Ocean Sci.* **2014**, *10*, 107–125. [CrossRef]

water MDPI

Article

Landscape, Soil, Lithology, Climate and Permafrost Control on Dissolved Carbon, Major and Trace Elements in the Ob River, Western Siberia

Iurii Kolesnichenko [1], Larisa G. Kolesnichenko [1], Sergey N. Vorobyev [1], Liudmila S. Shirokova [2,3], Igor P. Semiletov [4,5,6,7], Oleg V. Dudarev [4], Rostislav S. Vorobev [1], Uliana Shavrina [1], Sergey N. Kirpotin [1] and Oleg S. Pokrovsky [1,2,3,*]

1 BIO-GEO-CLIM Laboratory, Tomsk State University, Lenin pr., 36, 634050 Tomsk, Russia; vancansywork@mail.ru (I.K.); klg77777@mail.ru (L.G.K.); soil@green.tsu.ru (S.N.V.); tsu@erd.su (R.S.V.); ulyanashavrina@yandex.ru (U.S.); kirp@mail.tsu.ru (S.N.K.)
2 Geosciences and Environment Toulouse, UMR 5563 CNRS, 14 Avenue Edouard Belin, 31400 Toulouse, France; Liudmila.SHIROKOVA@Get.omp.eu
3 N. Laverov Federal Center for Integrated Arctic Research, Russian Academy of Sciences, 163069 Arkhangelsk, Russia
4 Laboratory of Arctic Research, V.I. Il'ichev Pacific Oceanological Institute, Far Eastern Branch of Russian Academy of Sciences, 43 Baltic Street, 690041 Vladivostok, Russia; ipsemiletov@alaska.edu (I.P.S.); dudarev@poi.dvo.ru (O.V.D.)
5 Tomsk Polytechnic University, 634050 Tomsk, Russia
6 Tomsk State University, 634050 Tomsk, Russia
7 Department of Chemistry, Moscow State University, 119192 Moscow, Russia
* Correspondence: oleg.pokrovsky@get.omp.eu

Citation: Kolesnichenko, I.; Kolesnichenko, L.G.; Vorobyev, S.N.; Shirokova, L.S.; Semiletov, I.P.; Dudarev, O.V.; Vorobev, R.S.; Shavrina, U.; Kirpotin, S.N.; Pokrovsky, O.S. Landscape, Soil, Lithology, Climate and Permafrost Control on Dissolved Carbon, Major and Trace Elements in the Ob River, Western Siberia. *Water* **2021**, *13*, 3189. https://doi.org/10.3390/w13223189

Academic Editor: Carlo Camporeale

Received: 14 October 2021
Accepted: 8 November 2021
Published: 11 November 2021

Abstract: In order to foresee possible changes in the elementary composition of Arctic river waters, complex studies with extensive spatial coverage, including gradients in climate and landscape parameters, are needed. Here, we used the unique position of the Ob River, draining through the vast partially frozen peatlands of the western Siberia Lowland and encompassing a sizable gradient of climate, permafrost, vegetation, soils and Quaternary deposits, to assess a snap-shot (8–23 July 2016) concentration of all major and trace elements in the main stem (~3000 km transect from the Tom River confluence in the south to Salekhard in the north) and its 11 tributaries. During the studied period, corresponding to the end of the spring flood-summer baseflow, there was a systematic decrease, from the south to the north, of Dissolved Inorganic Carbon (DIC), Specific Conductivity, Ca and some labile trace elements (Mo, W and U). In contrast, Dissolved Organic Carbon (DOC), Fe, P, divalent metals (Mn, Ni, Cu, Co and Pb) and low mobile trace elements (Y, Nb, REEs, Ti, Zr, Hf and Th) sizably increased their concentration northward. The observed latitudinal pattern in element concentrations can be explained by progressive disconnection of groundwaters from the main river and its tributaries due to a northward increase in the permafrost coverage. A northward increase in bog versus forest coverage and an increase in DOC and Fe export enhanced the mobilization of insoluble, low mobile elements which were present in organo-ferric colloids (1 kDa—0.45 μm), as confirmed by an in-situ dialysis size fractionation procedure. The chemical composition of the sampled mainstream and tributaries demonstrated significant ($p < 0.01$) control of latitude of the sampling point; permafrost coverage; proportion of bogs, lakes and floodplain coverage and lacustrine and fluvio-glacial Quaternary deposits of the watershed. This impact was mostly pronounced on DOC, Fe, P, divalent metals (Mn, Co, Ni, Cu and Pb), Rb and low mobile lithogenic trace elements (Al, Ti, Cr, Y, Zr, Nb, REEs, Hf and Th). The pH and concentrations of soluble, highly mobile elements (DIC, SO_4, Ca, Sr, Ba, Mo, Sb, W and U) positively correlated with the proportion of forest, loesses, eluvial, eolian, and fluvial Quaternary deposits on the watershed. Consistent with these correlations, a Principal Component Analysis demonstrated two main factors explaining the variability of major and trace element concentration in the Ob River main stem and tributaries. The DOC, Fe, divalent metals and trivalent and tetravalent trace elements were presumably controlled by a northward increase in permafrost, floodplain, bogs, lakes and lacustrine deposits on the watersheds. The DIC and labile alkaline-earth metals, oxyanions (Mo, Sb and W) and U were impacted by southward-dominating forest coverage, loesses and eluvial and fertile soils. Assuming that climate warming in the WSL will lead to a northward shift of the forest and permafrost boundaries, a "substituting space for time" approach predicts a future increase in the concentration of DIC and labile major and trace elements and a decrease of the transport of DOC and low soluble trace metals in the form of colloids in the main stem of the Ob River. Overall, seasonally-resolved transect studies of large riverine systems of western Siberia are needed to assess the hydrochemical response of this environmentally-important territory to on-going climate change.

Keywords: river; forest; bog; permafrost; carbon; major ions; iron; colloids; trace element

1. Introduction

Studies on hydrochemistry of large Arctic rivers are at the forefront of climate warming research due to their high importance in carbon (C) and greenhouse gases (GHG) regulation at the planetary scale and their high vulnerability to ongoing environmental changes [1,2]. Presently, researchers have achieved a satisfactory understanding of hydrological fluxes and river water hydrochemistry, including both suspended and particulate load, at the terminal (gauged) stations across the Arctic. This was possible thanks to the systematic work of the State Hydrological Surveys of the main Artic countries [3] and, more recently, concerted works of various international programs [4–9]. In contrast, the knowledge of spatial variations of major and trace components of the river water along the main stem of Arctic rivers and their tributaries, which is necessary for understanding the environmental controls and export mechanisms of riverine solutes, remains rather limited.

The Ob River, which is the largest Arctic river in terms of its watershed area (2,975,000 km^2), is an important vector of carbon, nutrients and major and trace element transfer to the Kara Sea [10,11]. It drains highly vulnerable discontinuous and sporadic permafrost (20% in average), which is extremely rich in organic C (OC) due to the dominance of peat soils [12]. Most recent hydrological studies demonstrated that, over past 80 years, the Ob River discharge increased by ca 7.7% [13] at a rate of 384 and 173 m^3/s (10 year^{-1}) in spring and winter, respectively, which was linked to a rapid increase in both warming and wetting of the territory [14]. This, together with its unique geographical situation and landscape setting, render the Ob River watershed among the key targeting regions for biogeochemical studies in the Arctic [2]. Indeed, the number of publications per year on the Ob River has increased from <10 prior to 1995 to 10–20 in 1996–2014 and 50–80 over the past 6 years.

Several detailed studies of Dissolved Organic Carbon (DOC) and major elements were conducted at the terminal gauging station of the river, near the Salekhard city [4–6]. These included DOC time-series observations by molecular-level techniques [15] and via remote sensing [16] and the quantification of particulate organic matter export [7]. In contrast, spatial coverage of the river main stem and its tributaries remains rather low, with just a few studies of the dissolved carbon and related CO_2 and CH_4 emissions [17,18] and one study of the molecular composition of DOC [19]. A large amount of data is available from the systematic State Rosgidromet monitoring of OC and major ions on four gauging stations of the Ob River (Salekhard, Belogor'e, Aleksandrovskoe and Kolpashevo) during 1970–2010 and measurements by Tomsk Politechnical University, as summarized in ref. [20]. The OC-controlled export of Hg by the Ob River has been considered by Mu et al. [21] and Sonke et al. [22]. A snapshot study of ^{137}Cs was performed at the scale of the Ob Basin [23]. Much less is known about other major and trace elements, especially their variations among the tributaries.

An important feature of the Ob River basin is the dominance of wetlands and mires, which contain huge amounts of OC and provide a sizable input of DOM and relevant metals, such as Fe, to the Ob River main stem and tributaries due to strong hydrological connectivity (i.e., [24–26]). As a result of this enhanced input of Fe and DOC to the Ob

River, its waters are likely to contain a high concentration of colloids. However, this aspect has never been tested for the Ob River, although organic and organo-mineral colloids can be important carriers of a number of trace elements, as is known in small rivers [27] and surface waters [28,29] across western Siberia.

The novelty of the present work is to acquire a snap-shot picture of the DOC and major and trace elements (TE), including their colloidal forms, in the main stem and several tributaries of the Ob River and to test, for the first time for this huge territory, a landscape control, including vegetation, type of soil and Quaternary deposits, on the chemical composition of the river water. Testing the landscape control on riverine solutes is now possible due to significant progress in digitalizing the available vegetation, permafrost and climate maps of large territories that allows straightforward landscape-based interpretation of river water chemistry (see examples in [30–32]). In this study, we hypothesized a dominant control of bog and permafrost coverage on the concentrations of DOC and major and trace elements in the main stem. In particular, in accord with previous studies of small WSL rivers [10,30,31,33–36], we expected a northward increase in DOC and a decrease in soluble alkaline-earth metal concentrations. In order to reveal the mechanisms of element transport in the Ob River and its tributaries, we assessed the colloidal and truly dissolved (low molecular weight) concentration of carbon and all major and trace elements in selected samples. We hypothesized a change in organic colloids after the confluence of the Ob and Irtysh rivers, following a recent study of DOM pattern in the Ob River main stem [19]. We anticipate that acquiring this new knowledge on riverine major and trace elements over the large climate, permafrost and vegetation gradient of the Ob River basin will allow empirical but straightforward prediction of future changes in river water chemistry, following a well-established "substituting space for time" approach.

2. Study Site and Methods

2.1. The Ob River of the WSL and Its Sampling

The Ob River, which delivers 15% of the total freshwater flow to the Arctic Ocean (404 $km^3 \cdot year^{-1}$), combines the features of southern rivers, draining steppe and forest-steppe regions, via its longest tributary, the Irtysh River, and the western Siberia Lowland rivers draining through a mixture of forest and mires. Further in the north, the Ob River is influenced by permafrost peatlands. The largest tributary of the Ob River is the Irtysh, which drains a territory of 1,643,000 km^2 and exhibits an annual discharge of 88.371 km^3, which is 37.5% of that of the Ob River upstream at their confluence at Khanty-Mansiisk. During the open water high flow period of 2016, the contribution of the Irtysh amounted to 49% of the Ob River discharge upstream of the confluence.

In this work, we used ship ('OM-341' vessel, see ref. [19]) route sampling and collected 23 main stem and 6 secondary channel water samples going from the north to the south between 8 July 2016 and 23 July 2016 (Figure 1). We covered, in total, 2952 km of the river length, which encompassed a sizable gradient in the latitude (from 66°47′19″ N in the most northern site, Salemal, to 56°54′24″ N in the most southern site, Kozjulyno, Supplementary Material Table S1). The 23 sampling points of the main stem are sufficiently representative to cover the full variability of hydrochemical parameters of the river during the studied season. This is supported by the highly homogeneous physio-geographical setting of the WSL, with minimal variations in runoff, lithology, vegetation and anthropogenic pressure over the quite large north–south gradient (i.e., [29–36]), as also reflected in the quite smooth pattern of the CO_2 concentration in the Ob River across the same latitudinal gradient [18]. In addition to the main stem, 11 tributaries of the Ob River were collected. The summer of 2016 was rather cold (−3.1° below normal (defined as prior to 2000) for July) and slightly humid (130% of the normal precipitation for July).

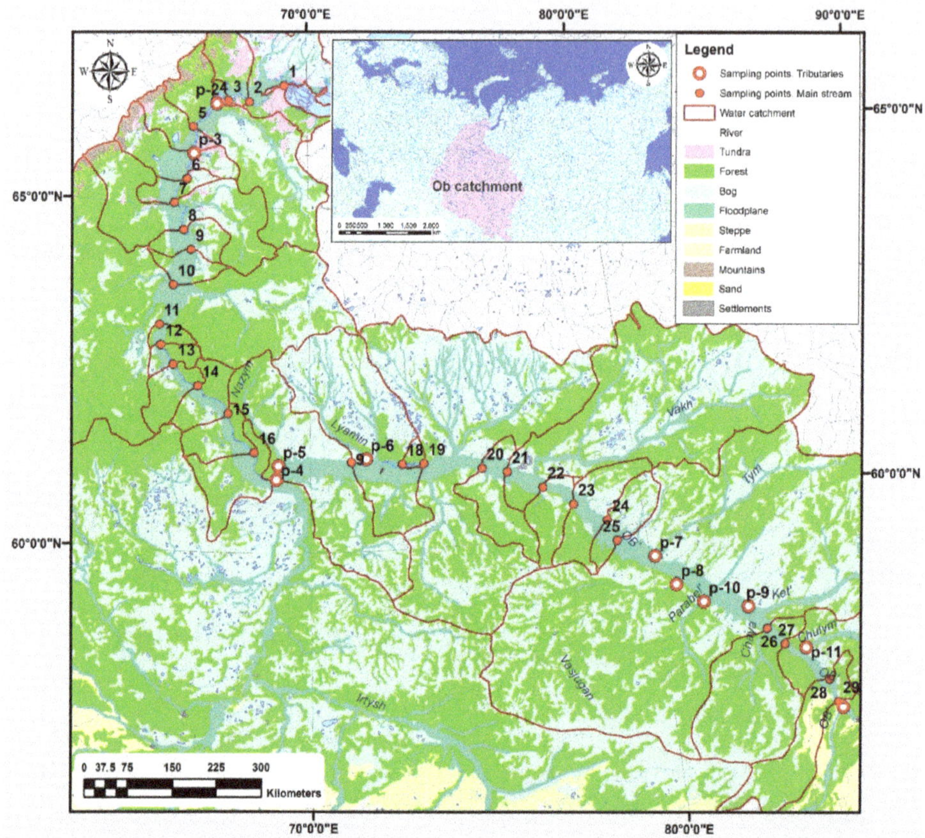

Figure 1. The Ob River and tributaries sampling points (solid and open circles, respectively) showing the dominant landscapes of the Ob River basin.

The river water was collected from the surface (0.5 m depth) in the central part of the main stem or a minimum of 500 m upstream of the tributary via a pre-cleaned polypropylene 1 L container and immediately filtered (<0.45 μm regenerated cellulose filter) using a pre-cleaned 250 mL polysulfone Nalgene filter unit, combined with a vacuum pump. First, 250 mL of MilliQ water was filtered, and the first portion of the river water filtrate (250 mL) was discarded. There was no decrease in the rate of filtration for the sampled volumes (typically less than 500 mL), so we did not expect any artifacts linked to filter clogging.

In addition to traditional 0.45 μm filtration, 10 selected samples of the main stem and tributaries were processed for 1 kDa (~0.0013 nm) dialysis. This technique allows one to quantify the nominal low molecular weight ($LMW_{<1kDa}$) and colloidal (1 kDa—0.45 μm) fractions. For this, large volumes of the river water were collected into a thoroughly cleaned 5 L plastic container via filtration through a 20 μm Nylon net, to avoid large particles, zooplankton and insects. Dialysis experiments were performed using 50 mL pre-cleaned dialysis bags placed in the river water over 3 to 5 days, as described elsewhere [28,37]. The plastic container was kept in darkness at a temperature similar to that of the river water and gently agitated due to ship movement and manually. The stability of the river water chemical composition during the full length of the dialysis procedure was verified by comparison of the dissolved (<0.45 μm) concentrations of all solutes before and after the exposure; the difference did not exceed 10% (at $p < 0.01$). As such, even if some microbial activity could occur during dialysis, it did not modify the colloidal composition of the river water.

2.2. Analytical Techniques

All filtered and dialyzed samples were stored in the refrigerator for 1 month prior to the analyses. Major anion concentrations (Cl^- and SO_4^{2-}) were measured by ion chromatography (Dionex 2000i) with an uncertainty of 2%. The dissolved organic carbon (DOC) and dissolved inorganic carbon (DIC) was determined using a Shimadzu TOC-VSCN Analyzer with an uncertainty of 3% and a detection limit of 0.3 mg/L [38]. In samples with low DOC (<10 mg/L), the DIC was also analyzed via a potentiometric alkalinity titration procedure with an uncertainty of $\pm 1\%$ and a detection limit of 5×10^{-5} mol L^{-1}; the difference between the results of the Shimadzu did not exceed 5%.

Major and trace elements were measured by quadrupole ICP-MS (7500ce, Agilent Technologies). Indium and rhenium were used as internal standards at their concentrations of approximately 3 μg L^{-1}. The international geostandard SLRS-5 (Riverine Water Reference Material for Trace Metals certified by the National Research Council of Canada) was used to check the validity and reproducibility of each analysis (see ref. [37] for analytical details). There was good agreement between our replicated measurements of SLRS-5 and the certified values (relative difference < 15%). Elements lacking the certified values in the SLRS sample or those having high intrinsic uncertainty of measurements (>20%) are not discussed in this work (Be, Sn and Te). For all major and most trace elements, analyzed by ICP MS, the concentrations in the blanks were below the analytical detection limits ($\leq$0.1–1 ng/L for Cd, Ba, Y, Zr, Nb, REE, Hf, Pb, Th and U; 1 ng/L for Ga, Ge, Rb, Sr and Sb; ~10 ng/L for Ti, V, Cr, Mn, Fe, Co, Ni, Cu, Zn and As). Further details of analyses are provided elsewhere [27,28,31].

2.3. Landscape Parameters of Tributaries and the Main Stem

The landscape parameters of the 9 tributaries and 23 points of the Ob River main stem were determined by digitalizing available soil, vegetation, lithological and geocryological maps (Figure 1, Table S2). The landscape and soil parameters were typified using the United States Database of Soil Resources (Available online: http://egrpr.soil.msu.ru/, accessed on 7 November 2021); the borders were verified and corrected according to available Landsat images. The permafrost parameters of the watershed were obtained from NCSCD data. The type of Quaternary deposits was taken from the State Geological Map of Russia with a resolution of 1:1,000,000 (Available online: http://www.geolkarta.ru/, accessed on 7 November 2021).

The Spearman rank order correlation coefficient (Rs) ($p < 0.05$) was used to determine the relationship between each major and trace element concentration and the latitude, climatic, lithological and landscape parameters of the watersheds. Further statistical treatment of element concentration drivers in river waters included a Principal Component Analysis with a variance estimation method and a scree test to minimize the number of governing factors. This analysis allowed us to test the effect of various environmental parameters (landscape, soil, vegetation, permafrost and type of Quaternary deposits) of the watershed on spatial variations of riverine solutes in both the Ob River main stem and the tributaries.

3. Results

3.1. Impact of the Latitude on the Element Concentration in the Main Stem of the Ob River

According to the major and trace element behavior along the water course of the main stem [39], from the south to the north transect, three main families were distinguished, as illustrated in Figures 2 and 3. The Dissolved Inorganic Carbon, alkaline-earth metals (Ca and Sr), Mo, W and U sizably (by a factor of 2.0) decreased their concentrations northward (Figure 2).

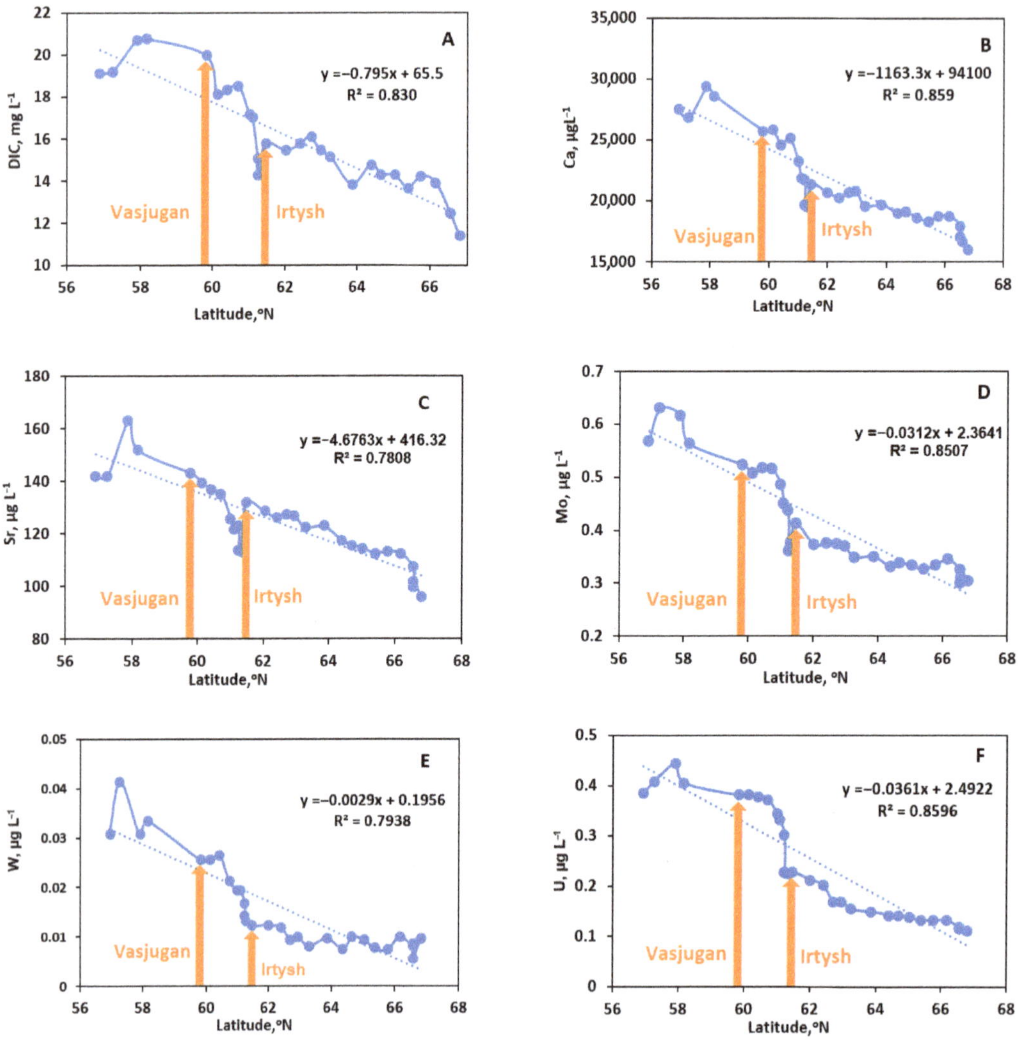

Figure 2. Latitudinal trends of labile major and trace elements decreasing their concentrations in the main stem of the Ob River from the south to the north: DIC (**A**), Ca (**B**), Sr (**C**), Mo (**D**), W (**E**) and U (**F**). The two main tributaries likely to control DOC and trace metal delivery, rivers Vasjugan and Irtysh, are shown by arrows.

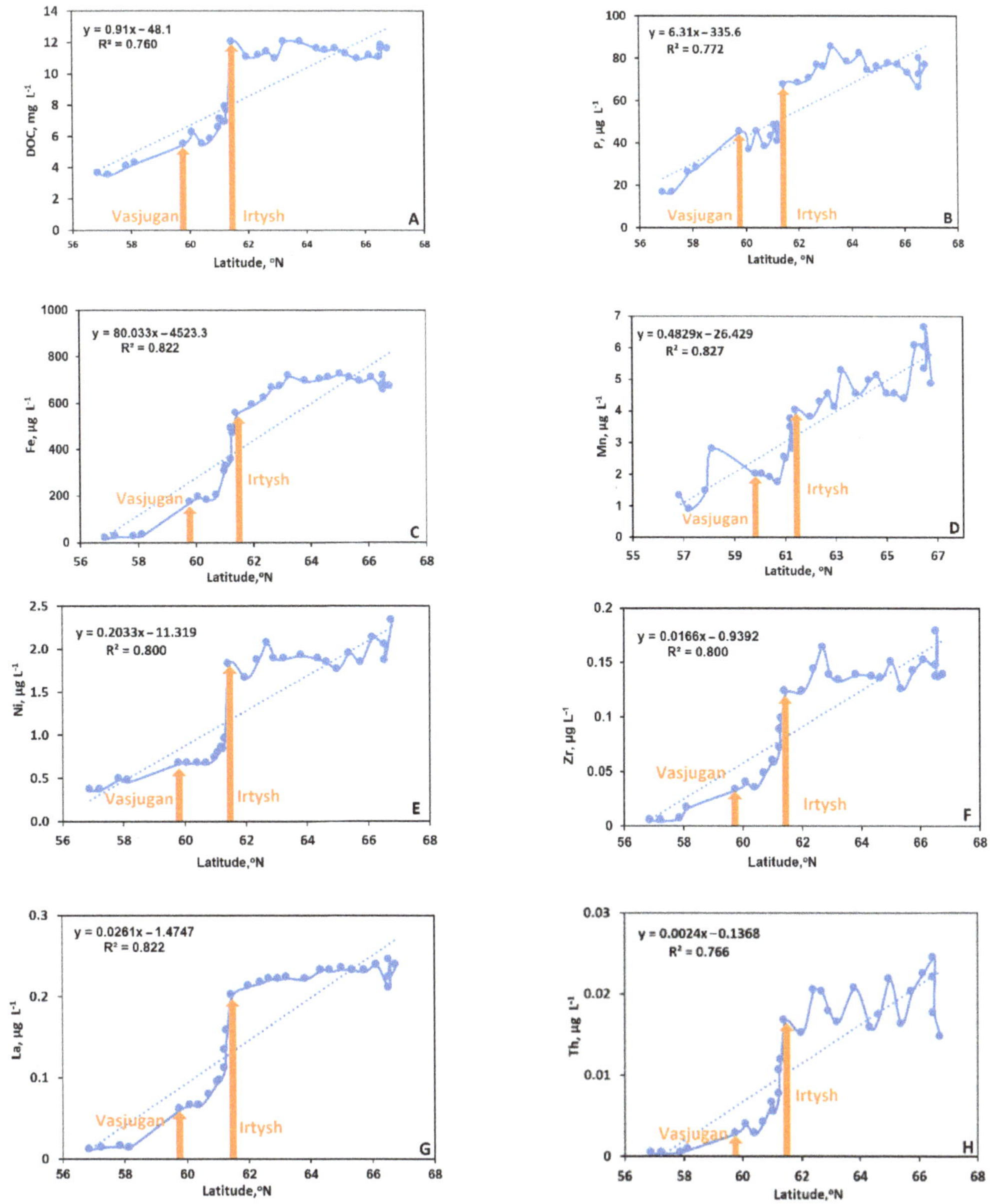

Figure 3. Latitudinal trends of DOC (**A**), nutrients (P (**B**), Fe (**C**), Mn (**D**) and Ni (**E**)) and insoluble trace elements (Zr (**F**), La (**G**) and Th (**H**)), showing a strong increase in concentration in the Ob main stem after the confluence with the Vasyugan and Irtysh River (shown by arrows).

The second group of elements included DOC, P, Mn, Al, Fe, Ti, Ni, Cu, Co, Rb, Pb, Y, Zr, Nb, REEs, Hf and Th, which strongly (by a factor of 2 to 10) increased their concentration from the south to the north (Figure 3). An increase in these element concentrations in the main stem of the Ob River occurred after the confluence, first, with Vasjugan and, then, with Irtysh; the impact of the Irtysh was especially seen for major anions (Cl and SO_4), alkali (Na and K) and Pb (Figure S1 of the Supplementary Material). Finally, a group of elements did not exhibit any sizable (>1.5 x) change in concentration over the main stem (Si, Li, V and Ba), or the evolution of their concentration did not follow any specific pattern (B, Mg, Cr, Zn, Cd, Ga, Ge, Cs and Tl), as illustrated in Figure S2. Note, some peaks in concentrations were detected for Zn, Cd and Pb, which could be tentatively linked to local pollution sources.

One potentially important tributary of the Ob River is the Vasyugan River. This tributary drains the largest world mire [40], and thus, it can substantially modify the Ob River chemical composition via delivering a number of DOM-bound elements. To assess this impact, the degree of element enrichment in the northern part of the Ob River (downstream of the Irtysh) was compared to its intermediate part (between Vasyugan and Irtysh) and the southern part (between Tom and Vasyugan). For this, we calculated the average concentrations of DOC and major and trace elements in these three segments of the main stem (Table 1). It can be seen that the elements mostly enriched (by a factor of 2 to 10) in the northern part of the river were DOC, Cl, P, Mn, Fe, some divalent metals (Mn, Ni and Pb) and insoluble lithogenic elements (Al, Ti, Nb, Y, REE, Ti, Zr, Hf and Th).

Table 1. Mean (±SD) concentration (μg L^{-1}) of major and trace elements in the three distinct parts of the Ob River main stem upstream and downstream of its confluence with Vasyugan and Irtysh.

Descriptions	Tom–Vasyugan (n = 4)	Vasyugan–Irtysh (n = 11)	Irtysh–Salemal (n = 16)
Cl	1740 ± 510	1660 ± 430	7240 ± 530
SO_4	7070 ± 1340	5106 ± 888	8480 ± 392
DOC	4600 ± 930	7200 ± 2100	11,500 ± 400
UV_{245} *	0.13 ± 0.45	0.28 ± 0.12	0.49 ± 0.02
DIC	18,500 ± 5300	16,690 ± 2180	15,100 ± 766
Li	2.63 ± 0.92	1.91 ± 0.17	2.74 ± 0.19
B	13.9 ± 2.5	10.0 ± 0.63	17.7 ± 1.64
Na	5866 ± 1471	4533 ± 296	8635 ± 566
Mg	5086 ± 803	4092 ± 308	4788 ± 236
Al	6.84 ± 1.96	11.56 ± 3.02	18.67 ± 2.28
Si	2987 ± 934	2905 ± 797	2440 ± 75
P	30.7 ± 14.7	49.030 ± 19.5	75.4 ± 5.7
K	1115 ± 183	972 ± 137	1571 ± 98
Ca	29,664 ± 3822	22,677 ± 2683	19,978 ± 914
Ti	3.27 ± 1.20	5.46 ± 1.84	8.80 ± 1.04
V	1.67 ± 0.3	1.37 ± 0.08	1.52 ± 0.03
Cr	0.14 ± 0.11	0.27 ± 0.12	0.30 ± 0.12
Mn	2.0 ± 1.48	18.0 ± 49.0	4.5 ± 0.5
Fe	59 ± 62	440 ± 475	664 ± 58
Co	0.04 ± 0.02	0.07 ± 0.1	0.07 ± 0.01
Ni	0.6 ± 0.2	0.9 ± 0.5	1.9 ± 0.1
Cu	1.7 ± 0.5	1.7 ± 0.2	2.3 ± 0.2
Zn	4.8 ± 3.2	3.0 ± 1.5	5.5 ± 3.2
Ga	0.01 ± 0.004	0.01 ± 0.002	0.01 ± 0.002
Ge	0.01 ± 0.003	0.007 ± 0.002	0.009 ± 0.001
As	1.4 ± 0.2	1.3 ± 0.5	1.5 ± 0.07

Table 1. *Cont.*

Descriptions	Tom–Vasyugan (n = 4)	Vasyugan–Irtysh (n = 11)	Irtysh–Salemal (n = 16)
Rb	0.6 ± 0.09	0.8 ± 0.5	1.2 ± 0.09
Sr	165 ± 33	126 ± 13	123 ± 6
Y	0.04 ± 0.03	0.1 ± 0.05	0.3 ± 0.01
Zr	0.02 ± 0.02	0.06 ± 0.03	0.1 ± 0.01
Nb	0.002 ± 0.002	0.007 ± 0.005	0.02 ± 0.002
Mo	0.7 ± 0.08	0.4 ± 0.1	0.4 ± 0.03
Cd	0.01 ± 0.01	0.009 ± 0.009	0.01 ± 0.006
Sb	0.1 ± 0.02	0.1 ± 0.02	0.09 ± 0.004
Cs	0.001 ± 0.00	0.001 ± 0.000	0.002 ± 0.000
Ba	24 ± 3	22 ± 0.9	24 ± 0.7
La	0.03 ± 0.02	0.1 ± 0.04	0.2 ± 0.01
Ce	0.04 ± 0.03	0.2 ± 0.08	0.4 ± 0.02
Pr	0.01 ± 0.01	0.03 ± 0.01	0.06 ± 0.003
Nd	0.03 ± 0.03	0.1 ± 0.04	0.2 ± 0.02
Sm	0.01 ± 0.01	0.03 ± 0.01	0.06 ± 0.004
Eu	0.004 ± 0.002	0.008 ± 0.002	0.015 ± 0.001
Gd	0.01 ± 0.01	0.03 ± 0.01	0.06 ± 0.004
Tb	0.001 ± 0.001	0.004 ± 0.001	0.008 ± 0.000
Dy	0.006 ± 0.004	0.02 ± 0.008	0.05 ± 0.003
Ho	0.001 ± 0.001	0.004 ± 0.002	0.009 ± 0.001
Er	0.004 ± 0.003	0.01 ± 0.004	0.03 ± 0.002
Tm	0.000 ± 0.000	0.002 ± 0.001	0.004 ± 0.000
Yb	0.003 ± 0.002	0.01 ± 0.004	0.02 ± 0.002
Lu	0.000 ± 0.000	0.002 ± 0.001	0.003 ± 0.000
Hf	0.002 ± 0.001	0.008 ± 0.004	0.02 ± 0.002
W	0.03 ± 0.01	0.02 ± 0.007	0.01 ± 0.002
Tl	0.002 ± 0.001	0.001 ± 0.000	0.002 ± 0.000
Pb	0.07 ± 0.1	0.2 ± 0.1	0.2 ± 0.02
Th	0.001 ± 0.001	0.007 ± 0.004	0.02 ± 0.002
U	0.4 ± 0.06	0.3 ± 0.1	0.2 ± 0.03

* units are cm^{-1}.

The distinction of the riverine solutes into three major groups depending on their latitudinal pattern was also confirmed by Spearman correlation coefficients between the element concentration and the latitude (Table S3 of the Supplementary Material). The elements of the first group exhibited quite strong (≤ -0.90, $p < 0.005$) negative correlations with latitude, whereas the elements of second group had $R_{Spearman} > 0.80$. The elements with a weak or non-systematic pattern showed statistically significant correlations ($p < 0.05$), but the $|R_{Spearman}|$ was typically below 0.8.

3.2. Major and Trace Elements in the Tributaries

The 11 tributaries sampled in this study exhibited highly contrasting behavior in both major and trace element concentrations. Similar to the main stem, this allowed the revealing of a distinct group of elements according to their latitudinal pattern. The southern tributaries (upstream of the Irtysh) were enriched in DIC, major anions, alkali and alkaline-earth metals and U, whereas the northern tributaries exhibited much higher concentrations of Mn, Fe and Co (>5 times) and were sizably enriched in DOC, Cr, P, Zn, Cd, Nb and trivalent and tetravalent low soluble trace elements (Figure S3).

The Irtysh River played a governing role in the concentration pattern of the main stem, as it presented a sizable addition of many elements, mostly labile anions and oxyanions (DIC, Cl, SO_4, B, Mo, Sb and W), alkalis and alkaline-earth metals (Li, Na, K, Rb, Mg, Ca, Sr and Ba), some divalent transition metals (Mn, Co and Ni), U and DOC to the main stem, given its high discharge (49% of the Ob River at the confluence in 2016) and, most importantly, elevated concentration of elements (Figure 4). Some elements, however, exhibited a higher concentration in the northern part of the Ob River compared to the Irtysh River: Al, Fe, Cr, Zn, Ga, Nb, Ti, Zr, Hf and REEs (Figure 4). Presumably, these elements were additionally delivered to the Ob River main stem via multiple small-size tributaries, which were not sampled in the present work.

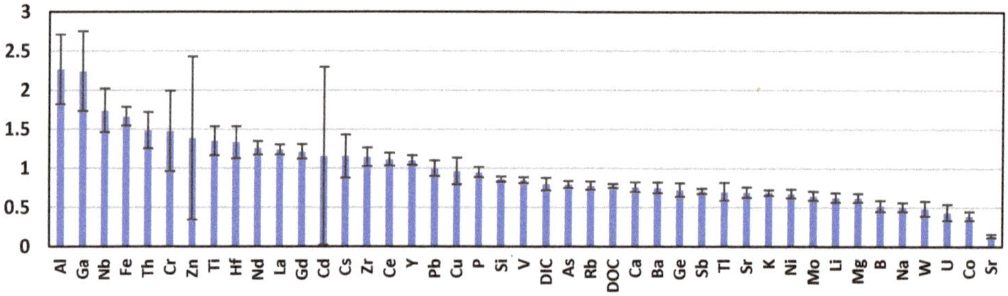

Figure 4. The ratio of the average (±s.d.) element concentrations of elements in the Ob River main stem downstream of the Irtysh (till the Ob mouth) to the concentration in the Irtysh River. Values above 1 indicate a sizable enrichment of the main stem in a given element by small tributaries in the north.

3.3. *Colloidal Status of Major and Trace Elements in the Ob River and Tributaries*

The dialysis procedure allowed a first-order assessment of the colloidal fraction (defined as the % of the element in the 1 kDa—0.45 μm fraction, divided by its total dissolved concentration in the <0.45 μm fraction). The percentage of colloidal fraction ranged from 0–10% (Cl, SO_4, DIC, Na, K, Li, B, Si, Mg, Ca and Mo) to 80–90% (trivalent and tetravalent hydrolysates), as listed in Table S4. In the northern part of the basin (downstream of the Irtysh), the distribution of the colloidal forms of elements among the tributaries and the main stem was quite homogeneous. This allowed identification of three main groups of solutes with respect to their colloidal status (Figure 5): (1) alkalis and alkaline-earth metals, major anions and Si and trace oxyanions (Mo, Sb and Ge) presenting between 0 and 20% of colloidal forms; (2) DOC, divalent transition metals (Ni, Cu and Cd), V, Cr, Cs, Tl, W and U, which were sizably impacted by colloids (20–70%) and (3) P, Fe, Mn, Co and all trivalent and tetravalent hydrolysates (Al, Ga, Y, REEs, Ti, Zr, Hf and Th), which were present essentially (>70–80%) in the colloidal form.

The DOC exhibited a remarkably homogeneous proportion of colloidal forms (42 ± 3%) in the main stem and most tributaries, including the Irtysh. However, the two most southern tributaries (Ket' and Tom') exhibited a much lower proportion of colloidal DOC, as well as of Fe, Co and Ni. Other colloid-affected elements (Al, Ti, P, V, Cr, As, Ga, Y and Zr) also demonstrated the lowest proportion of colloids in the southern tributaries (Irtysh, Ket and Tom). Uranium exhibited quite a particular colloidal pattern, with a gradual decreasing of the colloidal fraction from the north (70–90%) to the middle course, including the Irtysh (20 to 50%), and further decreasing in the two most southern tributaries, Ket' and Tom' (3%).

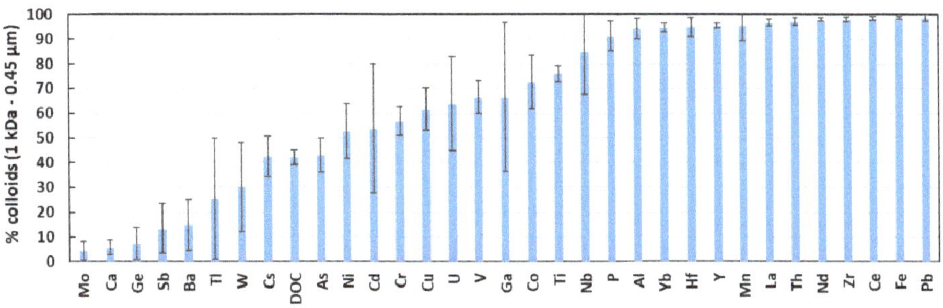

Figure 5. Proportion of the colloidal fraction of elements in the Ob River downstream of the Irtysh (4 points) and 3 tributaries (Pasaydeyakha, Poluy and Pitljar). Alkalis and alkaline-earth metals and oxyanions exhibited <10% of colloids and are not shown here.

It is noteworthy that the role of the largest tributary (Irtysh) in the colloidal pattern of the Ob River was not that significant: for most elements, their colloidal fraction in the Irtysh was not statistically different from that of the main stem and the downstream (northern) tributaries (Figure S4).

3.4. *Testing the Landscape, Soil and Quaternary Deposits Control on Element Concentration*

The main stem of the Ob River and its tributaries sampled in this study exhibited strong variability in the main landscape parameters, such as watershed size; mean annual air temperature (MAAT) and permafrost, forest, lake, bog and floodplain coverage, as well as the type of soil and Quaternary deposits. The main environmental parameters of the Ob River basin progressively evolved from the south to the north, which allowed testing of the impact of the climate and landscape on the element concentration in several selected points of the main stem. The latitude of the sampling point; the permafrost coverage and the proportion of bogs, lakes, floodplain, lacustrine and fluvio-glacial Quaternary deposits of the watershed strongly correlated ($p < 0.05$) with DOC, Fe, P, divalent metals (Mn, Co, Ni, Cu and Pb), Rb and low mobile lithogenic trace elements (Al, Ti, Cr, Y, Zr, Nb, REEs, Hf and Th) concentrations (Table S3A,B, Figures 6 and S5). Note that the impact of permafrost was especially strong in the northern part of the river basin. The pH and concentrations of soluble, highly mobile elements (DIC, SO_4, Ca, Sr, Ba, Mo, Sb, W and U) positively correlated with the proportion of forest, loess and fertile soils and eluvial, eolian, and fluvial Quaternary deposits on the Ob River watershed (Figures 6 and S5, where a number of the most important parameters, such as forest coverage (pH, Ca, Mo and U), floodplain area (DOC, P, Fe, Th) and watershed coverage by fluvio-glacial deposits (DOC, Al, Fe and La) are illustrated). Other landscape factors were of secondary importance for element control, yet they exhibited sizable correlations with particular elements. Examples are podzol soil coverage of the watershed that was positively linked to the concentration of Al, Cr, Ga, Th and saline soils that positively impacted the concentration of Cl, SO_4, Li, B, Na, K, As and Rb in the river water.

Pairwise correlations of the riverine element concentration with the landscape parameters of the watershed were further developed via a multi-parametric statistical approach (Principal Component Analysis). Considering both the main stem of the Ob River and its tributaries, two main factors were revealed, accounting for 41% and 13% of total variability, respectively (Figure 7). The F1 was presumably controlled by a northward increase in permafrost, floodplain, bogs, lakes and lacustrine deposits on the watersheds and acted on the DOC, Al, P, Ti, Fe, divalent metals, Rb, Cs, Nb and trivalent and tetravalent trace elements concentrations. The second factor (F2) included the Specific Conductivity, SO_4^{2-}, Li, B, Na, labile alkaline-earth metals (Mg, Sr and Ba), oxyanions (Mo, Sb and W) and oxyanions (As, Sb), which were impacted by southward-dominating forest coverage, loesses and eluvial and fertile soils. This factorial structure was found to be highly stable and preserved in the general features for both the Ob River main stem and its tributaries, if treated separately (Figure S6).

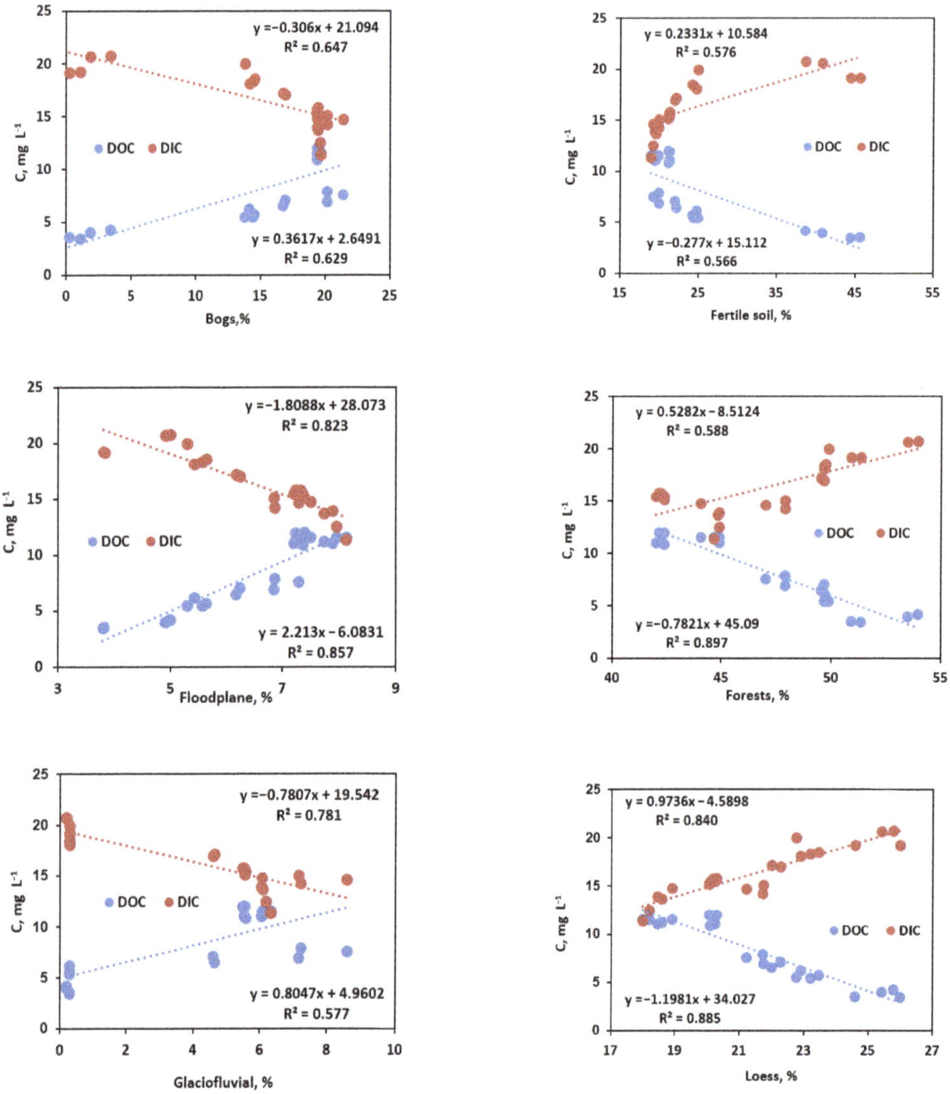

Figure 6. Examples of landscape factors, soil and Quaternary deposit control on the DOC and DIC concentrations in the Ob River. See examples of other factors and elements in Figure S5. A correlation matrix of environmental parameters of the watershed and riverine solutes is provided in Table S3.

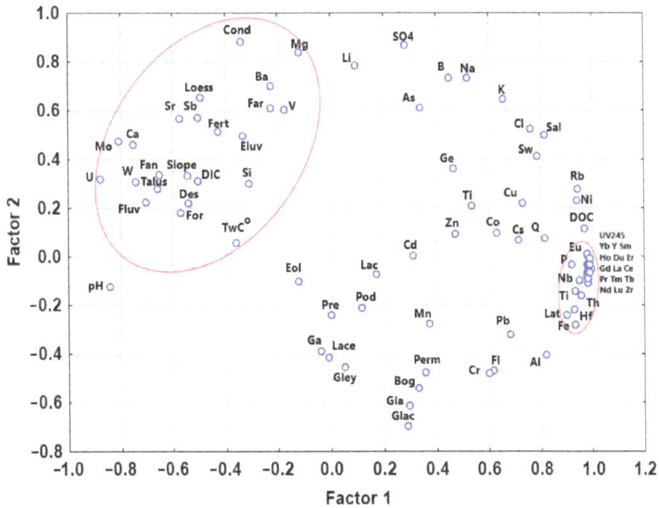

Figure 7. Results of the PCA treatment of all of the dataset (the Ob River main stem and tributaries), including the elementary composition of the river water and landscape parameters of the watersheds. The landscape parameters (% of watershed coverage) are abbreviated as follows: For, Forest; Fl, floodplain; Perm, permafrost; Far, farmland; Pre, pre-Quaternary deposits; Glac, glacial deposits; Slope, slopewash deposits, Des, deserptium (rock stream deposits); Lac, lacustrine and glacio-lacustrine deposits; Fan, fan-alluvial deposits; Gla, glacial deposits; Eol, eolian deposits; Pod, podsols; Sal, saline soils; Fert, fertile soils; TwC°, Water temperature; Cond, Specific Conductivity.

4. Discussion

4.1. The Contrasting Spatial Distribution of DOC and Major and Trace Elements between Northern and Southern River Segment Is Due to the Latitudinal Pattern of Landscape, Soil and Quaternary Deposits

The hydrochemical composition of the Ob River basin demonstrated several distinct groups of elements, defined according to the latitudinal patterns of their concentrations, similar to what was reported for small rivers of the WSL [31,36]. The DOC, organically bound metals (V, Cr, Mn, Fe, Co, Ni, Cu, Zn, Cd and Pb) and the majority of low soluble trace elements (Al, REEs, Nb, Ti, Zr, Hf and Th) exhibited a minimal concentration in the upper reaches of Ob and a maximal concentration in the northern, permafrost-affected zone. The northward increase in the concentration of these elements in both the main stem and tributaries may have originated from multiple factors. First, enhanced input of lithogenic low mobility elements in the northern part of the Ob River (permafrost zone) may have occurred due to a decrease of river connectivity, with deep and shallow groundwater and an increase in the surface flow. The surface drainage through forest litter together with surface runoff from surrounding bogs have led to river water enrichment in organic and organo-ferric and organo-aluminum (DOC, Fe, Al) colloids [30,34]. These colloids act as main carriers of divalent metals, insoluble trivalent and tetravalent trace elements, Pb, Cr and V in western Siberian rivers [27,28].

Another possible explanation for a northward increase in the concentrations of lithogenic and low mobile elements and a decrease in the concentrations of soluble labile elements lies in the dynamics of peat formation/decay across the territory, as recently suggested for small rivers of the WSL [36]. Because of ongoing recovery from the last glaciation, the WSL terrestrial ecosystems are not at the stationary stage: the growth of mires in the south leads to active accumulation of peat [40,41], whereas the permafrost peatlands in the north thaw and degrade [42–44]. Given that the WSL peat preferentially accumulates heavy metals (V, Cr, Zn and Pb) and trivalent (TE^{3+}) and tetravalent (TE^{4+})

trace elements (Al, Y, REEs, Ti, Zr, Hf and Th) and is depleted in alkalis and alkaline-earths metals, As and Mo [45], it is possible that the riverine concentrations of trace elements inherit the element cycling between the growing/decaying peat and surface hydrological network.

At the same time, the obtained results did not completely confirm the initial hypothesis. We expected a northward decrease of the concentration of ground-water originated soluble, highly mobile anions, alkalis and alkaline earth metals. However, this was not observed for Cl, SO_4, Na and K. "Presumably, these elements bear the influence of salt soils of the Irtysh River, which flows through partially salty steppe and forest-steppe landscapes. Further to the north, the Ob River drains through paleo-marine deposits containing salt minerals (based on sedimentary cores available from extensive drilling of the territory [46]). This may enrich the main stem in highly mobile Na, Cl and SO_4.

It is important to note that unlike it was reported for water pCO_2 and molecular composition of riverine DOM (i.e., [17–19], the Irtysh River does not play a regulatory role of the lower Ob River water chemistry in terms of the concentrations of other major and trace elements, including organo-mineral colloids. Instead, the hydrochemical conditions of the Ob River are shaped by the integration of multiple small tributaries, the chemical composition of which gradually changes along the permafrost and landscape gradient, as is known for other river basins of western Siberia, such as Taz and Pur [36].

The concentration of riverine solutes was found to be strongly controlled by the abundance of bogs and lakes on the watershed and its floodplain coverage, for both the main stem and tributaries. The elements which positively correlated with wetlands reflect a leading role of both organic-rich soils and sediments of wetlands (unified here as bogs, lakes and floodplains) in DOC, P, K, divalent metals and insoluble elements mobilization (notably trivalent and tetravalent hydrolysates) from the watershed to the river. This presents a prominent contrast to the very small boreal catchments described in Northern Sweden, where the presence of wetlands at the watershed decreased the fluxes of dissolved metals from boreal forests to downstream [47–49]. According to these authors, such a decrease occurred due to a combination of low weathering in peat soils and the accumulation of organophilic metals in peat. The contrast between the northern Scandinavian and western Siberian settings is consistent with a possibility of peat degradation, rather than accumulation in northern part of the Ob River basin (downstream of the confluence with the Irtysh). This would lead to the enhanced release of trace metals to the streams and rivers. It is also possible that the presence of permafrost in the northern part of the WSL greatly shortens the water and element pathways between wetlands and rivers, compared to that in the non-permafrost regions of northern Sweden. The transport of trace metals released from mineral soils under the forest towards the river is therefore strongly enhanced by allochthonous organic matter originating from peat bogs and wetlands of the region.

Another landscape parameter is the presence of podzol soils at the watershed. Similar to glacial quaternary deposits, these soils lead to enrichment of the river water with low mobile, lithogenic elements, such as Al, Cr, Ga, REE, Hf and Th (see Table S3). These elements mark the presence of primary silicate minerals, often developed on felsic moraine, as has been shown in European boreal regions (i.e., ref. [50]).

In contrast to metals and low mobile trace elements, the concentrations of the DIC, alkaline-earth metals, oxyanions (Mo, As and W) and U decreased northward and were positively affected by the presence of forest, loess, eluvial, fluvial and fertile soils. On the one hand, this could reflect a northward decrease of the connectivity between groundwaters (enriched in these elements; see ref. [51]) and surface waters due to an increase in the permafrost coverage (i.e., [33,34]). On the other hand, a southward increase in watershed coverage by soils which are rich in these labile elements (loess and eluvial and eolian soils, containing carbonate concretions and fauna) may explain the preferential enrichment in these elements in the southern part of the Ob River watershed. Note that the eolian solid aerosol deposits in western Siberia are also enriched in these soluble elements [52]. It has not been excluded that both factors (soils and groundwaters) were pronounced at the scale of such a large watershed.

Finally, the presence of saline soils in the southern part of the Ob basin produced a sizable enrichment of the river water in anions (Cl and SO_4), B, alkalis (Li, Na and K), which presumably originated from local surface salt deposits and evaporates.

4.2. Riverine Transport of the Trace Element Occurs in the Form of Organo-Ferric Colloids

All major and trace elements in the Ob River and its tributaries presented a colloidal pattern, which was similar to the one established for boreal and permafrost-affected zones of high latitudes [28,37,53–57]. Soluble, highly labile alkalis, anions and oxyanions were negligibly affected by colloids as they do not interact with organic matter or colloidal Fe/Al hydroxides and thus present in the form of simple ions or neutral molecules. The second group of colloidal-bound trace elements reflected their capacity to either complex with colloidal organic matter (divalent transition metals) or co-precipitate with organo-ferric colloids (trivalent and tetravalent hydrolysates).

A prominent feature of the major part of the Ob Basin is that the spatial variability of the colloid distribution among the tributaries and the main stem was rather low, and only two southern tributaries (Ket and Tym) presented a remarkable contrast to the rest of the Ob River basin. This presumably reflects the homogeneous physico-geographical and landscape parameters of the majority of the Ob River and its tributaries north of the Irtysh. The two southern tributaries exhibit less bogs and lakes in their watersheds and contain carbonate minerals in their base rocks (clays and silts with carbonate concretions, [30]). In accordance with previous studies of small rivers of the WSL [58], we hypothesize that the presence of carbonate rocks and decreasing the proportion of wetlands in the southern part of the Ob basin leads to (1) a decrease in the concentration of DOC and Fe, which can serve as main carriers of trace metals and (2) an increase in the DIC concentration and enhanced delivery of the ionic form of elements from the groundwater. The latter is facilitated by a much stronger connectivity between DOC-poor and DIC-rich groundwater and surface waters in the permafrost-free part of the Ob Basin [34].

An interesting and unexpected result was the extremely high proportion of colloidal Mn (83 to 100% in all samples except Ket (23%)). The typical colloidal fraction of Mn in the boreal and permafrost-affected surface waters (rivers, lakes and bog waters), including peat porewater, is between 80 and 20% [28,29], and the summer period usually accounts for the lowest proportion of colloidal Mn, due to the dominance of low molecular weight (<1 kDa) exometabolites that are capable of complexing divalent metals [27]. However, the lowest proportion of Mn colloids was observed only in the most southern tributary, which also exhibits a low proportion of colloidal OC. In these rivers, Mn is likely to be present as a divalent cation, originated from the groundwater, or complexed with autochthonous low molecular weight DOM. In contrast, the majority of Mn in all of the other tributaries and the main stem was tightly bound to colloids. We therefore hypothesize that the transport of Mn occurs in the form of high molecular weight, Fe-rich organic colloids in the main part of the Ob Basin. It is possible that a high proportion of bogs in the watersheds of all rivers north of the Vasyugan River provides such an enhanced colloidal transport. These Mn-rich colloids are likely to be generated at the interface of anoxic and oxic surface waters of the bogs and then delivered to the river via surface flow. Note that the enhanced riverine Mn transport in mire-affected regions has been reported in Northern Europe [55,59].

Another striking example of speciation control on element migration in the river water is that of the uranyl (UO_2^{2+}) ion. Uranium (VI) speciation in surface waters rich in DOM and Fe is largely controlled by high molecular weight organo-ferric colloids [28,60]. In the non-permafrost zone, these colloids are formed at the riparian/hyporheic zone of the river where the Fe(II) and U(VI)-rich groundwaters mix with oxygenated, organic-rich surface waters [37]. In the permafrost zone, these colloids originate from the surface flow over the forest floor and mire waters. Thus, in the northern part of the Ob River basin, elevated concentrations of both Fe and DOC provide suitable conditions for U transport in the form of colloids (70–90% in the 1 kDa—0.45 μm form). Mires, which are highly abundant in the Ob basin north of the Ket tributary, are also capable of mobilizing uranium to the river water, as is known from other boreal settings [61]. In contrast, in the southern part of the basin, the groundwater contacting with the carbonate rocks of the basement or the loess soils are enriched in HCO_3^-. Such a chemical environment renders uranyl into soluble, non-colloidal carbonate complexes, as is known for other boreal regions which are impacted by carbonate rocks [62]. As a result, the proportion of the colloidal form of $UO_2(VI)$ progressively decreases southward and becomes as low as 3% in the two most southern tributaries (Ket and Tom), which exhibited the highest DIC and Ca and the lowest DOC and Fe concentration. Overall, the progressive southward decrease of the colloidal status of U reflects an increase in the connectivity between the surface and groundwaters of the WSL (i.e., [34,63]) and a decrease in the bog coverage of the river watershed in the same direction.

4.3. Possible Impact of Landscape Changes on Element Concentration in the Ob River and Its Tributaries

The unique geographical position of the Ob River, which traverses, from the south to the north, several distinct landscape zones and encompasses a large permafrost, MAAT and vegetation gradient, allows one to use a substituting space for time approach (i.e., ref. [64]) for foreseeing possible future changes in the river water hydrochemistry based on the contemporary pattern of riverine solutes. This approach has been efficiently used for small rivers of the WSL [30,31,33–35,58] but, to the best of our knowledge, never attempted for the Ob River main stem. The restrictions of this approach are the following: a lack of accounting for the time scale, necessary for the northern ecosystem to reach the new "more southern" state; ignoring the possible shift in the structure of the vegetation and soil microbial community, change in hydrologic seasons and the properties of groundwater [65]. Considering these restrictions, only a preliminary, first-order assessment of possible changes could be made.

Given the rather high similarity of riverine solutes in the Ob River and tributaries across different permafrost zones, north of the Irtysh (excluding the most southern part, south of Vasyugan), the shift in permafrost boundaries or the increase in the thickness of the active layer [66–69] are not expected to sizably impact the hydrochemical composition of rivers in the Ob basin, unless the connectivity between the deep underground and surface waters is modified (i.e., 10, 30). The latter may lead to the enhanced concentration of soluble highly mobile elements in the north, as has been demonstrated for the case of small WSL rivers of the Pur and Taz watersheds [27].

At the same time, the elements affected by the possible adsorption on the clay minerals underlying the peat deposits (first of all, DOC; see [70–74]) may modify their concentration in the northern part of the Ob basin. Globally, wetlands (bogs, lakes and floodplain zones) were the main controlling factor of the element concentration in the main stem and tributaries across the studied spatial gradient. Therefore, we believe that future changes in the WSL landscape, such as forestation of the northern part and a decrease in the proportion of bogs, will be all induced by changes in atmospheric precipitation, terrestrial productivity and duration of open-water seasons, i.e., [75]. However, quantification of the impact of these factors requires extensive ecosystem-level regional modeling [76], which goes beyond the scope of this study.

5. Conclusions

Based on the 3000 km north–south sampling transect of the Ob River main stem and its 11 tributaries, performed during the end of the spring flood—beginning of the summer baseflow period, this snapshot study of a large Arctic river dissolved (<0.45 μm) load revealed distinct physio-geographical control of the hydrochemistry of the river water. The land cover approach allowed testing of the control of the main physio-geographical parameters of the Ob River watersheds on element concentration along the main stem and among the tributaries. Bogs, floodplains, lake coverage and the permafrost presence on the watershed were found to be the main factors of the northward increasing in the concentration of DOC and low-soluble trace elements, which were present in the form of organic and organo-mineral colloids (Fe, Mn, Al, V, Cr, Co, Ni, Cu, Zn, Cd and Pb) and lithogenic trivalent and tetravalent TE (Al, Ti, Zr, Hf and Th) and REEs. These elements were also positively affected by the presence of glacial, lacustrine and fluvio-glacial Quaternary deposits. In contrast, soluble highly mobile alkalis, alkaline-earth metals, DIC, SO_4, B, Mo, Sb, As, W and U were positively affected by the presence of forest developed on loesses; fertile and saline soils and eluvial, fluvial and eolian Quaternary deposits, given that these substrates could contain soluble carbonate minerals and salt inclusions.

We revealed a high homogeneity in both the element concentration and colloidal status in the main stem of the Ob River and its tributaries located north of the confluence with the Irtysh. At the scale of the whole basin, the distribution of wetlands (bogs, floodplains, lakes and lacustrine deposits) was the dominant factor defining the elementary pattern in waters of the Ob River. In the northern part of the basin, the permafrost coverage exhibited sizable control on riverine solutes. As such, in case of drastic environmental changes in the WSL territory (permafrost boundary shift, active layer depth increase, vegetation coverage and precipitation regime), the changes in distribution of bogs, lakes and forest, might become the governing factors of modifications in the river hydrochemical regime.

Supplementary Materials: The following are available online at https://www.mdpi.com/article/10.3390/w13223189/s1, Figure S1: Latitudinal dependence of Cl, SO_4, Na, K and Pb concentration in the main stem of the Ob River; Figure S2: Main stem concentrations of elements which do not show any particular pattern with the latitude: Si, Mg, Cu, V, As and Sb; Figure S3: A histogram of the elemental ratio in 2 northern tributaries (downstream of the Irtysh) to 8 southern tributaries (upstream of the Irtysh) of the Ob River; Figure S4: Proportion of the colloidal (1 kDa—0.45 μm) fraction of DOC, Ca, Fe, Mn, Al, Cu, U, La and Th in the main stem (blue box plot column) and the tributaries of the southern (blue circles) and northern (red circles) part of the Ob Basin; Figure S5. Examples of major and trace element concentration with landscape parameters of the Ob River main stem and tributaries; Figure S6. Results of PCA treatment of the solute data and watershed characteristics (separately, Ob main stem and tributaries). Table S1: List of the sampled sites at the main stem of the Ob River and its tributaries. Table S2: Main landscape parameters and genetic type of the Quaternary deposits (% of the watershed coverage) of the tributaries and several key points at the Ob River main stem; Table S3: A correlation matrix of the element concentration in the Ob main stem tributaries and landscape coverage (%) of the watersheds; Table S4: Proportion of the colloidal fraction of elements in the Ob River downstream of the Irtysh (4 points) and 7 tributaries.

Author Contributions: I.K. and O.S.P. designed the study and wrote the paper; I.K., L.G.K. and S.N.V. performed sampling, analysis of major cations and their interpretation; I.P.S. and O.V.D. were the leaders of the ship expedition and provided data interpretation; L.S.S. was in charge of DOC, DIC and anion measurements and their interpretation; L.G.K., I.K., U.S. and R.S.V. provided GIS-based interpretation, mapping and identification of river watersheds. S.N.K. provided landscape characterization of river watersheds. All authors have read and agreed to the published version of the manuscript.

Funding: This research was funded by the Russian Scientific Foundation, RSF grant No 18-17-00237_P and grant No 21-77-30001 to IS, as well as by the Russian Foundation for Basic Research, RFBR grants No 19-55-15002, 20-05-00729_a, and RSF grant No 18-77-10045 for field work and a Belmont Forum VULCAR-FATE grant for some laboratory analyses. The APC was funded by O.S. Pokrovsky.

Institutional Review Board Statement: Not applicable.

Informed Consent Statement: Not applicable.

Data Availability Statement: All obtained data are published in ref. [39].

Acknowledgments: The study was partly carried out using the research equipment of the Unique Research Installation "System of experimental bases located along the latitudinal gradient" TSU, with financial support from the Ministry of Education and Science of Russia (RF—2296.61321X0043, agreement No. 075-15-2021-672). Furthermore, this study was supported by the Ministry of Education and Science of Russia (grant No 121-021-500057-4).

Conflicts of Interest: The authors declare no conflict of interest.

References

1. Turetsky, M.R.; Abbott, B.W.; Jones, M.C.; Anthony, K.W.; Olefeldt, D.; Schuur, E.A.G.; Grosse, G.; Kuhry, P.; Hugelius, G.; Koven, C.; et al. Carbon release through abrupt permafrost thaw. *Nat. Geosci.* **2020**, *13*, 138–143. [CrossRef]
2. Vonk, J.E.; Tank, S.E.; Walvoord, M.A. Integrating hydrology and biogeochemistry across frozen landscapes. *Nat. Commun.* **2019**, *10*, 1–4. [CrossRef] [PubMed]
3. Gordeev, V.V.; Martin, J.-M.; Sidorov, I.S.; Sidorova, M.V. A reassessment of the Eurasian river input of water, sediment, major elements, and nutrients to the Arctic Ocean. *Am. J. Sci.* **1996**, *296*, 664–691. [CrossRef]
4. Cooper, L.W.; McClelland, J.W.; Holmes, R.M.; Raymond, P.A.; Gibson, J.J.; Guay, C.K.; Peterson, B.J. Flow-weighted values of runoff tracers ($\delta^{18}O$, DOC, Ba, alkalinity) from the six largest Arctic rivers. *Geophys. Res. Lett.* **2008**, *35*, L18606. [CrossRef]
5. McClelland, J.W.; Déry, S.J.; Peterson, B.J.; Holmes, R.M.; Wood, E.F. A pan-arctic evaluation of changes in river discharge during the latter half of the 20th century. *Geophys. Res. Lett.* **2006**, *33*, L06715. [CrossRef]
6. Holmes, R.M.; Coe, M.T.; Fiske, G.J.; Gurtovaya, T.; McClelland, J.W.; Shiklomanov, A.I.; Spencer, R.G.M.; Tank, S.E.; Zhulidov, A.V. Climate change impacts on the hydrology and biogeochemistry of Arctic Rivers. In *Climatic Changes and Global Warming of Inland Waters: Impacts and Mitigation for Ecosystems and Societies*; Goldman, C.R., Kumagi, M., Robarts, R.D., Eds.; John Wiley and Sons: Hoboken, NJ, USA, 2013; pp. 1–26.
7. McClelland, J.W.; Holmes, R.M.; Peterson, B.J.; Raymond, P.A.; Striegl, R.G.; Zhulidov, A.V.; Zimov, S.A.; Zimov, N.; Tank, S.E.; Spencer, R.G.M.; et al. Particulate organic carbon and nitrogen export from major Arctic rivers. *Glob. Biogeochem. Cycles* **2016**, *30*, 629–643. [CrossRef]
8. Lammers, R.B.; Shiklomanov, A.I. A regional Hydrometeorological Data Network for Russia. 2013. Available online: https://www.r-arcticnet.sr.unh.edu/v4.0/index.html (accessed on 7 November 2021).
9. Ahmed, R.; Prowse, T.; Dibike, Y.; Bonsal, B.; O'Neil, H. Recent trends in freshwater influx to the Arctic Ocean from four major Arctic-draining rivers. *Water* **2020**, *12*, 1189. [CrossRef]
10. Frey, K.E.; McClelland, J.W. Impacts of permafrost degradation on arctic river biogeochemistry. *Hydrol. Process.* **2009**, *23*, 169–182. [CrossRef]
11. Vorobyev, S.N.; Pokrovsky, O.S.; Kolesnichenko, L.G.; Manasypov, R.M.; Shirokova, L.S.; Karlsson, J.; Kirpotin, S.N. Biogeochemistry of dissolved carbon, major, and trace elements during spring flood periods on the Ob River. *Hydrol. Process.* **2019**, *33*, 1579–1594. [CrossRef]
12. Beilman, D.W.; MacDonald, G.M.; Smith, L.C.; Reimer, P.J. Carbon accumulation in peatlands of West Siberia over the last 2000 years. *Glob. Biogeochem. Cycles* **2009**, *23*, GB1012. [CrossRef]
13. Wang, P.; Huang, Q.; Pozdniakov, S.P.; Liu, S.; Ma, N.; Wang, T.; Zhang, Y.; Yu, J.; Xie, J.; Fu, G.; et al. Potential role of permafrost thaw on increasing Siberian river discharge. *Environ. Res. Lett.* **2021**, *16*, 034046. [CrossRef]
14. Xu, M.; Kang, S.; Wang, X.; Wu, H.; Hu, D.; Yang, D.Q. Climate and hydrological changes in the Ob River Basin during 1936–2017. *Hydrol. Process.* **2020**, *34*, 1821–1836. [CrossRef]
15. Kaiser, K.; Canedo-Oropeza, M.; McMahon, R.; Amon, R.M.W. Origins and transformations of dissolved organic matter in large Arctic rivers. *Sci. Rep.* **2017**, *7*, 13064. [CrossRef] [PubMed]
16. Huang, J.; Wu, M.; Cui, T.; Yang, F. Quantifying DOC and its controlling factors in major Arctic rivers during ice-free conditions using Sentinel-2 data. *Remote Sens.* **2019**, *11*, 2904. [CrossRef]
17. Pipko, I.I.; Pugach, S.P.; Savichev, O.G.; Repina, I.A.; Shakhova, N.E.; Moiseeva, Y.A.; Barskov, K.V.; Sergienko, V.I.; Semiletov, I.P. Dynamics of dissolved inorganic carbon and CO_2 fluxes between the water and the atmosphere in the main channel of the Ob River. *Dokl. Chem.* **2019**, *484*, 52–57. [CrossRef]
18. Karlsson, J.; Serikova, S.; Vorobyev, S.N.; Rocher-Ros, G.; Denfeld, B.; Pokrovsky, O.S. Carbon emission from Western Siberian inland waters. *Nat. Commun.* **2021**, *12*, 825. [CrossRef]
19. Perminova, I.V.; Shirshin, E.A.; Zherebker, A.; Pipko, I.I.; Pugach, S.P.; Dudarev, O.V.; Nikolaev, E.N.; Grigoryev, A.S.; Shakhova, N.; Semiletov, I.P. Signatures of molecular unification and progressive oxidation unfold in dissolved organic matter of the Ob-Irtysh River system along its path to the Arctic Ocean. *Sci. Rep.* **2019**, *9*, 19487. [CrossRef]
20. Savichev, O.G.; Mazurov, A.K.; Pipko, I.I.; Sergienko, V.I.; Semiletov, I.P. Spatial patterns of the evolution of the chemical composition and discharge of river water in the Ob River basin. *Dokl. Earth Sci.* **2016**, *466*, 59–63. [CrossRef]
21. Mu, C.; Zhang, F.; Chen, X.; Ge, S.; Mu, M.; Jia, L.; Wu, Q.; Zhang, T. Carbon and mercury export from the Arctic rivers and response to permafrost degradation. *Water Res.* **2019**, *161*, 54–60. [CrossRef]
22. Sonke, J.E.; Teisserenc, R.; Heimbürger-Boavida, L.E.; Petrova, M.V.; Marusczak, N.; Le Dantec, T.; Chupakov, A.V.; Li, C.; Thackray, C.P.; Sunderland, E.M.; et al. Eurasian river spring flood observations support net Arctic Ocean mercury export to the atmosphere and Atlantic Ocean. *Proc. Natl. Acad. Sci. USA* **2018**, *115*, E11586–E11594. [CrossRef]

23. Semenkov, I.N.; Miroshnikov, A.Y.; Asadulin, E.E.; Usacheva, A.A.; Velichkin, V.I.; Laverov, N.P. The Ob River Basin as source of Kara Sea Contamination with global fallout of cesium-137. *Dokl. Earth Sci.* **2015**, *463*, 704–706. [CrossRef]
24. Kharanzhevskaya, Y.; Maloletko, A.; Sonyutkina, A.; Gielczewski, M.; Kirschey, T.; Michałowski, R.; Mirosław-Świątek, D.; Okruszko, T.; Osuch, P.; Trandziuk, P. Assesing mire-river interaction in a pristine Siberian bog-dominated watershed—Case study of a part of the Great Vasyugan Mire, Russia. *J. Hydrol.* **2020**, *590*, 125315. [CrossRef]
25. Kharanzhevskaya, Y.; Voistinova, E.S.; Sinyutkina, A.A. Spatial and temporal variations in mire surface water chemistry as a function of geology, atmospheric circulation and zonal features in the south-eastern part of Western Siberia. *Sci. Total Environ.* **2020**, *733*, 139343. [CrossRef] [PubMed]
26. Moskovchenko, D.V.; Babushkin, A.G.; Pikinerov, P.V. Ecological and Hydrochemical Characteristics of Wetlands 'Upper Dvuobye'. Tyumen State University Herald. *Nat. Resour. Use Ecol.* **2017**, *3*, 8–21. [CrossRef]
27. Krickov, I.V.; Pokrovsky, O.S.; Manasypov, R.M.; Lim, A.G.; Shirokova, L.S.; Viers, J. Colloidal transport of carbon and metals by western Siberian rivers during different seasons across a permafrost gradient. *Geochim. Cosmochim. Acta* **2019**, *265*, 221–241. [CrossRef]
28. Pokrovsky, O.S.; Manasypov, R.M.; Loiko, S.V.; Shirokova, L.S. Organic and organo-mineral colloids in discontinuous permafrost zone. *Geochim. Cosmochim. Acta* **2016**, *188*, 1–20. [CrossRef]
29. Raudina, T.V.; Loiko, S.; Kuzmina, D.M.; Shirokova, L.S.; Kulizhsky, S.P.; Golovatskaya, E.A.; Pokrovsky, O.S. Colloidal organic carbon and trace elements in peat porewaters across a permafrost gradient in Western Siberia. *Geoderma* **2021**, *390*, 114971. [CrossRef]
30. Pokrovsky, O.S.; Manasypov, R.M.; Shirokova, L.S.; Loiko, S.V.; Krickov, I.V.; Kopysov, S.; Zemtzov, V.A.; Kulizhsky, S.P.; Vorobyev, S.N.; Kirpotin, S.N. Permafrost coverage, watershed area and season control of dissolved carbon and major elements in western Siberia rivers. *Biogeosciences* **2015**, *12*, 6301–6320. [CrossRef]
31. Pokrovsky, O.S.; Manasypov, R.M.; Loiko, S.; Krickov, I.A.; Kopysov, S.G.; Kolesnichenko, L.G.; Vorobyev, S.N.; Kirpotin, S.N. Trace element transport in western Siberia rivers across a permafrost gradient. *Biogeosciences* **2016**, *13*, 1877–1900. [CrossRef]
32. Vorobyev, S.N.; Kolesnichenko, Y.; Korets, M.A.; Pokrovsky, O.S. Testing landscape, climate and lithology impact on carbon, major and trace elements of the Lena River and its tributaries during a spring flood period. *Water* **2021**, *13*, 2093. [CrossRef]
33. Frey, K.E.; McClelland, J.W.; Holmes, R.M.; Smith, L.C. Impacts of climate warming and permafrost thaw on the riverine transport of nitrogen and phosphorus to the Kara Sea. *J. Geophys. Res.* **2007**, *112*, G04S58. [CrossRef]
34. Frey, K.E.; Siegel, D.I.; Smith, L.C. Geochemistry of west Siberian streams and their potential response to permafrost degradation. *Water Resour. Res.* **2007**, *43*, W03406. [CrossRef]
35. Frey, K.E.; Smith, L.C. Amplified carbon release from vast West Siberian peatlands by 2100. *Geophys. Res. Lett.* **2005**, *32*, L09401. [CrossRef]
36. Pokrovsky, O.S.; Manasypov, R.M.; Kopysov, S.; Krickov, I.V.; Shirokova, L.S.; Loiko, S.V.; Lim, A.G.; Kolesnichenko, L.G.; Vorobyev, S.N.; Kirpotin, S.N. Impact of permafrost thaw and climate warming on riverine export fluxes of carbon, nutrients and metals in western Siberia. *Water* **2020**, *12*, 1817. [CrossRef]
37. Vasyukova, E.V.; Pokrovsky, O.S.; Viers, J.; Oliva, P.; Dupré, B.; Martin, F.; Candaudaup, F. Trace elements in organic- and iron-rich surficial fluids of the boreal zone: Assessing colloidal forms via dialysis and ultrafiltration. *Geochim. Cosmochim. Acta* **2010**, *74*, 449–468. [CrossRef]
38. Prokushkin, A.S.; Pokrovsky, O.S.; Shirokova, L.S.; Korets, M.A.; Viers, J.; Prokushkin, S.G.; Amon, R.M.W.; Guggenberger, G.; McDowell, W.H. Sources and the flux pattern of dissolved carbon in rivers of the Yenisey basin draining the Central Siberian Plateau. *Environ. Res. Lett.* **2011**, *6*, 045212. [CrossRef]
39. Pokrovsky, O.O.S.; Kolesnichenko, L.G. Chemical Composition and Landscape Parameters of the Ob River and Its Tributaries, July 2016. Mendeley Database. 2021. Available online: https://data.mendeley.com//datasets/dkbypg976y/1 (accessed on 7 November 2021).
40. Kirpotin, S.N.; Joosten, H.; Tanneberger, F.; Elshehawi, S.; Semenova, N.M.; Volkova, I.I.; Volkov, I.V.; Feurdean, A.; Pokrovsky, O.S.; Antoshkina, O.A.; et al. Great Vasyugan Mire: How the world's largest peatland helps addressing the world's largest problems. *Ambio* **2021**, *50*, 2038–2049. [CrossRef]
41. Antipina, T.G.; Preis, Y.I.; Zenin, V.N. Dynamics of forest vegetation and climate in the southern taiga of Western Siberia in the Late Holocene according to spore–pollen analysis and AMS dating of the peat bog. *Russ. J. Ecol.* **2019**, *50*, 445–452. [CrossRef]
42. Fotiev, S.M. Arctic peatlands of the Yamal-Gydan province of Western Siberia. *Earth's Cryosphere* **2017**, *21*, 3–13. [CrossRef]
43. Loiko, S.; Raudina, T.; Lim, A.; Kuzmina, D.; Kulizhskiy, S.; Pokrovsky, O. Microtopography Controls of Carbon and Related Elements Distribution in the West Siberian Frozen Bogs. *Geosciences* **2019**, *9*, 291. [CrossRef]
44. Panova, N.K.; Trofimova, S.S.; Antipina, T.G.; Zinoviev, E.V.; Gilev, A.V.; Erokhin, N.G. Holocene dynamics of vegetation and ecological conditions in the southern Yamal Peninsula according to the results of comprehensive analysis of a relict peat bog deposit. *Russ. J. Ecol.* **2010**, *41*, 20–27. [CrossRef]
45. Stepanova, V.M.; Pokrovsky, O.S.; Viers, J.; Mironycheva-Tokareva, N.P.; Kosykh, N.P.; Vishnyakova, E.K. Major and trace elements in peat profiles in Western Siberia: Impact of the landscape context, latitude and permafrost coverage. *Appl. Geochem.* **2015**, *53*, 53–70. [CrossRef]

46. Liss, O.L.; Abramova, L.I.; Avetov, N.A.; Berezina, N.A.; Inisheva, L.I.; Kurnishnikova, T.V.; Sluka, Z.A.; Tolpysheva, T.Y.; Shvedchikova, N.K. *Wetland Systems of West Siberia and Their Importance for Nature Conservation*; Grifi K Publisher: Tula, Russia, 2001; p. 584. (In Russian)
47. Lidman, F.; Morth, C.M.; Laudon, H. Landscape control of uranium and thorium in boreal Streams—Spatiotemporal variability and the role of wetlands. *Biogeosciences* **2012**, *11*, 4773–4785. [CrossRef]
48. Kohler, S.J.; Lidman, F.; Laudon, H. Landscape types and pH control organic matter mediated mobilization of Al, Fe, U and La in boreal catchments. *Geochim. Cosmochim. Acta* **2014**, *135*, 190–202. [CrossRef]
49. Lidman, F.; Kohler, S.J.; Morth, C.-M.; Laudon, H. Metal Transport in the Boreal Landscape-The Role of Wetlands and the Affinity for Organic Matter. *Environ. Sci. Technol.* **2014**, *48*, 3783–3790. [CrossRef] [PubMed]
50. Vasyukova, E.V.; Oliva, P.; Viers, J.; Martin, F.; Dupre, B.; Pokrovsky, O.S. Chemical weathering of mafic rocks in boreal subarctic environment (North-West Russia) under influence of glacial moraine deposits. *Chem. Geol.* **2019**, *509*, 115–133. [CrossRef]
51. Pasechnik, E.Y.; Guseva, N.V.; Savichev, O.G.; Lgotin, V.A.; Balobanenko, A.A.; Domarenko, V.A.; Vladimirova, O.N. Trace elements composition of underground waters of the upper hydrogeodynamic zone in the basin of upper Ob as a factor of formation of their ecological-geochemical condition. *Bull. Tomsk Polytech. Univ. Geo Assets Eng.* **2020**, *331*, 54–63.
52. Shevchenko, V.P.; Pokrovsky, O.S.; Vorobyev, S.N.; Krickov, I.V.; Manasypov, R.M.; Politova, N.V.; Kopysov, S.G.; Dara, O.M.; Auda, Y.; Shirokova, L.S.; et al. Impact of snow deposition on major and trace element concentrations and fluxes in surface waters of Western Siberian Lowland. *Hydrol. Earth Syst. Sci.* **2017**, *21*, 5725–5746. [CrossRef]
53. Pokrovsky, O.S.; Schott, J. Iron colloids/organic matter associated transport of major and trace elements in small boreal rivers and their estuaries (NW Russia). *Chem. Geol.* **2002**, *190*, 141–179. [CrossRef]
54. Lyvén, B.; Hassellöv, M.; Turner, D.R.; Haraldsson, C.; Andersson, K. Competition between iron- and carbon-based colloidal carriers for trace metals in a freshwater assessed using flow field-flow fractionation coupled to ICPMS. *Geochim. Cosmochim. Acta* **2003**, *67*, 3791–3802. [CrossRef]
55. Ingri, J.; Widerlund, A.; Land, M. Geochemistry of major elements in a pristine boreal river system; Hydrological compartments and flow paths. *Aquat. Geochem.* **2005**, *11*, 57–88. [CrossRef]
56. Ingri, J.; Malinovsky, D.; Rodushkin, I.; Baxter, D.C.; Widerlund, A.; Andersson, P.; Gustafsson, Ö.; Forsling, W.; Öhlander, B. Iron isotope fractionation in river colloidal matter. *Earth Planet. Sci. Lett.* **2006**, *245*, 792–798. [CrossRef]
57. Ilina, S.M.; Poitrasson, F.; Lapitsky, S.A.; Alekhin, Y.V.; Viers, J.; Pokrovsky, O.S. Extreme iron isotope fractionation between colloids and particles of boreal and temperate organic-rich waters. *Geochim. Cosmochim. Acta* **2013**, *101*, 96–111. [CrossRef]
58. Krickov, I.V.; Lim, A.G.; Manasypov, R.M.; Loiko, S.V.; Vorobyev, S.N.; Shevchenko, V.P.; Dara, O.M.; Gordeev, V.V.; Pokrovsky, O.S. Major and trace elements in suspended matter of western Siberian rivers: First assessment across permafrost zones and landscape parameters of watersheds. *Geochim. Cosmochim. Acta* **2020**, *269*, 429–450. [CrossRef]
59. Björkvald, L.; Buffam, I.; Laudon, H.; Morth, C. Hydrogeochemistry of Fe and Mn in small boreal streams: The role of seasonality, landscape type and scale. *Geochim. Cosmochim. Acta* **2008**, *72*, 2789–2804. [CrossRef]
60. Bagard, M.L.; Chabaux, F.; Pokrovsky, O.S.; Prokushkin, A.S.; Viers, J.; Dupré, B.; Stille, P. Seasonal variability of element fluxes in two Central Siberian rivers draining high latitude permafrost dominated areas. *Geochim. Cosmochim. Acta* **2011**, *75*, 3335–3357. [CrossRef]
61. Porcelli, D.; Andersson, P.S.; Wasserburg, G.J.; Ingri, J.; Baskaran, M. The importance of colloids and mires for the transport of uranium isotopes through the Kalix River watershed and Baltic Sea. *Geochim. Cosmochim. Acta* **1997**, *61*, 4095–4113. [CrossRef]
62. Chupakov, A.V.; Pokrovsky, O.S.; Moreva, O.Y.; Shirokova, L.S.; Neverova, N.V.; Chupakova, A.A.; Kotova, E.I.; Vorobyeva, T.Y. High resolution multi-annual riverine fluxes of organic carbon, nutrient and trace element from the largest European Arctic river, Severnaya Dvina. *Chem. Geol.* **2020**, *538*, 119491. [CrossRef]
63. Ala-Aho, P.; Soulsby, C.; Pokrovsky, O.S.; Kirpotin, S.N.; Karlsson, J.; Serikova, S.; Vorobyev, S.N.; Manasypov, R.M.; Loiko, S.; Tetzlaff, D. Using stable isotopes to assess surface water source dynamics and hydrological connectivity in a high-latitude wetland and permafrost influenced landscape. *J. Hydrol.* **2018**, *556*, 279–293. [CrossRef]
64. Blois, J.L.; Williams, J.W.; Fitzpatrick, M.C.; Jackson, S.T.; Ferrier, S. Space can substitute for time in predicting climate-change effects on biodiversity. *Proc. Natl. Acad. Sci. USA* **2013**, *110*, 9374–9379. [CrossRef]
65. Zemtsov, V.A. The influence of physical and geographical conditions on the natural regulation of the river flow of the West Siberian Plain. In *Issues of the Geography of Siberia*; Zemtsov, A.A., Ed.; Tomsk State University: Tomsk, Russia, 1979; pp. 46–58. (In Russian)
66. Stendel, M.; Christensen, J.H. Impact of global warming on permafrost conditions in a coupled GCM. *Geophys. Res. Lett.* **2002**, *29*, 1632. [CrossRef]
67. Anisimov, O.; Kokorev, V.; Zhil'tsova, Y. Temporal and spatial patterns of modern climatic warming: Case study of Northern Eurasia. *Clim. Chang.* **2013**, *118*, 871–883. [CrossRef]
68. Romanovsky, V.E.; Drozdov, D.S.; Oberman, N.G.; Malkova, G.V.; Kholodov, A.L.; Marchenko, S.S.; Moskalenko, N.G.; Sergeev, D.O.; Ukraintseva, N.G.; Abramov, A.A.; et al. Thermal state of permafrost in Russia. *Permafr. Periglac. Proc.* **2010**, *21*, 136–155. [CrossRef]
69. Vasiliev, A.A.; Streletskaya, I.D.; Shirokov, R.S.; Oblogov, G.E. Evolution of cryolithozone of coastal zone of western Yamal during climate change. *Kriosf. Zemli* **2011**, *2*, 56–64.

70. Gentsch, N.; Mikutta, R.; Alves, R.J.E.; Barta, J.; Capek, P.; Gitte, A.; Hugelius, G.; Kuhry, P.; Lashchinskiy, N.; Palmtag, J.; et al. Storage and transformation of organic matter fractions in cryoturbated permafrost soils across the Siberian Arctic. *Biogeosciences* **2015**, *12*, 4525–4542. [CrossRef]
71. Kaiser, C.; Meyer, H.; Biasi, C.; Rusalimova, O.; Barsukov, P.; Richter, A. Conservation of soil organic matter through cryoturbation in arctic soils in Siberia. *J. Geophys. Res.* **2007**, *112*, 9–17. [CrossRef]
72. Kawahigashi, M.; Kaiser, K.; Kalbitz, K.; Rodionov, A.; Guggenberger, G. Dissolved organic matter in small streams along a gradient from discontinuous to continuous permafrost. *Glob. Chang. Biol.* **2004**, *10*, 1576–1586. [CrossRef]
73. Mergelov, N.S.; Targulian, V.O. Accumulation of organic matter in the mineral layers of permafrost-affected soils of coastal lowlands in East Siberia. *Eurasian Soil Sci.* **2011**, *44*, 249–260. [CrossRef]
74. Oosterwoud, M.R.; Temminghoff, E.J.M.; van der Zee, S.E.A.T.M. Quantification of DOC concentrations in relation with soil properties of soils in tundra and taiga of Northern European Russia. *Biogeosci. Discuss.* **2010**, *7*, 3189–3226. [CrossRef]
75. Ernakovich, J.; Hopping, K.A.; Berdanier, A.B.; Simpson, R.T.; Kachergis, E.J.; Steltzer, H.; Wallenstein, M.D. Predicted responses of arctic and alpine ecosystems to altered seasonality under climate change. *Glob. Chang. Biol.* **2014**, *20*, 3256–3269. [CrossRef]
76. Zemtsov, V. Ecoregions and the problem of surface water quality objectives indication in the Ob river basin (Siberia, Russia). In *Hydrology: Science & Practice for the 21st Century*; British Hydrological Society: London, UK, 2004; Volume II, pp. 226–231.

water

MDPI

Article

Major-Ion Chemistry and Quality of Water in Rivers of Northern West Siberia

Irina Ivanova [1,*], Oleg Savichev [1,2], Nikolay Trifonov [1], Yulia V. Kolubaeva [1] and Natalia Volkova [1]

1 Tomsk Branch of the Trofimuk Institute of Petroleum Geology and Geophysics, Siberian Branch of Russian Academy of Sciences, 634055 Tomsk, Russia; osavichev@mail.ru (O.S.); trifonovnik@mail.ru (N.T.); kolubaeva@inbox.ru (Y.V.K.); volkovana@ipgg.sbras.ru (N.V.)

2 School of Earth Sciences and Engineering, National Research Tomsk Polytechnic University, 634050 Tomsk, Russia

* Correspondence: ivanovais@ipgg.sbras.ru

Abstract: This study reports a synthesis of years-long hydrogeochemical monitoring in northern West Siberia, performed by the Russian Meteorological Service (Rosgidromet) and several academic institutions. Natural factors and intensive human economic activity lead to the disruption of the ecosystems of the northern territories of Western Siberia. The aim of this study is to estimate the background water chemistry parameters in the rivers of northern West Siberia in the beginning of the 21st century. The mean values hydrochemical and geochemical indicators were determined with STATISTICA software, which can be used as background values in assessing the actual and allowable anthropogenic impact on water bodies. We revealed four water chemistry provinces: western Ob Gulf and Ob estuary catchments (I); eastern Ob Gulf and Taz Gulf catchments, except for the Taz River and its tributaries (II); Taz River catchments (III); Yenisei River catchments, right bank (IV). The major-ion chemistry of the sampled river waters records a combination of geological, geomorphological, and hydrological conditions in the four provinces. The features typical of the northern West Siberian Plain are especially prominent in province II, which has the lowest average total of major ions (Σ_{mi}), the highest chemical oxygen demand (potassium dichromate COD), and the highest contents of Fe and phosphates. The Σ_{mi} value is the highest in province IV. The river waters from four provinces share similarity in quite high organic contents (both potassium dichromate and permanganate COD), as well as high NH_4^+ and Fe. The long-term average Σ_{mi} of the waters is predicted not to change much in the coming one or two decades, though it may decrease slightly in the winter season but increase in the fall and spring time.

Keywords: river water; major ions; background concentration; water discharge; northern West Siberia

Citation: Ivanova, I.; Savichev, O.; Trifonov, N.; Kolubaeva, Y.V.; Volkova, N. Major-Ion Chemistry and Quality of Water in Rivers of Northern West Siberia. *Water* **2021**, *13*, 3107. https://doi.org/10.3390/w13213107

Academic Editor: Maurizio Barbieri

Received: 24 September 2021
Accepted: 2 November 2021
Published: 4 November 2021

1. Introduction

The world is currently living through notable climate change evident in many regions, including the high latitudes and specifically the Kara Sea drainage basin [1–13]. The near-surface air temperatures in the northern territories are increasing faster than in the middle and tropic latitudes. Correspondingly, the warming causes prominent effects on permafrost, which appear as (i) permafrost thawing, especially near rivers and lakes; (ii) coastline erosion; (iii) inputs of highly saline waters from cryopegs into the water turnover; (iv) changes in the contours of surface watersheds and aquifers; (v) disturbed water-level phases [4,6,9,11,12,14–21]. The environment changes, probably, will entrain changes in the water and solid flow regimes [22], recharge patterns, and water–rock interaction, and thus affect the chemistry and quality of surface waters.

Huge reserves of fresh water are concentrated in Western Siberia, the ecological state of which has not been alarming until recently. However, the environmental problems of the Arctic territories are becoming more relevant, since natural factors and intensive human economic activity, which associated with the extraction of hydrocarbons, lead to the disruption of the ecosystems of the northern territories of Western Siberia. The river water quality is a very important factor concerning human health and aquatic ecosystems. The gradual decrease of river water quality is considered a serious concern because it threatens the sustainability of the aquatic ecosystem. In this respect, hydrogeochemical monitoring is an environmental issue of top priority [23]. Such monitoring can follow standard government programs by meteorological and environment agencies or make part of extended research programs with a large scope of parameters and conditions to check. In Russia the monitoring results are published on a regular basis in reports of the Ministry of Natural Resources and its departments [24–26]. The water quality and related environmental problems have had a large literature for the recent decades, but most of the available publications focus on specific localities and water bodies (commonly lakes), or on a limited number of parameters. Some publications deal with long rivers (most often estuaries), which straddle several climate zones, but zonal variations remain overlooked [1–3,7,11,12,15,20,27–36]. Thus, the modern state of rivers in the Kara Sea basin is yet poorly constrained.

The aim of this study is to estimate the background water chemistry parameters in the rivers of northern West Siberia in the beginning of the 21st century. For this purpose, additional water chemistry surveys were undertaken in the Kara basin by a joint team from the Tomsk Branch of the A.A. Trofimuk Institute of Petroleum Geology and Geophysics (Tomsk), Tomsk Polytechnical University, and Tomsk State University (Tomsk). The synthesis of our results and published evidence, including the reports of the Russian Meteorological Service (Rosgidromet), has provided the required information on the concentrations of dissolved elements, which can make basis for further assessment of climate change effects and anthropogenic loads associated with petroleum production.

2. Materials and Methods

2.1. Study Area

The territory of Northern West Siberia encompasses the Arctic, Subarctic, and temperate climate zones. The winter season generally lasts as long as seven to eight months, with monthly means of −23 to −29 °C, while the summer is within two months, and the mean July temperature is 10 to 14 °C. The amount of atmospheric precipitation increases from 300 to 550 mm/yr off the coast, out of which 40% is solid precipitation. The snow depth reaches 70–90 cm. The region is located in the West Siberian Plain, with a smooth terrain as low as 35–55 m above sea level. The dominant soil in the Taiga being podzols and sandy soils. Soils are represented mainly by Cryic Histosols, less often by Eutric Histosols, with thick organogenic horizons at the study area. The state of organogenic (peat) soil horizons significantly affects the depth of permafrost. This is a zone of continuous permafrost in the Yamal, Taz, and Gydan peninsulas and sporadic permafrost south of the peninsulas down to 64° N [37].

Geologically, the region belongs to the Late Paleozoic–Mesozoic West Siberian Plate, with Mesozoic-Cenozoic sediments upon a Paleozoic basement. Microcrystalline schists, phyllites and phyllitic schists of various compositions-muscovite, sericite, sericite-chlorite and chlorite, sometimes siliceous, are conventionally assigned to the Proterozoic (Rf_3-V_1). Lower Mesozoic (T-J) deposits are represented by alternating clays, marls, sandstones, and siltstones. Undivided Jurassic–Cretaceous deposits are represented by glauconite-quartz siltstones with interlayers of limestones and sandstones. Cretaceous deposits are represented by clays, interbedded with sandy–silty rock units. Undivided deposits (K-P_1) are represented by clays (opoka-like interlayers, with nodules of marl and siderite, small pyrite inclusions) and mica kaolinized sands (with lenses and clay interlayers). The sediments of the Paleocene and Eocene are represented mainly by clayey, clayey-siliceous and siliceous rocks. Oligene deposits are represented by alternating sands and silts with clay interlayers. Neogene deposits are represented by sandy-argillaceous lignite deposits. The greatest part of the territory is occupied by marine, glacial, and mixed marine–glacial plains, where Middle Quaternary silty and clayey sediments are overlain by Middle

Quaternary and Late Quaternary sand and silt deposited in lacustrine, fluvial, or mixed environments. The mostly peaty oligotrophic soils enclose thick organic horizons.

2.2. Sampling and Analytical Procedures

The sampled rivers with the catchment areas of $\leq$2000 km^2 and from 2000 to 50,000 km^2 [38] (Figure 1), at several sites: (1) tributaries of the Ob estuary and the Ob Gulf (Shchuchiya and Sob Rivers and Sob tributaries); (2) Poluy River, at Rosgidromet stations near Poluy community and Salekhard city; (3) tributaries of the eastern Ob Gulf catchment; (4) Nadym River and its tributaries; (5) Pur River, Pyaku-Pur River near Tarko-Sale town, and Pur tributaries; (6) Taz Gulf tributaries; (7) Taz River near Ratta community, and Taz tributaries; (8) streams in Turukhan River catchments, including Turukhan River near Yanov Stan community; (9) peat bogs and swampy areas (water-logged topographic lows with frozen organic-rich layer on top) [39]. The bog waters were sampled over the whole territory from the eastern slopes of the Urals (Ob tributaries) to the Bolshaya Kheta catchment. The sampling was from small and medium rivers, which reflect, respectively, local and zonal conditions, whereas large rivers that drain more than 50,000 km^2 across multiple climate zones (Ob, Yenisei, Nadym, Pur, and Taz Rivers) are unsuitable for characterizing the water chemistry of the northern Kara basin.

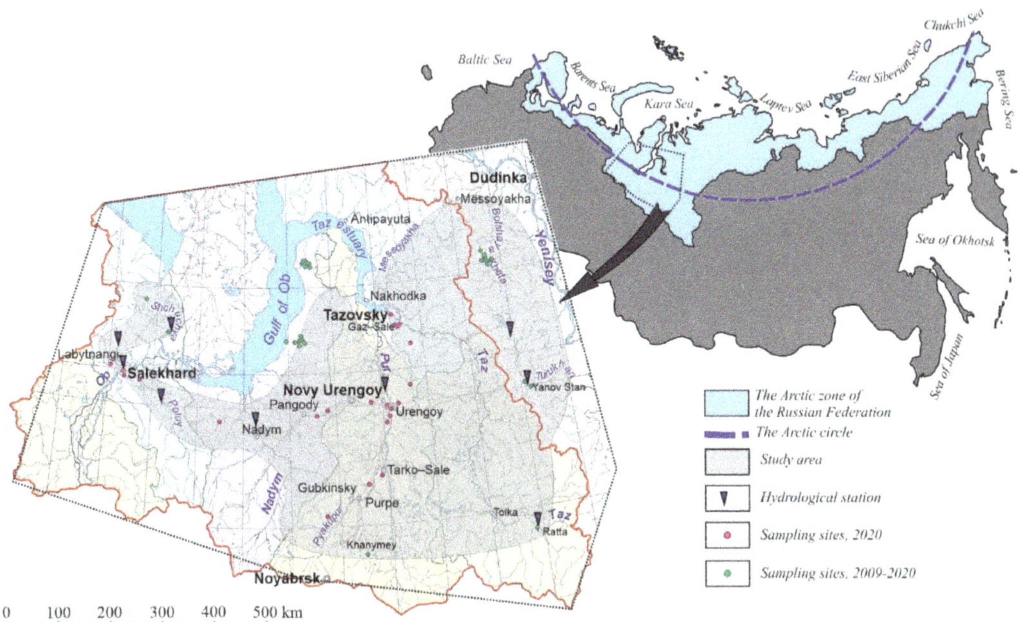

Figure 1. Location map of study area in northern West Siberia.

The database on river water chemistry includes the results, which obtained by authors in 2009 through 2020 and published elsewhere [14,31,32,36,37,40–42], as well as the latest data were obtained from 23 river water samples collected during the fieldtrip to the Yamal-Nenets district in summer 2020 (Figure 1). The choice of these rivers was restricted by the possibility by using road transportation. The water samples were obtained upstream of the bridge. The total number of analyzed samples amounted to 360.

The water temperature, pH, dissolved oxygen, and electric conductivity in the main stem and tributaries were measured directly in the field using Hanna portable instruments. The water was sampled in cleaned polypropylene bottle from 20–30 cm depth in 1 or 2 m offshore [43].

The concentrations of major ions and N, P, and Si compounds, concentrations of Fe, Cu, Zn, Pb, Cd, and Hg, pH, and chemical oxygen demand (both for potassium dichromate and permanganate) were estimated with reference to compatibility between our methods and the *Rosgidromet* practices [44]. Most of the river and bog water were analyzed in accredited laboratories. The samples collected in 2020 were analyzed at the Basic Research Laboratory of Hydrogeochemistry of Tomsk Polytechnic University (accreditation ROSS RU.0001.511901) by different methods, following the guidelines of *Rosgidromet* [44] and the World Meteorological Organization [45]: potentiometry for pH; titrimetry for chemical oxygen demand and the concentrations of CO_2, Ca^{2+}, Mg^{2+}, HCO_3^-, and Cl^-; turbidimetry for SO_4^{2-}; photometry and mass spectrometry with inductively coupled plasma (ICP-MS) on a *NexION 300D* spectrometer for Si, NH_4^+, NO_2^-, NO_3^-, phosphates, and iron; ion chromatography and atomic absorption spectrometry (AAS) for Na^+ and K^+; stripping voltammetry, AAS, and ICP-MS for Zn, Cu, Hg, Pb, Mn, Ni, and Al. The ICP-MC method was also applied to determine trace-element contents in filtered water samples (Sartorius filter, 0.45 μm pore size). The analytical work was performed at the certified laboratories of the Tomsk Polytechnical University and the Tomsk Regional Department of Natural Resources and Environment (*Oblkompriroda*).

2.3. Data Processing and Statistical Analysis

The data were processed by means of mathematical statistics using the *MS Excel* and *Statistica* software at a significance level value of 5%. The water compositions were classified into bicarbonate (HCO_3^-+CO_3^{2-}), sulfate (SO_4^{2-}), and chloride (Cl^-) ionic types according to major anions [46], and each type was further divided into calcic, magnesian, and sodic subtypes according to major cations.

The water chemistry studies consisted of several steps. The work began with data inventory and classification of rivers according to flow and recharge patterns [47,48]. Then statistics were calculated over specific parts of the Kara basin [49,50], and the rms errors were estimated as

$$\delta_A \approx \frac{\sigma}{\sqrt{N}}, \quad (1)$$

where σ is the rms error and N is the number of measurements (size of statistical sample). At the next step, the Fisher K_F (Equation (2)) and Student K_t (Equation (3)) distributions of the statistics were calculated at 5% significance level to test the variance [48–50].

$$K_F = \frac{\max\left(\sigma_x^2; \sigma_y^2\right)}{\min\left(\sigma_x^2; \sigma_y^2\right)}, \quad (2)$$

$$K_t = \frac{|A_x - A_y|}{\sqrt{N_x \cdot \sigma_x^2 + N_y \cdot \sigma_y^2}} \cdot \sqrt{\frac{N_x \cdot N_y \cdot (N_x + N_y - 2)}{(N_x + N_y)}}, \quad (3)$$

where A_x and A_y, σ_x and σ_y, N_x and N_y, are, respectively, the arithmetic means, rms errors, and sizes of the statistical samples x and y. If both Fischer and Student tests confirmed the specified low variance, the statistical samples were joined, and tested further in the same way.

Finally, the hydrochemical and hydrological parameters were correlated and the regression relationships were obtained separately for the statistical samples and for the whole data set, as recommended in [48]:

$$|r| \geq 0.70; \ |r| \geq \frac{2 \cdot (1 - r^2)}{\sqrt{N - 2}}; \ |k_r| \geq 2 \cdot \delta_{kr}, \quad (4)$$

where r is the correlation coefficient; k_r and δ_{kr} are the regression coefficient and its rms error.

3. Results and Discussion

The sampled rivers were first divided into eight groups: (1) tributaries of the Ob estuary and Ob Gulf: Shchuchiya, Sob (Rosgidromet station near Kharp community, and Sob tributaries), Ob Gulf tributaries; (2) Poluy River (Rosgidromet stations near Poluy community and Salekhard city); (3) tributaries of the Ob Gulf (right bank); (4) Nadym River (Rosgidromet station near Nadym community) and its tributaries; (5) Pur River (Rosgidromet station near Tarko-Sale community) and its tributaries; (6) Taz Gulf tributaries; (7) Taz River (Rosgidromet station near Ratta community) and its tributaries; (8) streams in the Turukhan (Rosgidromet station near Yanov Stan community) and Bolshaya Kheta catchments. Then the groups were joined into four provinces according to water chemistry similarity.

The rivers of group 1 flow within the western slope of the Urals, along major geological and hydrological boundaries: at the junction between the Ural orogenic belt and the West Siberian Basin filled with Paleozoic and Mesozoic sediments, and between the rivers of the mountainous-tundra zone and the plainland rivers of the West Siberian tundra and forest zones [47]. The rivers of groups 3–6 are located within the West Siberian Basin and its Paleogene, Neogene, and Quaternary sediments, and within the tundra and forest zones [47]. Groups 2 and 7 represent the conditions that are generally intermediate between the plainland rivers of the West Siberian Basin interior on one side and groups 1 and 8 on the other side [47,51].

For each group, we calculated the arithmetic means and rms errors of element concentrations, and variance in the Σ_{mi} series (Ca^{2+}, Mg^{2+}, Na^{+}, K^{+}, HCO_3^{-}, CO_3^{2-}, SO_4^{2-}, and Cl^{-}) was checked by means of Fisher (Table 1) and Student (Table 2) tests, at a 5% significance level. As a result, the Σ_{mi} data for groups 1 and 2 were jointed into one statistical sample corresponding to province I. Similar Σ_{mi} values were obtained for rivers in the Taz catchment. On the other hand, the rivers of groups 1, 2, and 7 jointly belong to province III, as they share the intermediate geological and geomorphological position between plainland and highland conditions, though the Taz catchment is located far from the eastern Ural slope. Groups 3–6 that include typically plainland rivers of the tundra, forest tundra, and northern taiga zones represent similar geological and geomorphological settings. The Σ_{mi} and variance for the Yenisei tributaries differ from the respective data for all other sampled rivers, including the Taz tributaries, as well as from bog waters. Bogs generally do not represent any specific climate zones and differ from rivers in both average Σ_{mi} and variance (Tables 1 and 2).

Table 1. Ratio between actual K_F (in formula 2) and critical K_F, 5% (5% significance) variance.

Area	1	2	3	4	5	6	7	8	9
1	0.56	**21.90**	**7.38**	**27.04**	**12.15**	**4.86**	**39.74**	**23.69**	**1.32**
2	**21.90**	0.39	**1.64**	0.48	0.69	**2.56**	0.70	0.42	**8.83**
3	**7.38**	**1.64**	0.55	**2.03**	0.91	0.86	**2.98**	**1.78**	**2.97**
4	**27.04**	0.48	**2.03**	0.63	0.86	**3.16**	0.93	0.26	**10.90**
5	**12.15**	0.69	0.91	0.86	0.39	**1.42**	**1.26**	0.75	**4.90**
6	**4.86**	**2.56**	0.86	**3.16**	**1.42**	0.57	**4.64**	**2.77**	**1.96**
7	**39.74**	0.70	**2.98**	0.93	**1.26**	**4.64**	0.36	0.38	**16.03**
8	**23.69**	0.42	**1.78**	0.26	0.75	**2.77**	0.38	0.23	**9.55**
9	**1.32**	**8.83**	**2.97**	**10.90**	**4.90**	**1.96**	**16.03**	**9.55**	0.53

Note: statistically significant difference is in bold. Numerals 1–9 stand for names of areas: 1 = rivers in the Yenisei catchment; 2 = rivers in the Nadym catchment; 3 = Poluy River; 4 = rivers in the Pur catchment; 5 = rivers in the Taz catchment; 6 = rivers in the western Ob Gulf catchment; 7 = rivers in the eastern Ob Gulf catchment; 8 = rivers in the Taz Gulf catchment; 9 = bogs and swampy areas.

Table 2. Ratio between actual K_t (in formula 3) and critical K_t, 5% (5% significance) variance, Student *t* test.

Area	1	2	3	4	5	6	7	8	9
1	0.00	**1.29**	**1.13**	**2.63**	**0.98**	**1.32**	**1.39**	0.95	0.64
2	**1.29**	0.00	**1.84**	0.19	**1.08**	**1.24**	0.86	0.34	**1.19**
3	**1.13**	**1.84**	0.00	**3.49**	0.79	0.38	**2.41**	**1.54**	0.54
4	**2.63**	0.19	**3.49**	0.00	**1.84**	**2.48**	0.95	0.31	**2.42**
5	0.98	**1.08**	0.79	**1.84**	0.00	0.39	**1.82**	**1.08**	0.73
6	**1.32**	**1.24**	0.38	**2.48**	0.39	0.00	**1.70**	**1.07**	0.76
7	**1.39**	0.86	**2.41**	0.95	**1.82**	**1.70**	0.00	0.34	**1.39**
8	0.95	0.34	**1.54**	0.31	**1.08**	**1.07**	0.34	0.00	0.92
9	0.64	**1.19**	0.54	**2.42**	0.73	0.76	**1.39**	0.92	0.00

Note: statistically significant difference is in bold. Numerals 1–9 stand for names of areas: 1 = rivers in the Yenisey catchment; 2 = rivers in the Nadym catchment; 3 = Poluy River; 4 = rivers in the Pur catchment; 5 = rivers in the Taz catchment; 6 = rivers in the western Ob Gulf catchment; 7 = rivers in the eastern Ob Gulf catchment; 8 = rivers in the Taz Gulf catchment; 9 = bogs and swampy areas.

The major-ion totals in the river waters are the highest in the Yenisei tributaries (province IV) and the lowest in the right side of the Ob Gulf and the Taz Gukf tributaries (province II). The contents of total iron and phosphates, as well as the chemical oxygen demand (dichromate COD) are, on the contrary, the highest in province II and lowest in province IV. The waters in provinces I–III are fresh and calcium bicarbonate of type 2, according to the classification of Alekin [46]. The compositions of this type commonly result from water interaction with sediments and weathered bedrocks. The waters of the Yenisei tributaries are likewise fresh and have a calcium bicarbonate chemistry, but belong to type 1 [46] corresponding to the interaction of waters with Na- and K-rich igneous rocks. These results agree with the local geology, i.e., with the presence of Cretaceous sediments in the eastern part of the area (Figure 2). However, the greatest part of the territory, between the Ob and Taz Rivers, is occupied by Upper Quaternary and Middle Quaternary sediments covered with peat.

The bog waters of the region are on average fresh and have calcium bicarbonate chemistry, like the river waters, and are intermediate between waters of types 1 and 2 in the contents of major ions. This composition may be due to thinning of water lenses and/or to accumulation of Na and K ions in the active layer above the permafrost. Compared to the river waters, the bog waters expectedly have higher dichromate COD due to peat formation, and Fe iron contents due to accumulation of dissolved compounds with organic acids.

In general, the major-ion totals are higher along the margins of the West Siberian Plain, while the organic compounds and their derivatives are more abundant in the swampy central areas. This fact has to be taken into account when estimating natural and anthropogenic effects on water bodies, e.g., using average values for provinces I–IV as background pH, COD, major ion concentrations, inorganic compounds of N, phosphates, Si, and Fe (Table 3).

Unfortunately, the available data on trace-element contents are insufficient to provide a reliable basis for distinguishing the water chemistry provinces, though some trends appear in lateral variations across the region: e.g., Ni and Cu increase slightly eastward. The waters were sampled for a larger number of elements during the trip of September 2020 by a team from the Tomsk branch of the Trofimuk institute of petroleum geology and geophysics of Siberian branch of Russian academy of sciences (Table 4). The regional average contents of Li, B, V, Co, Ni, Ge, As, Rb, Cd, Pb, La, Ce, and Pr we obtained differ for less than 50% from the previous estimates over the rivers from the Kara Sea drainage basin [36].

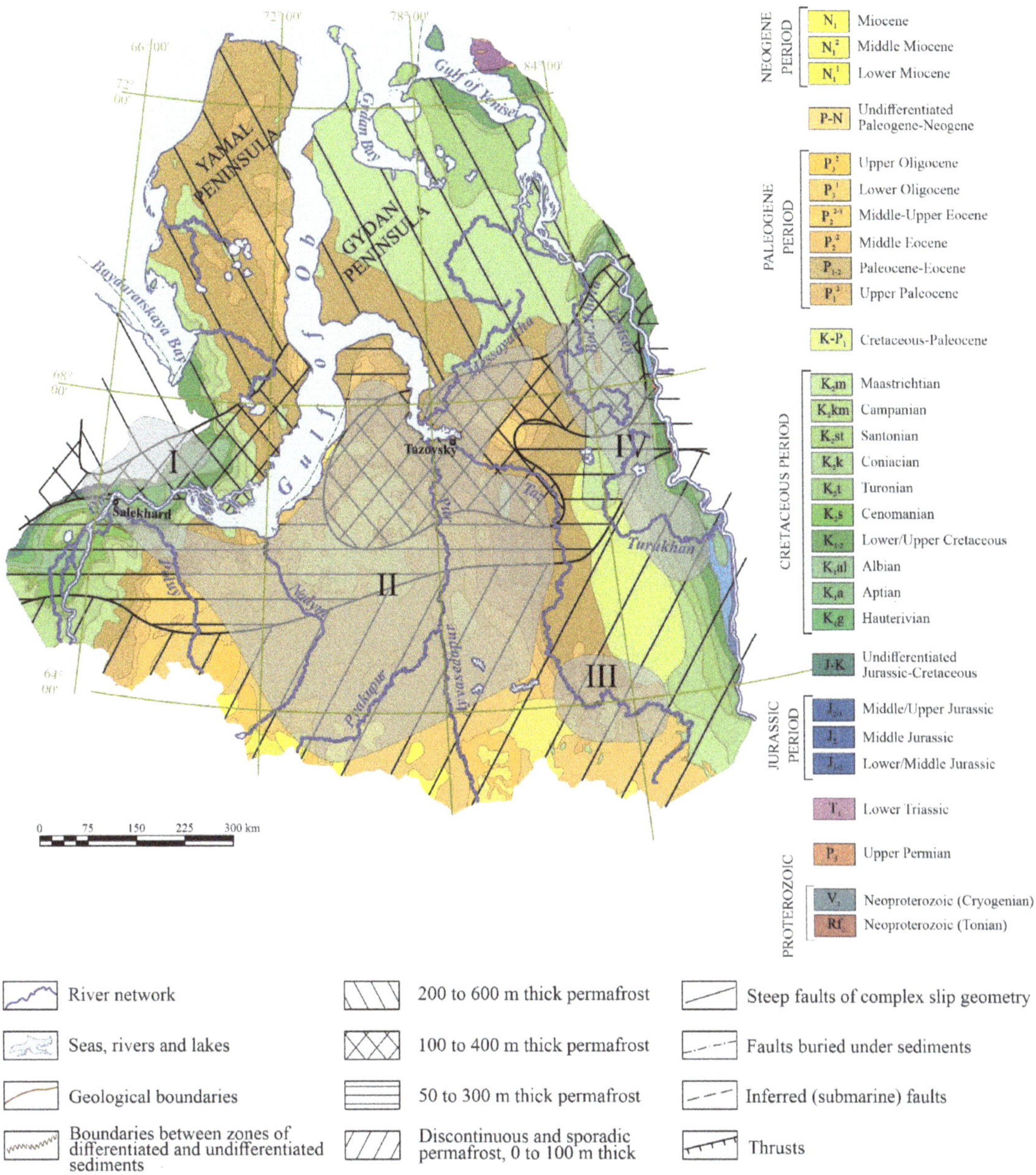

Figure 2. Map of northern West Siberia and local geology [52].

Table 3. Average values of hydrochemical parameters of river (four provinces) and bog (region as a whole) waters.

	I Ob Gulf Catchment (Western Part)			II Catchments of Ob Gulf (western part) and Taz Gulf			III Upper Taz Reaches and Taz Tributaries			IV Yenisei Triburaties			V Bogs		
	A	δ_A	*N*	*A*	δ_A	*N*	*A*	δ_A	*N*	*A*	δ_A	*N*	*A*	δ_A	*N*
pH,	7.05	0.07	56	6.17	0.07	134	7.03	0.14	19	7.04	0.11	32	5.64	0.17	41
Water classification after [46]															
Class/group/type	C/Ca/II			C/Ca/II			C/Ca/II			C/Ca/I			C/Ca/I(II)		
mgO/L															
PO	10.34	1.42	72	11.80	0.99	96	13.10	1.82	13	11.77	0.96	12	53.55	17.45	8
BO	20.55	1.82	65	30.62	2.30	108	27.68	2.70	17	20.50	1.61	23	167.85	41.35	8
mg/L															
CO_2	15.0	3.1	21	9.3	1.8	16	2.1	1.1	5	7.1	1.6	18	18.2	3.9	8
Σ_{mi}	72.0	5.3	84	36.8	2.2	126	62.0	8.0	19	136.8	23.1	49	98.9	16.5	41
Ca^{2+}	8.0	0.6	84	4.2	0.2	126	8.1	1.0	19	14.1	1.9	49	0.9	0.4	8
Mg^{2+}	3.8	0.3	84	1.8	0.1	126	3.3	0.5	19	6.9	1.0	49	0.5	0.2	8
Na^+	6.5	0.9	84	3.6	0.3	126	3.7	0.8	19	8.7	2.7	49	0.7	0.3	8
K^+	0.7	0.1	84	0.5	0.1	126	1.0	0.3	19	2.7	1.2	49	0.2	0.1	8
HCO_3^-	35.9	3.0	84	17.3	1.5	126	38.4	5.9	19	88.8	14.2	49	5.4	2.9	8
Cl^-	9.1	1.0	84	4.2	0.4	126	3.0	0.5	19	5.5	1.4	49	0.8	0.2	8
SO_4^{2-}	7.8	0.7	84	7.7	1.0	126	3.6	0.5	19	10.1	3.3	49	3.5	1.0	8
NO_3^-	0.22	0.13	4	0.28	0.06	70	0.81	0.39	10	1.21	0.31	27	0.07	0.03	8
NO_2^-	0.01	0.00	8	0.03	0.00	71	0.01	0.00	10	0.13	0.08	19	0.01	0.00	8
NH_4^+	0.53	0.20	4	0.35	0.05	74	0.31	0.06	11	0.57	0.09	17	2.55	0.47	8
Phosphates	0.06	0.04	4	0.14	0.04	54	0.06	0.03	5	0.02	0.01	8	0.05	0.02	8
Si	4.75	0.37	71	5.25	0.37	80	5.04	0.75	13	3.06	0.52	11	2.44	0.56	8
Fe	0.644	0.064	80	1.247	0.096	135	0.744	0.139	19	0.550	0.090	34	8.135	2.591	41
μg/L															
Ni	2.3	0.5	16	10.1	2.2	62	22.1	8.5	9	21.9	5.8	15	2.7	0.4	41
Cu	0.8	0.1	4	8.9	1.8	62	6.1	2.3	9	8.0	1.4	15	0.5	0.1	8
Zn	2.3	0.9	4	15.4	4.2	62	19.7	14.4	9	63.1	18.8	15	7.1	2.2	8
Cd	<0.01	<0.01	4	0.28	0.07	45	0.07	0.05	9	0.04	0.01	9	0.03	0.01	8
Pb	<0.1	<0.1	4	2.0	0.5	62	2.1	0.9	9	2.5	0.7	15	1.0	0.5	8

Note: *A* and δ_A are, respectively, arithmetic mean and its error found by Equation (1); *N* is the number of measurements; BO and PO are carbon oxygen demand for potassium bichromate and potassium permanganate, respectively; Σ_{mi} is the total of major ions (Ca^{2+}, Mg^{2+}, Na^+, K^+, HCO_3^-, CO_3^{2-}, SO_4^{2-}, and Cl^-).

Table 4. Average values of trace elements in rivers of the Kara basin, sampled in September 2020, µg/L.

Element	Element Concentration (September 2020)		A^* [36]
	A	δ_A	
Li	1.72	0.23	2.23
B	8.63	2.32	11.10
Al	169.5	74.4	27.8
Ti	5.77	3.89	0.39
V	0.80	0.28	1.01
Cr	0.96	0.16	0.27
Mn	15.93	3.51	34.00
Co	0.18	0.04	0.13
Ni	1.97	0.37	1.17
Cu	0.63	0.10	1.06
Zn	1.44	0.28	2.36
Ge	0.01	0.00	0.01
As	0.72	0.09	0.45
Rb	0.46	0.08	0.64
Sr	23.21	5.11	69.70
Y	0.311	0.044	0.051
Mo	0.040	0.012	0.670
Cd	0.003	0.001	0.004
Cs	0.008	0.005	0.002
Ba	5.55	0.66	22.40
Pb	0.105	0.047	0.090
La	0.217	0.042	0.166
Ce	0.475	0.096	0.310
Pr	0.064	0.011	0.046
U	0.031	0.019	0.300

Note: A is the arithmetic mean value; δ_A is the rms error found by Equation (1), at $N = 23$; A^* = average over the rivers of the Kara basin according to [36].

The total of major ions varies seasonally as a function of runoff patterns. The Σ_{mi} and the water discharge Q_t are commonly related as

$$_{\text{mi}} = a \cdot \frac{Q_t}{Q_a}^{-b} \quad (5)$$

See the general Σ_{mi} vs. Q_t/Q_a relationship for all four provinces in Figure 3, where Q_t is the water discharge (flow rate) at a certain time and Q_a is the respective average value; a and b are empirical constants. In the case of province IV, at very small discharge ratios ($Q_t/Q_a < 0.01$), mainly in the winter low-water season, Σ_{mi} can even increase slightly at higher Q_t/Q_a (Figure 4). This state may correspond to influx from saline water lenses in permafrost (cryopegs). The effect of saline water inputs smooths down as the surface runoff into the river network increases, and the major-ion total decreases in a stable way with increasing discharge at $Q_t/Q_a > 0.2$–0.3. The Σ_{mi} and the discharge correlate in an intricate way, being first directly and then inversely proportional. This reversal was attributed [53] to the water chemistry control jointly by slope, surface, and subsurface run-off components, while the run-off, in its turn, depends nonlinearly on the air temperature [53].

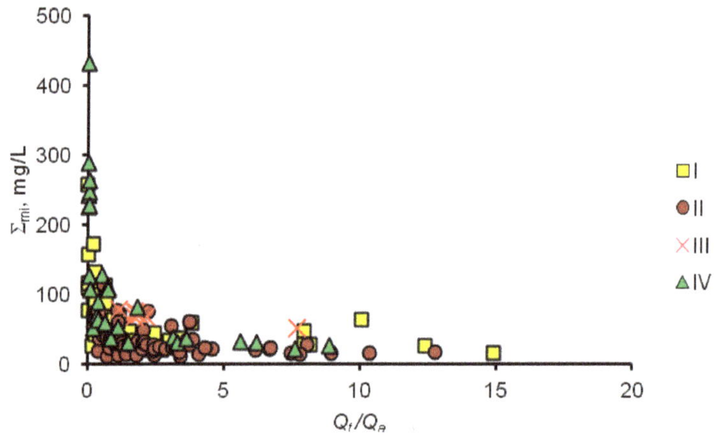

Figure 3. Σ_{mi} vs. Q_t/Q_a relationship for provinces I–IV (Table 3); (I) $\Sigma_{mi} = 50.121 \cdot \frac{Q_t}{Q_a}^{-0.200}$; (II) $\Sigma_{mi} = 35.140 \cdot \frac{Q_t}{Q_a}^{-0.337}$; (III) $\Sigma_{mi} = 71.723 \cdot \frac{Q_t}{Q_a}^{-0.092}$; (IV) $\Sigma_{mi} = 59.463 \cdot \frac{Q_t}{Q_a}^{-0.407}$.

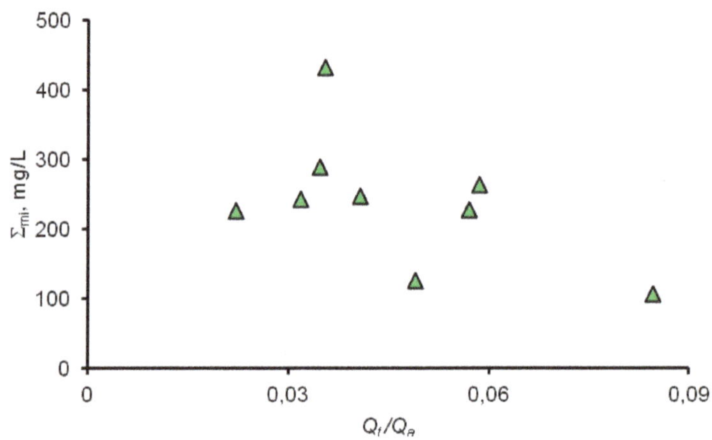

Figure 4. Σ_{mi} vs. Q_t/Q_a relationship for province IV (Table 3) at small Q_t/Q_a ratios.

The power–law relationship (Figure 3) can be used for preliminary estimation of possible climate-driven changes in the water chemistry of the Arctic rivers. Specifically, it was found out that the subsurface component of the river run-off in Northern Asia has undergone a statistically significant increase in the past two to four decades, and the duration of water-level seasons has changed [4,7,13,15]. Therefore, the major-ion total can be expected to increase slightly in the winter low-water time but may increase in the fall and spring seasons.

4. Conclusions

The analyzed statistical samples of total major-ion contents in the rivers of northern West Siberia reveal four water chemistry provinces: western Ob Gulf and Ob estuary catchments (I); eastern Ob Gulf and Taz Gulf catchments, except for the Taz River and its tributaries (II); Taz River catchment (III); and Yenisei River catchment (right bank) (IV). The compositions of river waters in different provinces record the hydrogeochemical conditions corresponding to the Ural eastern slope (province I) and to the junction between the West Siberian and Central Siberian plates (province IV), while the conditions in province III are intermediate between the two provinces. The conditions in province II are the most typical of the northern West Siberian Plain: the lowest average Σmi, the highest dichromate COD, and the highest contents of Fe and phosphates. The major-ion total is the highest in the rivers of province IV, as a result of its geological specificity and due to inputs of saline waters from cryopegs in watershed permafrost.

Some trace elements (e.g., Cu and Ni) show increasing trends from west to east. The river waters from all four provinces share similarity in quite high organic contents inferred from both dichromate and permanganate COD, as well as high NH_4^+ and Fe. Thus, the waters of the area do not comply with the quality requirements for the waters of fishery and household uses.

The total major-ion concentrations vary seasonally: they decrease at higher run-off and water levels and increase in the low-water season. In view of the revealed correlation between the water chemistry and water level (discharge) patterns, and in the presence of climate-induced changes in the subsurface run-off component and in the duration of hydrological seasons, it can be assumed that the long-term average Σ_{mi} values will hardly change much in the coming one or two decades, though they may decrease slightly in the winter season but increase in the fall and spring time.

Author Contributions: Sampling, conceptualization, I.I.; methodology, writing—soriginal draft preparation, O.S.; analyses, N.V.; interpretation of analytical results, Y.V.K.; visualization, N.T. All authors have read and agreed to the published version of the manuscript.

Funding: This research was funded by the Russian Science Foundation, grant No 20-77-10084.

Institutional Review Board Statement: Not applicable.

Informed Consent Statement: Not applicable.

Data Availability Statement: The data that support the findings of this study are available from the corresponding author upon reasonable request.

Conflicts of Interest: The authors declare no conflict of interest.

References

1. Bring, A.; Destouni, G. Hydrological and hydrochemical observation status in the pan-Arctic drainage basin. *Polar. Res.* **2009**, *28*, 327–338. [CrossRef]
2. Gordeev, V.V. *The River-Sea System and Its Role in the Oceanic Water Chemistry*; Shirshov Institute of Oceanology of the Russian Academy of Sciences: Moscow, Russia, 2009; 36p.
3. Holmes, R.M.; McClelland, J.W.; Peterson, B.J.; Tank, S.E.; Bulygina, E.; Eglinton, T.I.; Gordeev, V.V.; Gurtovaya, T.Y.; Raymond, P.A.; Repeta, D.J.; et al. Seasonal and annual fluxes of nutrients and organic matter from large rivers to the Arctic Ocean and surrounding seas. *Estuaries Coasts* **2012**, *35*, 369–382. [CrossRef]
4. Shiklomanov, A.I.; Lammers, R.B.; Lettenmaier, D.P.; Polischuk, Y.M.; Savichev, O.G.; Smith, L.G. Hydrological changes: Historical analysis, contemporary status, and future projections. In *Environmental Changes in Siberia: Regional Changes and Their Global Consequences*; Groisman, P.Y., Gutman, G., Eds.; Springer: Berlin, Germany, 2013; pp. 111–154. Available online: http:link.springer.com/chapter/10.1007/978-94-007-4569-8_4 (accessed on 24 September 2021).
5. Bradshaw, R.H.M.; Sykes, M.T. *Ecosystem Dynamics: From the Past to the Future*; John Wiley & Sons: Chichester, UK, 2014; 321p.
6. Reykhard, L.; Shulga, N.; Novichkova, Y.A.; Kozina, N.; Dara, O.; Sapozhnikov, P.; Gordeev, V.; Belyaev, N.; Boev, A. Biomineral indicators of hydrological, geological and climatic processes in the Arctic. *Acta. Crystal. Found. Adv.* **2018**, *74*, e250. [CrossRef]
7. Dzhamalov, R.G.; Safronova, T.I. Effect of permafrost rocks on water resources formation in Eastern Siberia: Case study of some rivers in Eastern Siberia. *Water Resour.* **2018**, *45*, 455–465. [CrossRef]
8. Song, C.; Wang, G.; Mao, T.; Dai, J.; Yang, D. Linkage between permafrost distribution and river runoff changes across the Arctic and the Tibetan Plateau. *Sci. China Earth Sci.* **2020**, *62*, 292–302. [CrossRef]
9. Gusev, E.M.; Nasonova, O.N.; Shurkhno, E.A.; Dzhogan, L.Y. Scenario forecasting of changes in water balance components in the Ob–Irtysh basin in the context of possible climate change. *Water Resour.* **2019**, *46*, 647–658. [CrossRef]
10. Kiyo, T.; Campbell, F.; Lantz, T.C.; Fraser, R.H.; Hogan, D. High Arctic vegetation change mediated by hydrological conditions. *Ecosystems* **2021**, *24*, 106–121. [CrossRef]
11. Manasypov, R.M.; Lim, A.G.; Krickov, I.V.; Shirokova, L.S.; Vorobyev, S.N.; Kirpotin, S.N.; Pokrovsky, O.S. Spatial and seasonal variations of C, nutrient, and metal concentration in thermokarst lakes of Western Siberia across a permafrost gradient. *Water* **2020**, *12*, 1830. [CrossRef]

12. Pokrovsky, O.S.; Manasypov, R.M.; Kopysov, S.G.; Krickov, I.V.; Shirokova, L.S.; Loiko, S.V.; Lim, A.G.; Kolesnichenko, L.G.; Vorobyev, S.N.; Kirpotin, S.N. Impact of permafrost thaw and climate warming on riverine export fluxes of carbon, nutrients and metals in Western Siberia. *Water* **2020**, *12*, 1817. [CrossRef]
13. Shiklomanov, A.; Déry, S.; Tretiakov, M.; Yang, D.; Magritsky, D.; Georgiadi, A.; Tang, W. River freshwater flux to the Arctic Ocean. In *Arctic Hydrology, Permafrost and Ecosystems*; Yang, D., Kane, D., Eds.; Springer: Berlin, Germany, 2021; pp. 703–738. [CrossRef]
14. Savichev, O.G.; Mazurov, A.K.; Pipko, I.I.; Sergienko, V.I.; Semiletov, I.P. Spatial patterns of the evolution of the chemical composition and discharge of river water in the Ob River Basin. *Dokl. Earth Sci.* **2016**, *466*, 47–51. [CrossRef]
15. Savichev, O.G.; Moiseeva, Y.A. A model of intra-annual flow distribution with scanty observational data. *ENVIROMIS2018 IOP Conf. Ser. Earth Environ. Sci.* **2018**, *211*, 1–6. [CrossRef]
16. Kopysov, S.G.; Erofeev, A.A. The influences of climate on runoff: A case study of four catchments in Western Siberia. *IOP Conf. Ser. Earth Environ. Sci.* **2019**, *381*, 012046. [CrossRef]
17. Krickov, I.V.; Lim, A.G.; Manasypov, R.M.; Loiko, S.V.; Vorobyev, S.N.; Pokrovsky, O.S.; Shevchenko, V.P.; Dara, O.M.; Gordeev, V.V. Major and trace elements in suspended matter of Western Siberian rivers: First assessment across permafrost zones and landscape parameters of watersheds. *Geochim. Cosmochim. Acta* **2020**, *269*, 429–450. [CrossRef]
18. Pokrovsky, O.S.; Viers, J.; Chupakov, A.V.; Shirokova, L.S.; Gordeev, V.V.; Shevchenko, V.P. Trace elements in the form of organo-mineral colloids in the mixing zone of the Arctic river. In *Biogeochemistry of Trace Elements*; Chemical Research and Application: New York, NY, USA, 2018; pp. 245–289.
19. Victorov, A.S.; Orlov, T.V.; Trapeznikova, O.N.; Kapralova, V.N.; Arkhipova, M.V. Regularities of the distribution of lake areas on thermokarst plains with fluvial erosion. *Dokl. Earth Sci.* **2020**, *491*, 282–284. [CrossRef]
20. Moiseenko, T.I.; Dinu, M.I.; Gashkina, N.A.; Kremleva, T.A.; Khoroshavin, V.Y. Geochemical features of elements distributions in the lake waters of the Arctic region. *Geochem. Intern.* **2020**, *58*, 613–623. [CrossRef]
21. Lim, A.G.; Loiko, S.V.; Kuzmina, D.M.; Krickov, I.V.; Shirokova, L.S.; Kulizhsky, S.P.; Vorobyev, S.N.; Pokrovsky, O.S. Dispersed ground ice of permafrost peatlands: Potential unaccounted carbon, nutrient and metal sources. *Chemosphere* **2021**, *266*, 128953. [CrossRef]
22. Schneider, C.; Laize, C.L.R.; Acreman, M.C.; Florke, M. How will climate change modify river flow regimes in Europe? *Hydrol. Earth Syst. Sci.* **2013**, *17*, 325–339. [CrossRef]
23. Hasan, N.-E.-A.; Salam, R.; Paul, R.; Ambade, I. Health Risk and Water Quality Assessment of Surface Water in an Urban River of Bangladesh. *Sustainability* **2021**, *13*, 6832. [CrossRef]
24. Shiklomanov, I.A. (Ed.) *Water Resources of Russia*; State Hydrological Institute: St. Petersburg, Russia, 2008; 600p. (In Russian)
25. Water, C. *Surface and Groundwater Resources: Uses and Water Quality, Year 2018*; Rosgidromet: St. Petersburg, Russia, 2019; 153p. (In Russian)
26. Rybaiskiy, N.G. (Ed.) *Water Resources of the Russian Federation in 2018: State and Uses*; Government Report; NIA-Priroda: Moscow, Russia, 2019; 290p. (In Russian)
27. Shvartsev, S.L.; Savichev, O.G. The ecological and geochemical state of the major tributaries of the Ob in its middle course. *Water Resour.* **1997**, *24*, 707–713.
28. Shvartsev, S.L.; Savichev, O.G.; Vertman, G.G.; Zarubina, R.F.; Nalivaiko, N.G.; Trifonova, N.G.; Turov, Y.P.; Frizen, L.F.; Yankovskii, V.V. The ecological and geochemical state of the water in the middle course of the Ob River. *Water Resour.* **1996**, *23*, 673–682.
29. Savenko, V.S. Chemical composition of sediment load carried by rivers. *Geochem. Intern.* **2007**, *45*, 816–824. [CrossRef]
30. Savichev, O.G. Influence of large tributaries on water chemistry in the Middle Ob river. *Bull. Tomsk State Univ.* **2010**, *340*, 222–228. (In Russian)
31. Savichev, O.G.; Kolesnichenko, L.G.; Saifulina, E.V. The ecologo-geochemical state of water bodies in the Taz-Yenisei interfluve. *Geogr. Nat. Resour.* **2011**, *32*, 333–336. [CrossRef]
32. Guseva, N.V.; Kopylova, Y.G.; Khvashchevskaya, A.A. Geochemical types of natural waters in the interfluve of the Enzor'yakhi and Yun'yakhi (Eastern Slope of the Polar Urals). *Water Resour.* **2013**, *40*, 417–425. [CrossRef]
33. Shirokova, L.S.; Pokrovsky, O.S.; Kirpotin, S.N.; Desmukh, C.; Pokrovsky, B.G.; Audry, S.; Viers, J. Biogeochemistry of organic carbon, CO_2, CH_4, and trace elements in thermokarst water bodies in discontinuous permafrost zones of Western Siberia. *Biogeochemistry* **2013**, *113*, 573–593. [CrossRef]
34. Gordeev, V.V.; Lein, A.Y.; Belyaev, N.A.; Ivanov, M.V. The distribution of heavy metals in the near-bottom particulate matter, fluffy layer, and bottom sediments over the Yenisei profile in the Kara sea. *Dokl. Earth Sci.* **2014**, *456*, 720–725. [CrossRef]
35. Moskovchenko, D.V.; Babushkin, A.G.; Yurtaev, A.A. The impact of the Russian oil industry on surface water quality (a case study of the Agan River catchment, West Siberia). *Environ. Earth Sci.* **2020**, *79*, 355. [CrossRef]
36. Savenko, A.V.; Savenko, V.S.; Pokrovsky, O.S. New data on the concentrations of dissolved trace elements in waters of Russian Arctic rivers. *Dokl. Earth Sci.* **2020**, *491*, 257–263. [CrossRef]
37. Griva, G.I. *Local Geology and Environment Conditions in Gas Fields of the Yamal Peninsula*; Tomsk State University: Tomsk, Russia, 2005; 352p. (In Russian)
38. State Standard. *Working Document GOST 19179-73. Land Hydrology*; Terms Definitions; Gosstandart USSR: Moscow, Russia, 1988; 47p. (In Russian)
39. State Standard. *Working Document GOST 21123-85. Peat. Terms and Definitions*; 01.07.1986. Izd; Standartov: Moscow, Russia, 1985; 85p. (In Russian)

40. Savichev, O.G.; Moiseeva, Y.A. Surface water stability to the anthropogenic effect in tundra and forest tundra in Western Siberia. *Bull. Arct. Fed. Univ. Ser. Nat. Sci.* **2016**, *4*, 36–46. [CrossRef]
41. Zemtsov, V.A.; Savichev, O.G. Resources, regime and quality of surface waters in the Ob River basin: History, current state and problems of research. *Intern. J. Environ. Stud.* **2015**, *72*, 386–396. [CrossRef]
42. Guseva, N.V.; Kopylova, Y.G. Chemistry of Natural Waters in the Yuniyakha—Enzoryakha interfluve, Eastern Slope of the Polar Urals. *Bull. Tomsk State Univ.* **2009**, *327*, 224–228. (In Russian)
43. Wilde, F.D.; Radtke, D.B. Guidelines for field-measured water-quality properties. In *National Field Manual for the Collection of Water-Quality Data*; Wilde, F.D., Ed.; U.S. Geological Survey Techniques of Water-Resources Investigations: Reston, VA, USA, 2008; Book 9, Chapter A6. Available online: https://pubs.er.usgs.gov/publication/twri09A6.0 (accessed on 24 September 2021).
44. Organization of Pollution Monitoring in Land Surface Waters. *Implemented since 2017-04-03*; Working Document RD 52.24.309–2016; Ministry of Natural Resources and Environment of the Russian Federation (Rosgidromet): Moscow, Russia, 2016; 135p. (In Russian)
45. Hydrology—From Measurement to Hydrological Information. *Guide to Hydrological Practices*, 6th ed.; Book I; World Meteorological Organization WMO 168: Geneva, Switzerland, 2008; 296p.
46. Alekin, O.A. *Fundamentals of Hydrochemistry*; Gidrometeoizdat: Leningrad, Russia, 1970; 444p. (In Russian)
47. Kuzin, P.S.; Babkin, V.I. *Geographic Atterns of the River Flow Regime*; Gidrometeoizdat: Leningrad, Russia, 1979; 200p. (In Russian)
48. State Standard. *Working Document SP 33-101-2003. Guide to Determination of Basic Hydrological Parameters*; Gosstroy Rossii: Moscow, Russia, 2004; 72p. (In Russian)
49. Schmidt, T.S. (Ed.) *Calculation of Hydrological Parameters. A Manual*; Gidrometeoizdat: Leningrad, Russia, 1984; 448p. (In Russian)
50. *Procedure for Calculating Background Element Concentrations in Water for Standardization of Wastewater Discharge*; Working Document RD 52.24.622–2017; 14.06.2017, Rostov-na-Donu, FGBU «SCHI»russia; Rosgidromet: Moscow, Russia, 2017; 96p. (In Russian)
51. Nudner, V.A. (Ed.) *Hydrogeology of the USSR*; West Siberian Plain (Tyumen, Omsk, Novosibirsk, and Tomsk Regions); Nedra: Moscow, Russian, 1970; Volume 16, 368p. (In Russian)
52. *Geological Map of the Yamal-Nenets Autonomous District and Krasnoyarsk Region*; 1:2500000, as of 01.09.2020; VSEGEI: St. Petersburg, Russia, 2020. (In Russian)
53. Savichev, O.G.; Guseva, N.V. Methodology of management of river basins geochemical balance in Western Siberia. *Bull. Tomsk. Polytech. Univ. Geo. Assets. Eng.* **2020**, *331*, 28–45. [CrossRef]

water

MDPI

Article

Testing Landscape, Climate and Lithology Impact on Carbon, Major and Trace Elements of the Lena River and Its Tributaries during a Spring Flood Period

Sergey N. Vorobyev [1], Yuri Kolesnichenko [1], Mikhail A. Korets [2] and Oleg S. Pokrovsky [3,4,*]

1 BIO-GEO-CLIM Laboratory, Tomsk State University, 35 Lenina, 634050 Tomsk, Russia; soil@green.tsu.ru (S.N.V.); vancansywork@mail.ru (Y.K.)
2 V.N. Sukachev Institute of Forest of the Siberian Branch of Russian Academy of Sciences—Separated Department of the KSC SB RAS, 660036 Krasnoyarsk, Russia; mik@ksc.krasn.ru
3 Geosciences and Environment Toulouse, UMR 5563 CNRS, University of Toulouse, 14 Avenue Edouard Belin, 31400 Toulouse, France
4 N. Laverov Federal Center for Integrated Arctic Research, Russian Academy of Sciences, 23 Nab. Northern Dvina, 163002 Arkhangelsk, Russia
* Correspondence: oleg.pokrovsky@get.omp.eu

Citation: Vorobyev, S.N..; Kolesnichenko, Y.; Korets, M.A..; Pokrovsky, O.S.. Testing Landscape, Climate and Lithology Impact on Carbon, Major and Trace Elements of the Lena River and Its Tributaries during a Spring Flood Period. *Water* **2021**, *13*, 2093. https://doi.org/10.3390/w13152093

Academic Editor: Bommanna Krishnappan

Received: 14 June 2021
Accepted: 28 July 2021
Published: 30 July 2021

Abstract: Transport of carbon, major and trace elements by rivers in permafrost-affected regions is one of the key factors in circumpolar aquatic ecosystem response to climate warming and permafrost thaw. A snap-shot study of major and trace element concentration in the Lena River basin during the peak of spring flood revealed a specific group of solutes according to their spatial pattern across the river main stem and tributaries and allowed the establishment of a link to certain landscape parameters. We demonstrate a systematic decrease of labile major and trace anion, alkali and alkaline-earth metal concentration downstream of the main stem of the Lena River, linked to change in dominant rocks from carbonate to silicate, and a northward decreasing influence of the groundwater. In contrast, dissolved organic carbon (DOC) and a number of low-soluble elements exhibited an increase in concentration from the SW to the NE part of the river. We tentatively link this to an increase in soil organic carbon stock and silicate rocks in the Lena River watershed in this direction. Among all the landscape parameters, the proportion of sporadic permafrost on the watershed strongly influenced concentrations of soluble highly mobile elements (Cl, B, DIC, Li, Na, K, Mg, Ca, Sr, Mo, As and U). Another important factor of element concentration control in the Lena River tributaries was the coverage of the watershed by light (for B, Cl, Na, K, U) and deciduous (for Fe, Ni, Zn, Ge, Rb, Zr, La, Th) needle-leaf forest (pine and larch). Our results also suggest a DOC-enhanced transport of low-soluble trace elements in the NW part of the basin. This part of the basin is dominated by silicate rocks and continuous permafrost, as compared to the carbonate rock-dominated and groundwater-affected SW part of the Lena River basin. Overall, the impact of rock lithology and permafrost on major and trace solutes of the Lena River basin during the peak of spring flood was mostly detected at the scale of the main stem. Such an impact for tributaries was much less pronounced, because of the dominance of surface flow and lower hydrological connectivity with deep groundwater in the latter. Future changes in the river water chemistry linked to climate warming and permafrost thaw at the scale of the whole river basin are likely to stem from changes in the spatial pattern of dominant vegetation as well as the permafrost regime. We argue that comparable studies of large, permafrost-affected rivers during contrasting seasons, including winter baseflow, should allow efficient prediction of future changes in riverine 'inorganic' hydrochemistry induced by permafrost thaw.

Keywords: river; hydrochemistry; permafrost; forest; landscape; lithology; carbonate rocks; trace element; major element

1. Introduction

The climate warming, permafrost thaw, change of hydrological connectivity and northward shift of tree line currently observed in northern Eurasia and throughout the Arctic and subarctic regions [1,2] are expected to result in massive carbon (C), nutrients, major and trace elements mobilization from permafrost soils and groundwaters to the rivers, and further to the Arctic Ocean [3,4]. Large Arctic rivers are the main players in element delivery from the land to the ocean, which becomes especially important during climate warming and permafrost thaw because such rivers integrate large areas of essentially pristine permafrost and forested regions [5]. As a result, assessment of lateral riverine export fluxes of organic and inorganic solutes in permafrost-affected regions is needed for defining adequate models of ecosystem functioning under various climate change scenarios [6].

The Lena River of Central and Eastern Siberia is located almost entirely within a permafrost zone; it exhibits the highest seasonal variation in water flow, and demonstrates an annual discharge increase over recent decades [7–15]. Among all the solutes transported by the Lena River water, organic carbon has received most attention [16–22] although information on general hydrochemistry is also available [23–29]. Most recently, novel isotopic approaches were applied for nutrients [30] and trace metals such as Li [31] and Fe [32]. However, the majority of these studies were performed during summer-autumn baseflow or the end of spring flood (July–September) whereas the peak of spring flood, when the main part of annual riverine transport of solutes in high latitude occurs [33–36], has not been sufficiently studied. The exception is regular (monthly to weekly) monitoring of the Lena River terminal gauging station at Kysyr, performed by the Russian Hydrological Survey and within the PARTNERS/ARCTIC GRO projects [34,37]. Furthermore, in contrast to this extensive research on carbon and major element transport by the Lena River main stem, its tributaries remain very poorly known. Thus, no detailed hydro-chemical studies at the peak of spring flood have been performed and the understanding of various contrasting tributaries of the Lena River remains quite poor. Taken together, our knowledge of environmental factors controlling the major and trace element concentration in the Lena River basin during the most crucial hydrological period of the year remains limited and thus deserves special attention. In particular, distinguishing the role of groundwater feeding and surface influx (vegetation, organic topsoil and mineral soil leaching) to the river remains one of the main challenges in understanding the mechanisms of solute flux formation in high latitude rivers. Here we suggest a landscape approach to unravel the role of different environmental factors for the Lena River basin, taking advantage of recent progress in GIS-based mapping of Siberian territory [38,39]. We anticipate that this information should allow, for the first time, testing the impact of main physio-geographical factors (vegetation, soil, lithology, climate and permafrost coverage) on the spatial pattern of major and minor solutes at the scale of a large Siberian river.

Based on the bulk of information acquired by previous chemical and isotopic work on major and trace elements in boreal and permafrost-affected rivers [40–47], we hypothesize that, during spring flood, the impact of lithology (which is reflected via deep subsoil waters and groundwater feeding) will be quite low and the river water chemical composition will be dominated by surficial flow originating from water leaching of organic topsoil and vegetation litter. In order to test this hypothesis, we sampled the main stem of the Lena River and its tributaries at the peak of the spring flood and we analyzed dissolved (<0.45 μm) major and trace components of the river water. We aimed at identifying environmental factors controlling element transport in the Lena River basin in order to use this information to foresee future changes in elemental export from the land to the Arctic Ocean.

2. Study Site, Materials and Methods

2.1. The Lena River and Its Tributaries

The sampled Lena River main stem and its 20 tributaries are located along a 2400 km latitudinal transect from SW to NE and includes watersheds of distinct sizes, geomorphology, permafrost extent, lithology, climate and vegetation (Figure 1 and Figure S1; Table S1). The total watershed area of the rivers sampled in this work is about 1.66 million km^2, representing 66% of the entire Lena River basin. Permafrost is mostly continuous except for some discontinuous and sporadic patches in the southern and south-western parts of the Lena basin [48]. The mean annual air temperatures (MAAT) along the transect range from −5 °C in the southern part of the Lena basin to −9 °C in the central part of the basin. For the studied part of the main stem, the MAAT is −8.76 ± 2.95 °C, whereas the range of MAAT for the tributaries is from −4.7 to −15.9 °C. The mean annual precipitation ranges from 350–500 mm y^{-1} in the southern and south-western part of the basin to 200–250 mm y^{-1} in the central and northern parts [48]. The mean precipitation of the studied Lena River basin and its 20 tributaries is 418 ± 77 mm y^{-1}. The lithology of the Siberian Platform which is drained by the Lena River is highly diverse and includes Archean and Proterozoic magmatic and metamorphic rocks, late Proterozoic, Cambrian and Ordovician dolostones and limestones, volcanic rocks of Permo-Triassic age and essentially terrigenous silicate sedimentary rocks of the Phanerozoic. Note that salt deposits are abundant in the SW part of the basin, where the carbonate rocks dominate the lithology of the watersheds. Further description of the Lena River basin landscapes, vegetation and lithology can be found in numerous works [18,20,23–25,29,49].

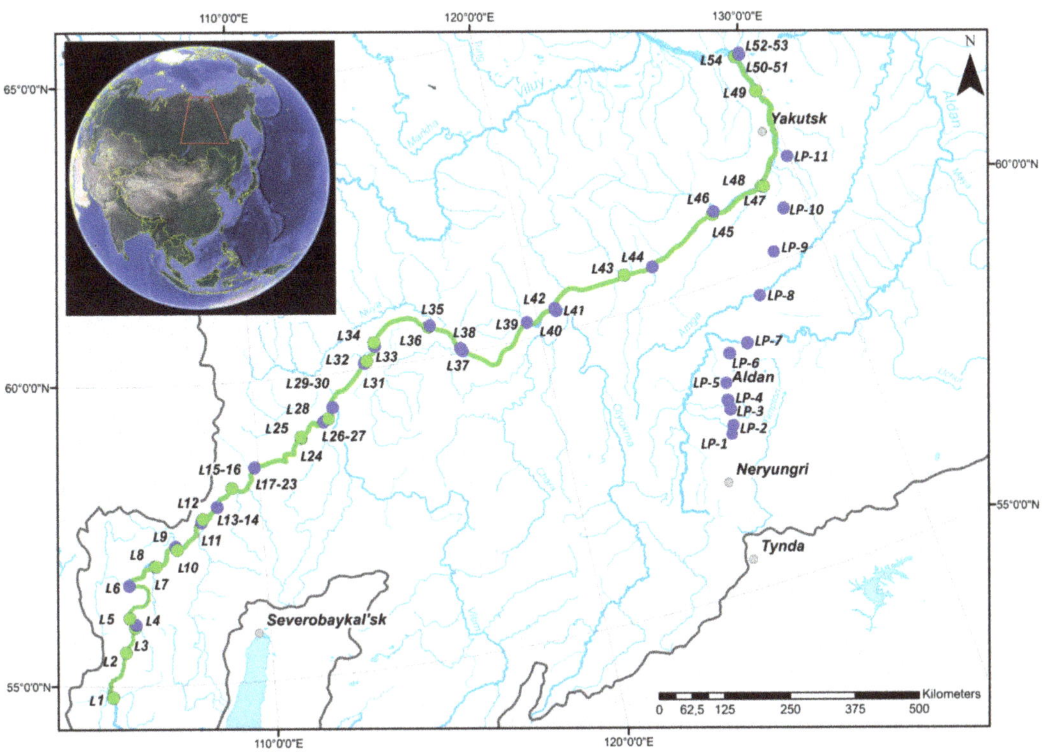

Figure 1. Map of the studied part of the Lena River basin with sampling points of the main stem (green circles) and tributaries (blue circles). The positions of tributaries are shown in Figure S1. See Table S1 for tributaries and river watersheds (LP) parameters.

The peak of annual discharge depends on the latitude and occurs in May in the south (Ust-Kut) and in June in the middle and low reaches of the Lena River (Yakutsk, Kuysyr). The peak of discharge at the Kuysur station is 140,000–160,000 m^3/s, the winter baseflow discharge is 2000–3000 m^3/s, and the summer baseflow discharge is 20,000–40,000 m^3/s [7–15]. From 29 May to 17 June 2016, we moved downstream the Lena River with an average speed of 30 km h^{-1} [50]. As such, we followed the progression of spring and moved from the southwest (Zhigalovo) to the northeast (Yakutsk), thus collecting river water at approximately the same stage of maximal discharge. Note that route sampling is a common way to assess river water chemistry in extreme environments [23,51], and generally a single sampling during high flow season provides the best agreement with time-series estimates [52]. Regular stops each 80–100 km at the middle of the river allowed sampling of major hydro-chemical parameters of the main stem. We also moved 500–1500 m upstream of selected tributaries where we sampled the tributary for hydrochemistry.

In addition to 20 tributaries, evenly distributed over the boat route on the main stem, between 29–31 May 2016, we sampled 10 rivers belonging to the Lena River watershed (mostly the middle part and upper reaches of Aldan). The choice of these 10 rivers of the Lena watershed was restricted by the possibility of ground access. For this, we used road transportation and we moved at least 500 m upstream of each river where it crossed the bridge to grab a water sample 2 m offshore (Figure 1).

2.2. Hydrochemical Measurements

The water temperature, pH, dissolved oxygen, and electric conductivity in the main stem and tributaries were measured directly in the field using Hanna and WTW portable instruments. The water was sampled in pre-cleaned polypropylene bottle from 20–30 cm depth in the middle of the river and immediately filtered through disposable single-use sterile Sartorius filter units (0.45 μm pore size). The first 50 mL of the filtrate was discarded. Filtered river waters were processed using the analytical approaches employed by the GET Laboratory (Toulouse) to analyze DOM-rich waters from boreal and permafrost-bearing settings [53,54]. Filtered solutions for cations and trace element analyses were acidified (pH = 2) with ultrapure double-distilled HNO_3 and stored in HDPE bottles previously washed with 1 M HCl and rinsed with MilliQ deionized water. Filtered water samples for anions were not acidified and were stored in HDPE bottles previously washed according to the above-described procedure for cations. The major anion concentrations (Cl^- and SO_4^{2-}) were analyzed by ion chromatography (Dionex 2000i), with an uncertainty of 2%. The DOC and Dissolved Inorganic Carbon (DIC) were determined by a Shimadzu TOC-VSCN Analyzer with an uncertainty of 3% and a detection limit of 0.1 mg/L. The major and trace elements were measured by quadrupole ICP-MS (7500ce, Agilent Technologies, Santa Clara, California, USA). Indium and rhenium were used as internal standards. The international geo-standard SLRS-5 (Riverine Water Reference Material for Trace Metals, certified by the National Research Council of Canada) was used to check the validity and reproducibility of analyses. Good agreement existed between our replicated measurements of SLRS-5 and the certified values (relative difference < 15%).

2.3. Landscape Parameters of the Lena River Basin and Data Treatment

The physio-geographical characteristics of the 20 Lena tributaries and the other 10 rivers of the Lena River basin sampled in this study and the two points of the Lena main stem (upstream and downstream r. Aldan Table S1 of the Supplementary Materials) were determined by digitalizing available soil, vegetation, lithological, and geocryological maps. The landscape parameters were typified using the TerraNorte Database of Land Cover of Russia ([38], http://smiswww.iki.rssi.ru/default.aspx?page=356, assessed on 29 July 2021) with original resolution of 230 × 230 m pixel size. This included various types of forest (evergreen, deciduous, needle-leaf/broadleaf), grassland, tundra, wetlands, water bodies and urban area. The climate and permafrost parameters of watershed were obtained from CRU grids data (1950–2016) and NCSCD data, respectively, whereas the biomass and soil organic carbon content was obtained from BIOMASAR2 and NCSCD databases. The permafrost extent type layer was taken from NCSCDv2 (The Northern Circumpolar Soil Carbon Database, ref. link: http://su.diva-portal.org/smash/record.jsf?pid=diva2%3A637770&dswid=1526,

assessed on 29 July 2021) in vector GIS format with a scale of 1:1,000,000 (which is about 1 × 1 km pixel resolution in rasterized format). The lithology layer was taken from the GIS version of the Geological map of the Russian Federation (scale 1:5,000,000 in vector format which is about 5 × 5 km pixel resolution in rasterized format). To test the effect of carbonate rocks on dissolved C parameters, we distinguished felsic plutonic, terrigenous silicate rocks (sedimentary siliciclastic of Archean, early Proterozoic and Cainozoic age) and dolostones and limestones of upper Proterozoic, Cambrian and Ordovician age.

The Pearson rank order correlation coefficient (Rs) ($p < 0.05$) was used to determine the relationship between each major and trace element concentration and climatic, lithological and landscape parameters of the Lena River tributaries. Further statistical treatment of element concentration drivers in river waters included a Principal Component Analysis using a variance estimation method, which allowed test of the effect of various environmental factors of the watershed on behavior of riverine solutes in both the Lena River main stem and its tributaries. All graphics and figures were created using MS Excel 2010, MS Visio Professional 2016 and GS Grapher 11 package. Statistical treatment was performed using STATISTICA-7 (http://www.statsoft.com, assessed on 29 July 2021).

3. Results

3.1. Element Concentration in the Main Stem of the Lena River

All measured hydro-chemical parameters of the Lena River and tributaries are available from the Mendeley database repository [55]. The concentrations of major and trace elements in the main stem averaged over 2600 km distance are listed in Table 1. For a number of elements, the greatest changes occurred between first 0–800 (±200) km (upstream of Kirenga/Chaika) and the remaining 800–2600 km downstream to the Aldan River. These two parts of the river transect reflect sizable change in the lithology of rocks, landscape and climate and distinguish upper reaches and the middle course of the Lena River.

Table 1. Major and trace element concentration (average ± s.d.) in the Lena River main stem. The distances are along the river, starting from Zhigalovo (the Lena River headwaters).

	Upper Reaches, 0–804 km	Middle Course, 804–2600 km
T_{air}, °C	21.9 ± 1.3	19.6 ± 0.65
T_{water}, °C	11.9 ± 0.4	9.6 ± 0.3
O_2, mg L^{-1}	6.1 ± 0.1	6.3 ± 0.05
pH	8.0 ± 0.1	7.64 ± 0.03
S.C., µS cm^{-1}	226 ± 17	99.7 ± 5.9
Depth, m	3.8 ± 0.554	7.5 ± 0.63
Cl mg L^{-1}	13.7 ± 0.89	9.2 ± 1.1
SO_4 mg L^{-1}	17.9 ± 2.0	6.42 ± 0.36
DOC, mg L^{-1}	11.0 ± 1.0	9.36 ± 0.34
DIC, mg L^{-1}	15.3 ± 1.8	5.3 ± 0.19
Li, µg L^{-1}	2.87 ± 0.32	1.08 ± 0.07
B, µg L^{-1}	8.9 ± 0.93	3.2 ± 0.39
Na, µg L^{-1}	10,400 ± 595	6770 ± 738
Mg, µg L^{-1}	7203 ± 858	2331 ± 109
Al, µg L^{-1}	69.3 ± 8.6	117 ± 5.8
Si, µg L^{-1}	1970 ± 40	1804 ± 18.5
P, µg L^{-1}	11.6 ± 2.2	6.38 ± 0.43
K, µg L^{-1}	651 ± 46	483 ± 8.6

Table 1. *Cont.*

	Upper Reaches, 0–804 km	Middle Course, 804–2600 km
Ca, μg L^{-1}	24,800 ± 3090	8440 ± 264
Ti, μg L^{-1}	7.95 ± 0.97	10.2 ± 0.67
V, μg L^{-1}	1.17 ± 0.10	0.689 ± 0.06
Cr, μg L^{-1}	0.33 ± 0.075	0.45 ± 0.05
Mn, μg L^{-1}	8.54 ± 1.2	4.79 ± 0.23
Fe, μg L^{-1}	80.0 ± 7.6	85.7 ± 3.8
Co, μg L^{-1}	0.059 ± 0.004	0.058 ± 0.003
Ni, μg L^{-1}	0.41 ± 0.08	0.54 ± 0.05
Cu, μg L^{-1}	1.2 ± 0.13	1.1 ± 0.05
Zn, μg L^{-1}	3.38 ± 0.36	4.6 ± 0.5
Ga, μg L^{-1}	0.012 ± 0.0016	0.0164 ± 0.0009
Ge, μg L^{-1}	0.0073 ± 0.0005	0.0076 ± 0.0003
As, μg L^{-1}	0.338 ± 0.03	0.171 ± 0.005
Rb, μg L^{-1}	0.35 ± 0.058	0.66 ± 0.019
Sr, μg L^{-1}	254 ± 39	66.8 ± 3.6
Y, μg L^{-1}	0.372 ± 0.038	0.422 ± 0.005
Zr, μg L^{-1}	0.189 ± 0.02	0.167 ± 0.007
Nb, μg L^{-1}	0.022 ± 0.003	0.038 ± 0.0026
Mo, μg L^{-1}	0.416 ± 0.02	0.251 ± 0.018
Cd, μg L^{-1}	0.0093 ± 0.0008	0.0099 ± 0.001
Sb, μg L^{-1}	0.0242 ± 0.0014	0.0155 ± 0.0007
Cs, μg L^{-1}	0.0019 ± 0.0003	0.003 ± 0.0003
Ba, μg L^{-1}	33.68 ± 3.906	10.29 ± 0.488
La, μg L^{-1}	0.402 ± 0.085	1.02 ± 0.07
Ce, μg L^{-1}	0.39 ± 0.07	1.02 ± 0.04
Pr, μg L^{-1}	0.09 ± 0.02	0.17 ± 0.003
Nd, μg L^{-1}	0.36 ± 0.06	0.629 ± 0.01
Sm, μg L^{-1}	0.0867 ± 0.01	0.119 ± 0.002
Eu, μg L^{-1}	0.018 ± 0.002	0.019 ± 0.0004
Gd, μg L^{-1}	0.0763 ± 0.0099	0.105 ± 0.0012
Tb, μg L^{-1}	0.0115 ± 0.0014	0.014 ± 0.00016
Dy, μg L^{-1}	0.0662 ± 0.0075	0.0785 ± 0.001
Er, μg L^{-1}	0.0361 ± 0.0041	0.0414 ± 0.00049
Tm, μg L^{-1}	0.00474 ± 0.00056	0.00575 ± 0.0001
Yb, μg L^{-1}	0.0312 ± 0.0035	0.0374 ± 0.00063
Lu, μg L^{-1}	0.00446 ± 0.00049	0.00537 ± 0.00010
Hf, μg L^{-1}	0.0326 ± 0.0047	0.0259 ± 0.0009
Pb, μg L^{-1}	0.065 ± 0.011	0.0807 ± 0.0066
Th, μg L^{-1}	0.0243 ± 0.004	0.059 ± 0.0043
U, μg L^{-1}	0.310 ± 0.02	0.256 ± 0.004

According to element behavior along the Lena River main stem, from Zhigalovo to Aldan (2600 km downstream), three groups of solutes could be distinguished: (i) Cl, SO_4, DIC, Li, B, Na, Mg, K, Ca, As, Sr, Mo, Sb, Ba and U, decreasing the concentration from SW to NE (Figure 2); (ii) Al, Ti, Cr, Fe, Ga, Rb, Y, Zr, Nb, Cs, REEs, Ce, Hf, Th, increasing their concentration from SW to NE (Figure 3); and finally (iii) DOC, Si, P, V, Mn, Co, Ni, Cu, Zn, Ge, Cd, Hf and Pb, which did not exhibit any statistically significant dependence ($r < 0.3$, $p < 0.05$) on the river distance (Figure 4). The Mann-Whitney test demonstrated significant ($p < 0.001$) difference in most element concentration between the SW part of the Lena main stem (upper reaches, 0–800 km) and its middle course (800 km—Aldan River), as listed in Table S2 of the Supplementary Materials. The exceptions are DOC, P, Cr, Fe, Co, Ni, Cu, Zn, Ge, Y, Zr, Cd, heavy REE, Hf and Pb which did not show statistically significant differences.

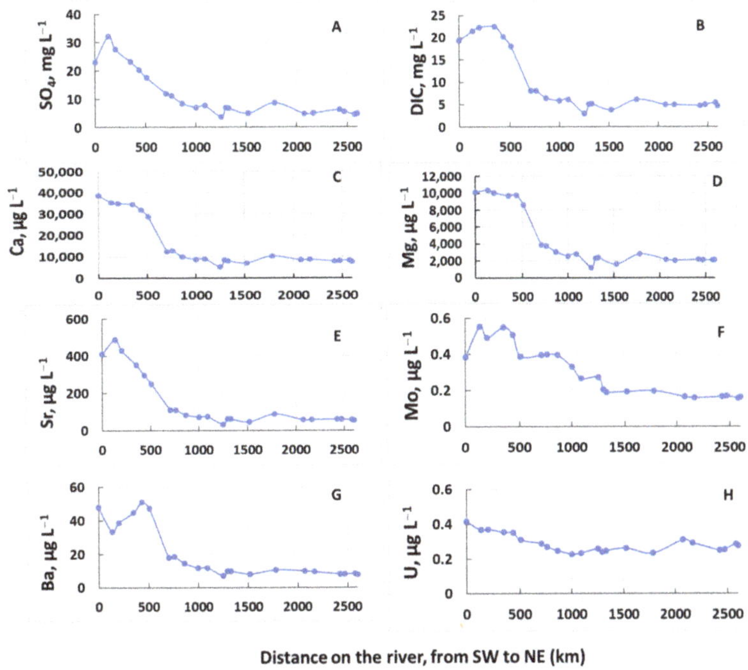

Figure 2. Examples of elements decreasing their concentration in the main stem of the Lena River from SW to NE: SO_4 (**A**), DIC (**B**), Mg (**C**), Ca (**D**), Sr (**E**), Mo (**F**) Ba (**G**) and U (**H**) The distance is km of river from Zhigalovo (0 km, upper reaches) to the Aldan River (~2600 km).

A linear (Pearson) pairwise correlation between solutes in the main stem demonstrated three potential interlinked carriers and sources of dissolved elements in the river water (Table S3 of the Supplementary File). Firstly, this is DOC, which positively correlated with B, Mg, P, K, Ca, V, Ni, Cu, As, Zr, Sb, Ba, Hf and U. The second potential carrier and source tracer is Fe which correlated with the largest number of trace elements (Si, P, Ti, V, Mn, Co, Ga, Y, Zr, Nb, Cs, REEs, Hf and Th). Only a few elements were significantly ($p < 0.05$) linked to both DOC and Fe (P, V, Zr, Hf). Aluminum most strongly correlated with Ti, Ga, Rb, Y, Nb, Cs, all REE, and Th.

3.2. *Dissolved Elementary Composition of Tributaries of the Lena Basin: Impact of Permafrost, Climate, Vegetation, Soil and Landscape on Major and Trace Element Concentration*

The Lena River tributaries exhibited strong variability of dissolved elementary composition, independent of river size and location within the main river basin. A Pearson correlation of solutes in the tributaries demonstrated three groups of elements depending on their link to potential carriers (or common source) in the river water (Table S3 of the Supplementary Materials). Similar to the main stem, DOC positively correlated ($R_{Pearson} > 0.5$) with Si, P, K, V, Cr, Ni, Cu, As, Y, Zr, Sb, HREE, Hf, and Th. Dissolved Fe strongly correlated with Si, P, Ti, Cr, Co, Ni, Cu, Ge, As, Rb, Y, Zr, Cd, Sb, La, heavy REE, Hf and Th, but preferentially (higher than DOC) correlated with P, Ti, Cr, Ni, Ge, Th. Finally, Al most strongly correlated with Ti, Cr, Ga, Nb, Cs, REE, (from Ce to Er) and Th. Similarly in the main stem, a number of elements were significantly ($p < 0.05$) linked to both DOC and Fe (Si, P, Ti, Cr, Cu, As, Y, Zr, Sb, HREE, and Th).

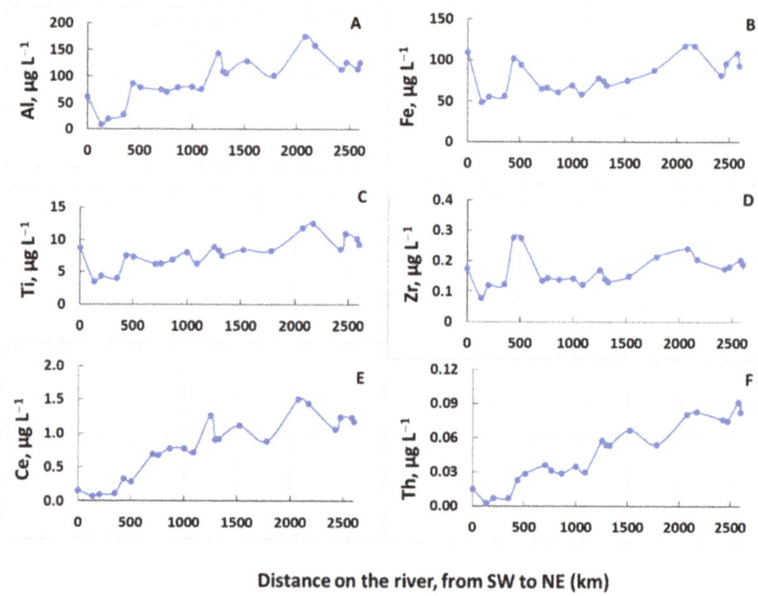

Figure 3. Examples of elements increasing their concentration in the main stem of the Lena River from SW to NE: Al (**A**), Fe (**B**), Ti (**C**), Zr (**D**), Ce (**E**) and Th (**F**). The distance is km of river from Zhigalovo (0 km, upper reaches) to the Aldan River (2600 km).

To test the impact of physio-geographical parameters on the hydrochemistry of the Lena River tributaries, a correlation matrix of element concentration with climate, permafrost, vegetation coverage and lithology was constructed (Table S4). These correlations revealed several environmental factors that are likely to control the elemental composition of various rivers of the Lena Basin. The first important factor was the vegetation, namely the coverage of watershed by light (Li, B, Cl, SO_4, Na, V, Sr, U) and deciduous (Si, Ni, Ge, Zr, Hf, Th) needle-leaf forest and broadleaf forest (DOC, Cl, SO_4, Li, Na, Si, Ca, Ti, V, Co, Ge, Sr, Cd, some heavy REE (Er, Yb), Hf and U) as illustrated in Figure 5. The humid grassland coverage of the watershed increased the concentrations of low-mobile elements such as Al, Ga, light REE, and also relatively labile Cs, Ba, Pb, U. Among other environmental variables, the proportion of sporadic permafrost strongly controlled the concentration of soluble highly mobile elements such as Cl, B, Na, Sr, Mo, and U, whereas the presence of carbonate rocks of the watershed provided elevated concentration of DIC, Mo, Ba and U (Figure 6). It is important to note that other potentially important landscape factors of the river watershed (riparian vegetation, tundra, water bodies, peatlands, recent burns, proportion of peatland and bogs, tundra coverage, total aboveground phyto-mass, OC stock in the upper 0–100 cm of soil, mean annual temperature and precipitation) did not significantly correlate with major and trace element concentration in the Lena River tributaries (Table S4).

The relationships between river water chemistry and landscape parameters of the watersheds were further examined using multi-parametric statistics. The PCA demonstrated the presence of two factors capable of explaining only 32% of total variability in element concentration and landscape parameters of the watersheds (Figure S2). The first factor was positively linked to broadleaf forest and humid grassland coverage of the watershed and included Al, Fe, Ti, Cr, Ni, Ga, Y, Zr, Nb, Zr, REEs, Hf and Th. As for the second factor, the presence of light needle-leaf, mixed forest, percentage of sporadic permafrost coverage and carbonate rock on the watershed controlled the distribution of mobile elements such as Li, B, Si, Na, Mg, Ca, K, V, Sr, Ba, Mo, As, Sb and U.

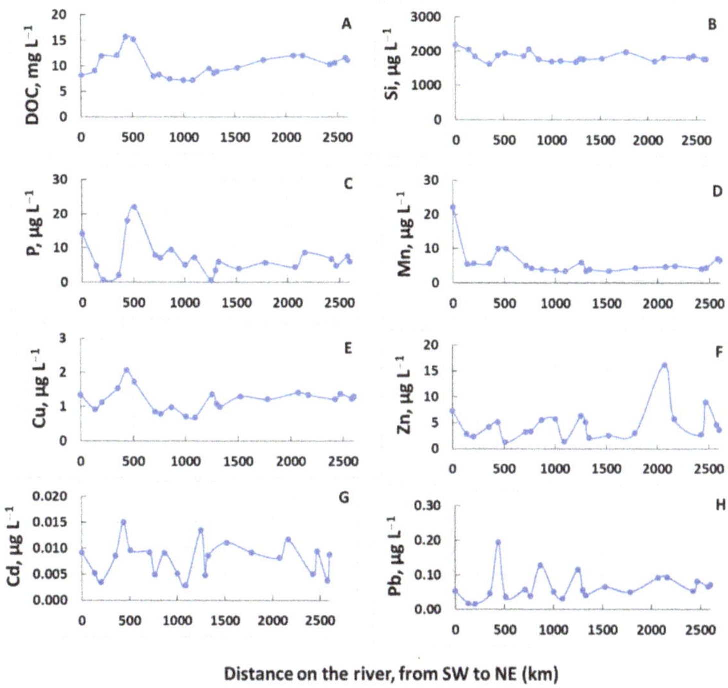

Figure 4. Examples of elements not exhibiting any concentration trend in the main stem of the Lena River from SW to NE: DOC (**A**), Si (**B**), P (**C**), Mn (**D**), Cu (**E**), Zn (**F**), Cd (**G**) and Pb (**H**). The distance is measured in km of river from Zhigalovo (0 km, upper reaches) to the Aldan River (2600 km).

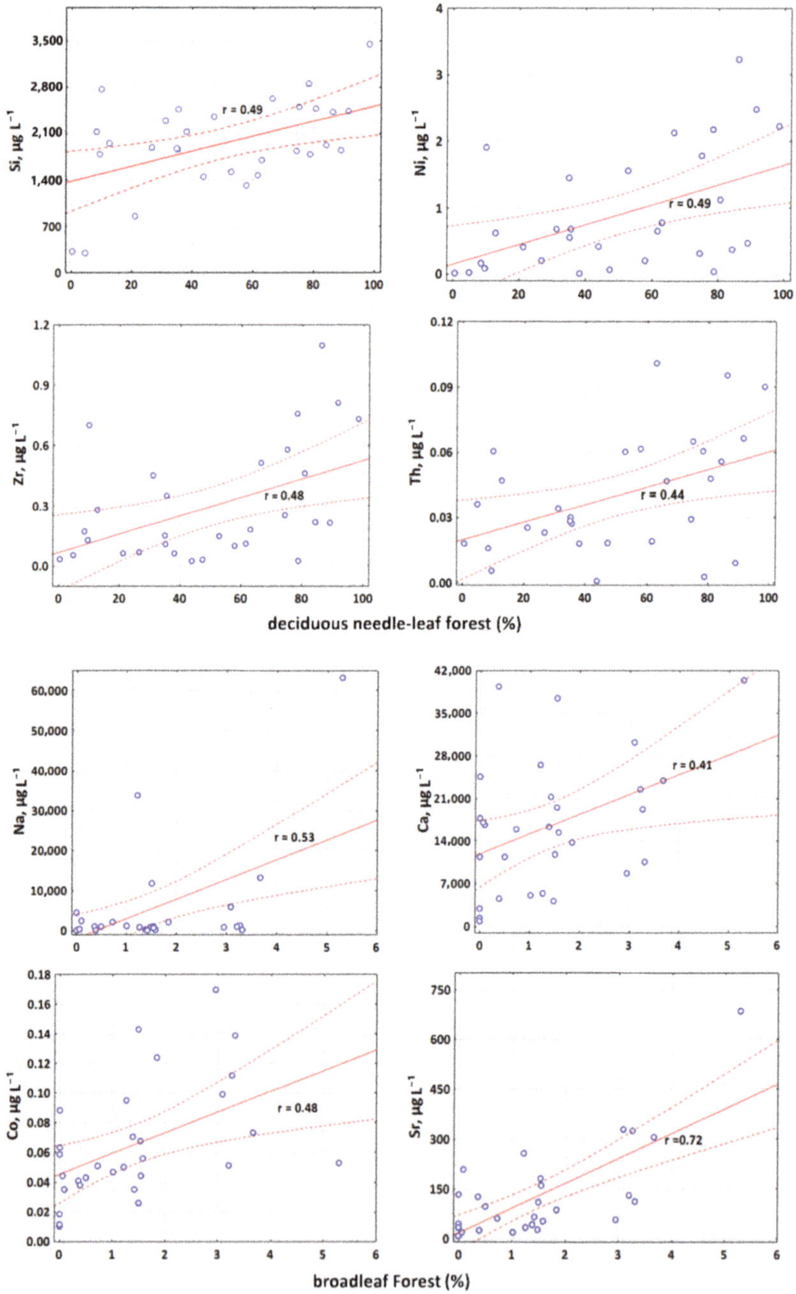

Figure 5. Significant ($p < 0.05$) positive control of landscape parameters—percentage of deciduous needle-leaf forest (Si, Ni, Zr, Th) and broadleaf forest (Na, Ca, Co, Sr) on element concentration in the Lena River tributaries.

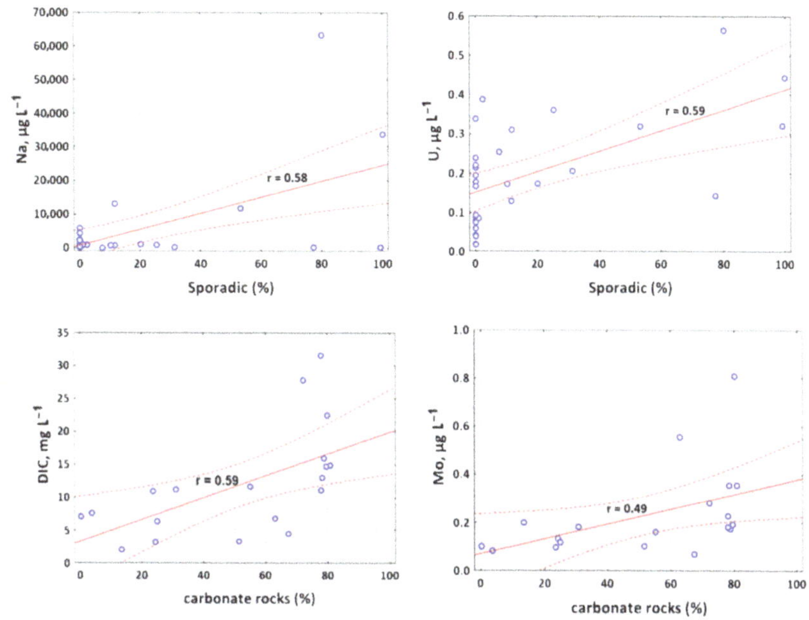

Figure 6. Significant ($p < 0.05$) positive control of permafrost distribution—percentage of sporadic permafrost coverage (Na and U) and lithology of rocks—percentage of carbonate rocks (DIC, Mo) on element concentration in the Lena River tributaries.

4. Discussion

4.1. Comparison with Other Data on Element Concentration in the Lena River and Its Tributaries

As for the Lena River main stem, there are several occasional measurements of the major river solutes in the middle course obtained during summer baseflow. During spring flood in June, we observed generally lower concentrations of Si (less than a factor of 2) compared to the July–August period [26], and also lower (less than × 2) concentrations of Ca, Mg, Sr, SO_4 but comparable concentrations of Na and Cl to those reported in the Yakutks-Kyusur region [28]. The Kyusur gauging state average Ba concentration 18 µg L^{-1}, [56]) is comparable with our data for the Lena main stem in the Yakutsk region. However, in general, the regular monitoring of the Lena River hydrochemistry at its terminal gauging station of Kyusur by Russian Hydrological Survey [57,58], and more recently, via PARTNERS and ARCTIC GRO projects [34,37], cannot be directly compared with our most northern sampling point (Yakutsk and the vicinity of the Aldan River), because there are substantial changes in the Lena River flow and landscape context over its last 800 km, from Yakutsk to the Lena delta. The most recent and complete data set on the upper and middle course of the Lena River, southwest of Aldan, is provided by the work of the group of Porcelli and Andersson [30–32]. There was full consistency in Li, Na, K, Si, Al and Fe dissolved concentrations measured in this study in June and those reported by these authors in July-August (range of 0.5–2, 10 to 1000, 500 to 900, 1700 to 2300, 70 to 140, and 50 to 130 µg L^{-1}, respectively). Overall, this comparison demonstrates rather stable concentrations over a large spatial and temporal range, with maximal variability of a factor of 2 to 3. Such a similarity of hydro-chemical composition of the main stream during quite contrasting seasons (spring flood and summer baseflow) is noteworthy, given the five-fold variation in the discharge. A likely cause of elevated Al, Fe and insoluble trace element concentration during high flow period is that these elements transport in the form of organic and organo-mineral colloids as is known for other permafrost regions [41]. The availability of DOM which stabilizes these Al, Fe-rich colloids and relevant trace elements is the highest during the high flow period, because mobilization of soil OM to the river is most efficient during freshet [36]. As such, despite sizable dilution during the high flow period, the concentration of these elements remains comparable with that during the baseflow. It thus can be concluded that such a transport enhancement of insoluble elements compensates for their source limitation due to dilution. Given that the first principal component acting on insoluble elements was linked to broadleaf forest and humid grassland vegetation, an empirical conclusion is that this type of vegetation is most efficient for mobilization of Fe and Al from the topsoil and forest litter to the river at the scale of the Lena Basin.

For the Lena river tributaries, the most comprehensive data set on major ions, Si and Sr was acquired in July-August of 1991–1996 by Huh and Edmond's group [23–25]. The following rivers could be quantitatively compared with those sampled in this work: Buotama (No 18), Tuolba (No 16), Siniaya (No 17), Olekma (No 14), Chuya (No 7), Orlinga (No 1), Tayura (No 3), Nuya (No 11), Vitim (No 8), Bolshoi Patom (No 12), Bolshoi Nimnyr (LP 3), and Amga (LP 8). For most of these rivers, the concentrations of Ca, Mg, and DIC were a factor of 3 to 15 lower during spring flood compared to summer baseflow. The concentrations of Na, Cl and SO_4 were typically 3 to 10 times lower, especially in the upper reaches (SW part) of the Lena basin, where the underground waters can bear the signature of Cambrian and Ordovician salt deposits. These waters are mostly pronounced in summer baseflow and only partially during spring flood, for example, Ichera, exhibiting 10 times higher NaCl concentration compared to other rivers (see database [55]). The impact of salty springs on the river water chemistry in SW part of the Lena basin (the Northern Baikal region) is fairly well known [58,59]. Noteworthy rather similar (<30% difference) concentration of K in all rivers sampled during spring flood (this study) and summer baseflow [23–25], despite significant dilution. A likely cause could be K leaching from terrestrial vegetation and silicate river suspended matter; both sources are most pronounced during spring flood, as is known from other permafrost Siberian rivers [41,60,61]. Enhanced coastal abrasion and extensive surface flux of melted snow are both responsible for efficient mobilization of K to the river during this period.

Another interesting observation is much higher (a factor of 3 to 20) Si concentrations during spring flood compared to summer baseflow. This is observed for a number of small rivers in the upper reaches of the Lena basin (Nuya, Tayura, Bolshaya Tira, Siniya and Buotama) as follows from the comparison of the data collected by Huh and Edmond group [23–25] and the results of the present study. It may reflect an uptake of this essential nutrient by diatomous periphyton growing in clearwater rivers which drain crystalline rocks of the Siberian Platform. Such an uptake during summer baseflow is known for many Arctic settings [40,62,63]. Another possible cause could be the fact that the mobilization of Si from subsurface soil is lower in summer compared to spring when the water mainly passes through the topsoil. Note that such a difference in Si concentration between two seasons is not observed for the main stem, presumably due to lower transparency of the water column and lower availability of solid surfaces, required for siliceous plankton and periphyton development in the summer. A nutrient limitation of the big river relatively to headwater streams may also contribute to lower uptake of Si in the main steam during the baseflow [16,28,30]. Additional sources of dissolved Si in the main stem could be dissolution of river suspended matter, which is highest during the high flow, as is known from other permafrost settings [41].

4.2. Possible Carries of Trace Elements in the Lena River Basin Based on Elementary Correlations and PCA

In boreal and permafrost-affected aquatic environments, the most important colloidal carriers of trace elements are dissolved organic matter (DOM) and Fe/Al hydroxides stabilized by DOM, as follows from ex-situ and in-situ fractionation results [53,60,64–68]. Based on correlations between element concentration in both the main stem and Lena's tributaries (Table S3), three main group of trace solutes were distinguished. First, these correlations revealed possible control of organic matter on transport of V, Ni, Cu, As and heavy REE. Note that strong complexation of DOM with divalent transition metals, notably Cu and Ni and HREE, is fairly well known ([53,68]). The second major carrier of trace metals in boreal waters is high molecular weight Fe oxy-(hydr)oxide colloids stabilized by organic matter. By virtue of their dual organic and mineral nature, these colloids are responsible for transport of the largest number of elements such as P, Cr, Ni, Zn, Ge, Sb, Y, Zr, La, Hf. Finally, Al-rich colloids and sub-colloidal particles are likely to have controlled typically lithogenic trace elements (Ti, Cr, Ga, Nb, Cs, REE, Th) and could be produced during riverine DOM leaching of alumo-silicate river suspended matter (i.e., Ref. [61]). Intensive mobilization of lithogenic elements via desorption from alumo-silicate material of the river suspended matter into soluble (<0.45 µm) form is consistent with strong correlations of these elements with Al, in both main stem and the tributaries (Table S3).

The PCA generally confirmed the presence of the group of low-soluble elements (Al, trivalent and tetravalent trace elements, Nb), which were affected by the first factor, and the group of labile elements—alkalis, alkaline earth metals (Li, Na, Mg, K, Ca, Sr), anions or neutral molecules (B, Si, V, Mo, As, Sb) and uranyl ion—which were controlled by the second factor (Figure S2 of the Supplementary Materials). The first factor was presumably linked to organic and organo-ferric colloids which usually carry low-mobile elements originating from silicate minerals. The second factor reflected migration of elements in truly dissolved (ionic or molecular) forms, not linked to any chemical carrier and originating from water–carbonate rock interaction in deep underground reservoirs.

4.3. Landscape Factors Controlling Spatial Pattern of Elementary Composition of Riverwater in the Lena Basin (Main Stem and Tributaries)

A decrease in concentrations of soluble highly mobile elements such as alkali and alkaline-earth metals, Cl, SO_4, B, Mo, As, Sb and U between the upper reaches of the Lena River (first 0–800 km of the transect) and the middle course of the main stem (second part, 800 km—Aldan) may reflect a decrease of connectivity between the river water and the deep underground waters. The latter are located within carbonate rocks and affected by salt deposits in the SW part of the Lena Basin compared to its NE part. We hypothesize that sporadic and isolated permafrost, frequently occurring in the SW part of the basin, facilitates the exchange between groundwater reservoirs and surface waters. In contrast, in the rest of the Lena Basin, essentially continuous permafrost prevents any impact of underground waters on river water chemistry, especially during the high flood period sampled in this work.

Another group of elements whose concentration systematically evolved over the river transect is the that of the lithogenic low-soluble elements (essentially trivalent and tetravalent hydrolysates). The concentration of these elements increased from the SW to the NE. On the one hand, this may reflect a change in dominant rocks: from carbonates and evaporites in the upper reaches of the Lena River basin to metamorphic and igneous silicates and Phanerozoic (silicate) sedimentary rocks in the middle course of the river. On the other hand, there is a general increase in soil OC stock from the SW to the NE part of the Lena River as follows from GIS assessment of the Lena basin parameters. We hypothesize that a coupled 'source' (abundance of silicate rocks) and 'transport' (organic colloids, stabilizing insoluble TE in solution) enhancement mechanisms are responsible for elevated concentrations of lithogenic insoluble elements in the NE part of the Lena River main stem relative to its SW part.

To sum up, in the Lena River basin, there are three major types of rock capable of affecting river water hydrochemistry: sedimentary carbonates (limestones and dolostones), Precambrian igneous and Phanerozoic sedimentary silicates, and salt deposits (evaporates) of early Phanerozoic age, distributed in the upper reaches of the Lena basin. The carbonate rocks play a visible role in the upper reaches of the Lena River and head waters. Enhanced connectivity between surface waters and deep groundwaters in the SW part of the basin, where the evaporate deposits are present, can provide highly mobile soluble elements (alkalis, alkaline-earth metals, oxyanions) to the river water. Silicate rocks which dominate the middle course of the Lena River act as an important source of lithogenic elements (trivalent, tetravalent hydrolysates), whose export from the soil to the river is facilitated by abundant DOM from coniferous vegetation.

The impact of environmental factors on major and trace element spatial pattern was further tested based on GIS-based landscape parameters of the Lena River tributaries. The proportion of sporadic and isolated permafrost strongly controlled concentrations of soluble highly mobile elements (Cl, B, DIC, Na, K, Mg, Ca, Mo, Ba and U). At the same time, the proportion of carbonate rocks at the watershed (per se) impacted only a very limited number of elements (Mg and Mo). A lack of direct impact of the rock lithology is consistent with low connectivity of deep groundwaters to the rivers during the high flow spring flood period. Another reason could be the rather small size (relative to the Lena River main stem) of the studied rivers and streams, so that the link between the river hyporheic zone and the underground reservoirs, especially under permafrost conditions, is not pronounced.

Another important landscape parameter controlling surficial and shallow subsurface flux to the river, as is known from adjacent permafrost territories covered by larch forest [41,42,69], is ground vegetation. Specifically, an important factor of element concentration control in the Lena River tributaries was the coverage of river watershed by light (Cl, SO_4, Na, Li, B, V, Sr, U) and deciduous (Si, Ni, Ge, Zr, Hf, Th) needle-leaf forest (dominance of *Pinus sylvestris* and *Larix*, respectively). The first group of elements may reflect their intense recycling with pine, which grow on carbonate-rich rocks of the SW part of the basin. In contrast, Si (and presumably Ge) are known to be concentrated in larch [40,69] and can be massively leached from their needles during the spring. A part of the second group of elements (usually low mobile tetravalent hydrolysates) could be affected by the DOC-mediated transport in the NW part of the basin. The NW part is dominated by silicate rocks and continuous permafrost, compared to carbonate rock-dominated and groundwater-affected SW part of the Lena basin. Enhanced mobility of Zr, Hf and Th in forested soils developed on silicate rocks is known in various boreal settings [68,69].

The PCA generally confirmed the linear correlations between the landscape parameters and the solute concentrations, although the explanation capacity of the two potential factors was rather low, at 21% and 11% of overall variability. This could be linked to relatively weak landscape control on elementary composition in the river water during the spring high flow season. Furthermore, the response of the hydrochemistry of small rivers of the Lena basin to key environmental parameters of the watersheds such as permafrost coverage and underground rock lithology was less pronounced than that of the main stem. A likely reason for this is weaker connectivity of small rivers with deep underground and subsurface waters during the studied period of high-flow spring flood. In contrast, given that the main input to small and medium size rivers in permafrost regions occurs via surface and shallow subsurface flow in June, essentially through plant litter and topsoil [42], the vegetation of the watershed exhibited discernable control on elementary composition of river waters.

Summarizing the analysis of the spatial pattern of solutes in the Lena River basin, we conclude that the working hypothesis of this study, regarding the low impact of rock lithology on elementary composition of the river water during spring flood, is verified only in the case of the Lena tributaries. The hydrochemistry of the main stem of the Lena clearly reflected the control of carbonate rocks and the impact of groundwater via isolated/sporadic permafrost in the southwestern part, contrasting with a lack of groundwater connection and dominance of silicate rocks in the northeastern part. However, another important factor increasing mobility of low soluble trivalent and tetravalent trace metals from SW to NE is dissolved organic carbon which, in turn, depends on the soil organic pool and the presence of various forests. At the same time, we do not exclude that the resolution of our GIS mapping resolution and spatial sampling are not sufficiently fine, so they cannot catch fine-scale landscape patches that are known to drive solute generation in small Arctic rivers ([70]).

Within a climate warming and permafrost thaw scenario, the main source of elements during spring flood will remain the surface flow, given that, for the majority of the Lena River basin, the permafrost is continuous and unlikely to become isolated or sporadic. This strongly contrasts with the scenario of permafrost thaw in other Siberian regions such as the Western Siberia Lowland (WSL), where the permafrost is mostly discontinuous to sporadic, and, instead of silicate crystalline and carbonate sedimentary rocks as mineral substrates in Eastern Siberia, the rivers of the WSL drain through thick, partially frozen peat deposits (i.e., Ref. [54,71,72]. During spring flood, surface flow rather than underground input regulates the pattern of riverine solutes in the Lena River basin. As such, the change in dominant forest species and the forestation of tundra [73,74], which typically requires several decades [75], rather than an increase in active layer depth, will determine the overall pattern of river water chemistry during high water flow. At the same time, the response of the Lena River main stem to the thawing of permafrost (leading to enrichment of the river water in soluble, highly mobile elements) might be more significant than that of its tributaries and other small rivers of the basin. The main reason is that the hydrological connectivity between river water and shallow subsurface or deep underground water is sizably higher in a large river compared to small tributaries. In this regard, comparable studies of large, permafrost-affected rivers of Eastern Siberia during most contrasting seasons, including notably winter baseflow, when the connection of the groundwater with surface waters is at its maximum, should allow efficient prediction of future changes in riverine 'inorganic' hydrochemistry induced by changes in the permafrost regime.

5. Conclusions

While seasonal and annual export fluxes (yields) of carbon (C) and inorganic solutes are fairly well known for all large Arctic rivers, spatial variations in elementary concentration along the river length and among its tributaries remain poorly understood. Moreover, the landscape factors controlling riverine element concentration in permafrost-affected regions are still poorly constrained. This is especially true for the largest river of Eastern Siberia, the Lena River, which drains through continuous permafrost zones with highly variable lithology and vegetation. In this work we measured dissolved carbon, major and trace elements over a 2600-km transect of the Lena River main stem (upper and middle reaches) including its 30 tributaries and watershed rivers, at the peak of the spring flood. There were two main group of solutes in the main stem depending on their spatial pattern: (i) elements that decreased their concentrations downstream, from SW to NE (Cl, SO_4, DIC, Li, B, Na, Mg, K, Ca, As, Sr, Mo, Sb, Ba and U), which reflected a decrease in the proportion of carbonate rocks in the watershed and the degree of groundwater feeding, and (ii) elements that increased their concentrations downstream (Al, Ti, Cr, Fe, Ga, Rb, Y, Zr, Nb, Cs, REEs, Hf and Th), which was tentatively linked to an increase in organic C stock in soils, larch forest coverage and enhanced mobilization of lithogenic elements from silicate minerals.

In contrast to the main stem, the chemical composition of the tributaries was only partially controlled by soils, rock lithology and permafrost. In particular, the type of permafrost distribution impacted mostly labile elements, whereas the role of carbonate rocks in the watersheds of the Lena River tributaries has not been explicitly pronounced. Furthermore, the watershed coverage by needle-leaf, broadleaf forest and humid grassland exhibited positive correlation with concentration of labile elements (Cl, SO_4, Li, B, Na, Si, Ca, V, Sr, U), but also some low-soluble traces (Al, Ga, REEE, Nb, Zr, Hf, Th). In accord with previous observations in permafrost-affected forested regions of Siberia, we believe that, during the high flow period of the spring flood, the main control on river water chemistry is exerted by surface flow. The latter often occurs over still frozen ground and thus primarily depends on the type of forest. As such, within the climate warming scenario, future changes in dominant ground vegetation and greening of tundra will mostly impact the river water chemistry, whereas the thawing of continuous permafrost and increase in the active layer depth might have a subordinate influence.

Supplementary Materials: The following are available online at https://www.mdpi.com/article/10.3390/w13152093/s1, Figure S1: Landscape map of the Lena River and tributaries (RLC based); Figure S2: Results of PCA treatment of the Lena River tributaries chemical composition, including landscape parameters. Table S1: Physico−geographical parameters of the studied Lena River tributaries; Table S2: Mann−Whitney test of the difference in element concentration in the upper (0−800 km) and middle (800−2600 km) course of the Lena River main stem; Table S3: Pearson correlation coefficients ($p < 0.05$) of major and trace elements with three main potential carriers—DOC, Fe and Al in the main stem of the Lena River and its tributaries; Table S4: Pearson correlation matrix of landscape parameters and river water chemistry.

Author Contributions: S.N.V. and O.S.P. designed the study and wrote the paper; Y.K. performed sampling of 10 tributaries and mapping; S.N.V. and O.S.P. performed sampling, analysis and their interpretation; M.A.K. provided landscape estimation of the Lena River tributaries. Each co-author has seen and approved the final paper and contributed to writing the manuscript. All authors have read and agreed to the published version of the manuscript.

Funding: This research was funded by RSF, grant number 18-17-00238-P and by RFBR, grants No 19-55-15002, 20-05-00729_a.

Institutional Review Board Statement: Not applicable.

Informed Consent Statement: Not applicable.

Data Availability Statement: The original data of element concentration in the Lena River and tributaries are available from: Pokrovsky, O.S. Lena River and tributaries water chemistry, Mendeley Data, 2021, doi:10.17632/h76pzm6vd3.1.

Acknowledgments: The reported study was carried out using the research equipment of the Unique Research Installation "System of experimental bases located along the latitudinal gradient" TSU.

Conflicts of Interest: The authors declare no conflict of interest.

References

1. Schuur, E.A.G.; McGuire, A.D.; Schädel, C.; Grosse, G.; Harden, J.W.; Hayes, D.J.; Hugelius, G.; Koven, C.D.; Kuhry, P.; Lawrence, D.M.; et al. Climate change and the permafrost carbon feedback. *Nature* **2015**, *520*, 171–179. [CrossRef]
2. Biskaborn, B.K.; Smith, S.L.; Noetzli, J.; Matthes, H.; Vieira, G.; Streletskiy, D.A.; Schoeneich, P.; Romanovsky, V.E.; Lewkowicz, A.G.; Abramov, A.; et al. Permafrost is warming at a global scale. *Nat. Commun.* **2019**, *10*, 264. [CrossRef]
3. Guo, L.; Ping, C.-L.; Macdonald, R.W. Mobilization pathways of organic carbon from permafrost to arctic rivers in a changing climate. *Geophys. Res. Lett.* **2007**, *34*. [CrossRef]
4. Vonk, J.E.; Tank, S.E.; Bowden, W.B.; Laurion, I.; Vincent, W.F.; Alekseychik, P.; Amyot, M.; Billet, M.F.; Canário, J.; Cory, R.M.; et al. Reviews and syntheses: Effects of permafrost thaw on Arctic aquatic ecosystems. *Biogeosciences* **2015**, *12*, 7129–7167. [CrossRef]
5. Vonk, J.E.; Tank, S.E.; Walvoord, M.A. Integrating hydrology and biogeochemistry across frozen landscapes. *Nat. Commun.* **2019**, *10*, 5377. [CrossRef] [PubMed]
6. Chadburn, S.E.; Krinner, G.; Porada, P.; Bartsch, A.; Beer, C.; Belelli Marchesini, L.; Boike, J.; Ekici, A.; Elberling, B.; Friborg, T.; et al. Carbon stocks and fluxes in the high latitudes: Using site-level data to evaluate Erath system models. *Biogeosciences* **2017**, *14*, 5143–5169. [CrossRef]
7. Yang, D.Q.; Kane, D.L.; Hinzman, L.D.; Zhang, X.B.; Zhing, T.J.; Ye, H.C. Siberian Lena River hydrological regime and recent change. *J. Geophys. Res. Atmos.* **2002**, *107*, 4694. [CrossRef]
8. Ye, B.S.; Yang, D.Q.; Kane, D.L. Changes in Lena River streamflow hydrology: Human impacts versus natural variations. *Water Resour. Res.* **2003**, *39*, 1200. [CrossRef]
9. Ye, B.; Yang, D.; Zhang, Z.; Kane, D.L. Variation of hydrological regime with permafrost coverage over Lena basin in Siberia. *J. Geophys. Res.* **2009**, *114*, D07102. [CrossRef]
10. Berezovskaya, S.; Yang, D.; Hinzman, L. Long-term annual water balance analysis of the Lena River. *Glob. Planet. Chang.* **2005**, *48*, 84–95. [CrossRef]
11. Smith, L.C.; Pavelksky, T.M. Estimation of river discharge, propagation speed, and hydraulic geometry from space: Lena River, Siberia. *Water Resour. Res.* **2008**, *44*, W03427. [CrossRef]
12. Gelfan, A.; Gustafsson, D.; Motovilov, Y.; Arheimer, B.; Kalugin, A.; Krylenko, I.; Lavrenov, A. Climate change impact on the water regime of two great Arctic rivers: Modelling and uncertainty issues. *Clim. Chang.* **2017**, *414*, 499–515. [CrossRef]
13. Gautier, E.; Depret, T.; Costard, F.; Virmoux, C.; Fedorov, A.; Grancher, D.; Konstantinov, P.; Brunstein, D. Going with the flow: Hydrologic response of middle Lena River (Siberia) to the climate variability and change. *J. Hydrol.* **2018**, *557*, 475–488. [CrossRef]

14. Suzuki, K.; Matsuo, K.; Yamazaki, D.; Ichii, K.; Iijima, Y.; Papa, F.; Yanagi, Y.; Hiyama, T. Hydrological variability and changes in the Arctic circumpolar tundra and the three largest Pan-Arctic river basins from 2002 to 2016. *Remote Sens.* **2018**, *10*, 402. [CrossRef]
15. Ahmed, R.; Prowse, T.; Dibike, Y.; Bonsal, B.; O'Neil, H. Recent trends in freshwater influx to the Arctic Ocean from four major Arctic-draining rivers. *Water* **2020**, *12*, 1189. [CrossRef]
16. Lara, R.J.; Rachold, V.; Kattner, G.; Hubberten, H.W.; Guggenberger, G.; Annelie, S.; Thomas, D.N. Dissolved organic matter and nutrients in the Lena River, Siberian Arctic: Characteristics and distribution. *Mar. Chem.* **1998**, *59*, 301–309. [CrossRef]
17. Raymond, P.A.; McClelland, J.W.; Holmes, R.M.; Zhulidov, A.V.; Mull, K.; Peterson, B.J.; Striegl, R.G.; Aiken, G.R.; Gurtovaya, T.Y. Flux and age of dissolved organic carbon exported to the Arctic Ocean: A carbon isotopic study of the five largest arctic rivers. *Glob. Biogeochem. Cycles* **2007**, *121*, GB4011. [CrossRef]
18. Semiletov, I.P.; Pipko, I.I.; Shakhova, N.E.; Dudarev, O.V.; Pugach, S.P.; Charkin, A.N.; McRoy, C.P.; Kosmach, D.; Gustafsson, Ö. Carbon transport by the Lena River from its headwaters to the Arctic Ocean, with emphasis on fluvial input of terrestrial particulate organic carbon vs. carbon transport by coastal erosion. *Biogeosciences* **2011**, *8*, 2407–2426. [CrossRef]
19. Goncalves-Araujo, R.; Stedmon, C.A.; Heim, B.; Dubinenkov, I.; Kraberg, A.; Moiseev, D.; Brachler, A. From fresh to marine waters: Characterization and fate of dissolved organic matter in the Lena River delta region, Siberia. *Front. Mar. Sci.* **2015**, *2*, 108. [CrossRef]
20. Kutscher, L.; Mörth, C.-M.; Porcelli, D.; Hirst, C.; Maximov, T.C.; Petrov, R.E.; Andersson, P.S. Spatial variation in concentration and sources of organic carbon in the Lena River, Siberia. *J. Geophys. Res. Biogeosci.* **2017**, *122*, 1999–2014. [CrossRef]
21. Griffin, C.G.; McClelland, J.W.; Frey, K.E.; Fiske, G.; Holmes, R.M. Quantifying CDOM and DOC in major Arctic rivers during ice-free conditions using Landsat TM and ETM+ data. *Remote Sens. Environ.* **2018**, *209*, 395–409. [CrossRef]
22. Cauwet, G.; Sidorov, I. The biogeochemistry of Lena River: Organic carbon and nutrients distribution. *Mar. Chem.* **1996**, *53*, 211–227. [CrossRef]
23. Huh, Y.; Tsoi, M.Y.; Zaitsev, A.; Edmond, J.M. The fluvial geochemistry of the rivers of eastern Siberia: I. Tributaries of the Lena River draining the sedimentary platform of the Siberian Craton. *Geochim. Cosmochim. Acta* **1998**, *62*, 1657–1676. [CrossRef]
24. Huh, Y.; Panteleyev, G.; Babich, D.; Zaitsev, A.; Edmond, J.M. The fluvial geochemistry of the rivers of Eastern Siberia: II. Tributaries of the Lena, Omoloy, Yana, Indigirka, Kolyma, and Anadyr draining collisional/accretionary zone of the Verkhoyansk and Cherskiy ranges. *Geochim. Cosmochim. Acta* **1998**, *62*, 2053–2075. [CrossRef]
25. Huh, Y.; Edmond, J.M. The fluvial geochemistry of the rivers of Eastern Siberia: III. Tributaries of the Lena and Anabar draining the basement terrain of the Siberian Craton and the Trans-Baikal Highlands. *Geochim. Cosmochim. Acta* **1999**, *63*, 967–987. [CrossRef]
26. Kuzmin, M.I.; Tarasova, E.N.; Bychinskii, V.A.; Karabanov, E.B.; Mamontov, A.A.; Mamontova, E.A. Hydrochemical regime components of Lena water. *Water Resour.* **2009**, *36*, 418–430. [CrossRef]
27. Pipko, I.I.; Pugach, S.P.; Dudarev, O.V.; Charkin, A.N.; Semiletov, I.P. Carbonate parameters of the Lena River: Characteristics and distribution. *Geochem. Internat.* **2010**, *48*, 1131–1137. [CrossRef]
28. Georgiadi, A.G.; Tananaev, N.I.; Dukhova, L.A. Hydrochemical conditions at the Lena River in August 2018. *Oceanology* **2019**, *59*, 797–800. [CrossRef]
29. Juhls, B.; Stedmon, C.A.; Morgenstern, A.; Meyer, H.; Holemann, J.; Heim, B.; Povazhnyi, V.; Overduin, P.P. Identifying drivers of seasonality in Lena River biogeochemistry and dissolved organic matter fluxes. *Front. Environ. Sci.* **2020**, *8*, 53. [CrossRef]
30. Sun, X.; Mörth, C.-M.; Porcelli, D.; Kutscher, L.; Hirst, C.; Murphy, M.J.; Maximov, T.; Petrov, R.E.; Humborg, C.; Schmitt, M.; et al. Stable silicon isotopic compositions of the Lena River and its tributaries: Implications for silicon delivery to the Arctic Ocean. *Geochim. Cosmochim. Acta* **2018**, *241*, 120–133. [CrossRef]
31. Murphy, M.; Porcelli, D.; Pogge von Strandmann, P.; Hirst, K.; Kutscher, L.; Katchinoff, J.; Morth, C.-M.; Maximov, T.; Andresson, P. Tracing silicate weathering processes in the permafrost-dominated Lena River watershed using lithium isotopes. *Geochim. Cosmochim. Acta* **2018**, *245*, 154–171. [CrossRef]
32. Hirst, C.; Andersson, P.; Kooijman, E.; Kutscher, L.; Maximov, T.; Moth, C.-M.; Porcelli, D. Iron isotopes reveal the sources of Fe-bearing particles and colloids in the Lena River basin. *Geochim. Cosmochim. Acta* **2020**, *269*, 678–692. [CrossRef]
33. Bowling, L.C.; Kane, D.L.; Gieck, R.E.; Hinzman, L.D.; Lettenmaier, D.P. The role of surface storage in a low-gradient Arctic watershed. *Water Resour. Res.* **2003**, *39*, 1087. [CrossRef]
34. Holmes, R.M.; McClelland, J.W.; Peterson, B.J.; Tank, S.E.; Bulygina, E.; Eglinton, T.I.; Gordeev, V.V.; Gurtovaya, T.Y.; Raymond, P.A.; Repeta, D.J.; et al. Seasonal and annual fluxes of nutrients and organic matter from large rivers to the Arctic Ocean and surrounding seas. *Estuaries Coasts* **2012**, *35*, 369–382. [CrossRef]
35. Pokrovsky, O.S.; Viers, J.; Dupré, B.; Chabaux, F.; Gaillardet, J.; Audry, S.; Prokushkin, A.S.; Shirokova, L.S.; Kirpotin, S.N.; Lapitsky, S.A.; et al. Biogeochemistry of carbon, major and trace elements in watersheds of Northern Eurasia drained to the Arctic Ocean: The change of fluxes, sources and mechanisms under the climate warming prospective. *C.R. Acad. Sci. Geosci.* **2012**, *344*, 663–677. [CrossRef]
36. O'Donnell, J.; Douglas, T.; Barker, A.; Guo, L. Changing Biogeochemical Cycles of Organic Carbon, Nitrogen, Phosphorus, and Trace Elements in Arctic Rivers. In *Arctic Hydrology, Permafrost and Ecosystems*; Yang, D., Kane, D.L., Eds.; Springer Nature: London, UK, 2021; pp. 315–348. [CrossRef]

37. McClelland, J.W.; Tank, S.E.; Spencer, R.G.M.; Shiklomanov, A.I. Coordination and sustainability of river observing activities in the Arctic. *Arctic* **2015**, *68*, 59. [CrossRef]
38. Bartalev, S.; Egorov, V.; Loupian, E.; Khvostikov, S. A new locally-adaptive classifi-cation method LAGMA for large-scale land cover mapping using remote-sensing data. *Remote Sens. Lett.* **2014**, *5*, 55–64. [CrossRef]
39. Mavromatis, V.; Prokushkin, A.S.; Korets, M.A.; Chmeleff, J.; Mounic, S.; Pokrovsky, O.S. Weak impact of landscape parameters and rock lithology on Mg isotope composition of the Yenisey River and its tributaries. *Chem. Geol.* **2020**, *540*, 119547. [CrossRef]
40. Pokrovsky, O.S.; Reynolds, B.C.; Prokushkin, A.S.; Schott, J.; Viers, J. Silicon isotope variations in Central Siberian rivers during basalt weathering in permafrost-dominated larch forests. *Chem. Geol.* **2013**, *355*, 103–116. [CrossRef]
41. Bagard, M.L.; Chabaux, F.; Pokrovsky, O.S.; Viers, J.; Prokushkin, A.S.; Stille, P.; Rihs, S.; Schmitt, A.D.; Dupre, B. Seasonal variability of element fluxes in two Central Siberian rivers draining high latitude permafrost dominated areas. *Geochim. Cosmochim. Acta* **2011**, *75*, 3335–3357. [CrossRef]
42. Bagard, M.L.; Schmitt, A.D.; Chabaux, F.; Pokrovsky, O.S.; Viers, J.; Stille, P.; Labolle, F.; Prokushkin, A.S. Biogeochemistry of stable Ca and radiogenic Sr isotopes in a larch-covered permafrost-dominated watershed of Central Siberia. *Geochim. Cosmochim. Acta* **2013**, *114*, 169–187. [CrossRef]
43. Mavromatis, V.; Prokushkin, A.S.; Pokrovsky, O.S.; Viers, J.; Korets, M.A. Magnesium isotopes in permafrost-dominated Central Siberian larch forest watersheds. *Geochim. Cosmochim. Acta* **2014**, *147*, 76–89. [CrossRef]
44. Mavromatis, V.; Rinder, T.; Prokushkin, A.S.; Pokrovsky, O.S.; Korets, M.A.; Chmeleff, J.; Oelkers, E.H. The effect of permafrost, vegetation, and lithology on Mg and Si isotope composition of the Yenisey River and its tributaries at the end of the spring flood. *Geochim. Cosmochim. Acta* **2016**, *191*, 32–46. [CrossRef]
45. Hindshaw, R.S.; Teisserenc, R.; Le Dantec, T.; Tananaev, N. Seasonal change of geochemical sources and processes in the Yenisei River: A Sr, Mg and Li isotope study. *Geochim. Cosmochim. Acta* **2019**, *255*, 222–236. [CrossRef]
46. Chupakov, A.V.; Pokrovsky, O.S.; Moreva, O.Y.; Shirokova, L.S.; Neverova, N.V.; Chupakova, A.A.; Kotova, E.I.; Vorobyeva, T.Y. High resolution multi-annual riverine fluxes of organic carbon, nutrient and trace element from the largest European Arctic river, Severnaya Dvina. *Chem. Geol.* **2020**, *538*, 119491. [CrossRef]
47. Brown, J.; Ferrians, O.J., Jr.; Heginbottom, J.A.; Melnikov, E.S. *Circum-Arctic Map of Permafrost and Ground Ice Conditions. National Snow and Ice Data Center/World Data Center for Glaciology*; Digital Media: Boulder, CO, USA, 2002.
48. Chevychelov, A.P.; Bosikov, N.P. Natural Conditions. In *The Far North*; Troeva, E.I., Isaev, A.P., Cherosov, M.M., Karpov, N.S., Eds.; Springer Nature: London, UK, 2010; p. 123.
49. Rachold, V.; Alabyan, A.; Hubberten, H.-W.; Korotaev, V.N.; Zaitsev, A.A. Sediment transport to the Laptev Sea-hydrology and geochemistry of the Lena River. *Polar Res.* **1996**, *15*, 183–196.
50. Gureyev, D. Tomsk State University: The Expedition on the River Lena from the Beginnings to the Aldan River 2016. Available online: https://www.youtube.com/watch?v=7IEiO4bgxc8 (accessed on 27 July 2021).
51. Spence, J.; Telmer, K. The role of sulfur in chemical weathering and atmospheric CO_2 fluxes: Evidence from major ions, $\delta^{13}C_{DIC}$, and $\delta^{34}S_{SO4}$ in rivers of the Canadian Cordillera. *Geochim. Cosmochim. Acta* **2005**, *69*, 5441–5458. [CrossRef]
52. Qin, J.; Huh, Y.; Edmond, J.M.; Du, G.; Ran, J. Chemical and physical weathering in the Min Jiang, a headwater tributary of the Yangtze River. *Chem. Geol.* **2006**, *227*, 53–69. [CrossRef]
53. Pokrovsky, O.S.; Manasypov, R.M.; Loiko, S.V.; Shirokova, L.S. Organic and organo-mineral colloids of discontinuous permafrost zone. *Geochim. Cosmochim. Acta* **2016**, *188*, 1–20. [CrossRef]
54. Pokrovsky, O.S.; Manasypov, R.M.; Loiko, S.; Krickov, I.A.; Kopysov, S.G.; Kolesnichenko, L.G.; Vorobyev, S.N.; Kirpotin, S.N. Trace element transport in western Siberia rivers across a permafrost gradient. *Biogeosciences* **2016**, *13*, 1877–1900. [CrossRef]
55. Pokrovsky, O.S. *Lena River and tributaries water chemistry, Mendeley Data*; Elsevier: Amsterdam, The Netherlands, 2021; Volume 1. [CrossRef]
56. Cooper, L.W.; McClelland, J.W.; Holmes, R.M.; Raymond, P.A.; Gibson, J.J.; Guay, C.K.; Peterson, B.J. Flow-weighted values of runoff tracers ($\delta^{18}O$, DOC, Ba, alkalinity) from the six largest Arctic rivers. *Geophys. Res. Lett.* **2008**, *35*, L18606. [CrossRef]
57. Gordeev, V.V.; Sidorov, I.S. Concentrations of major elements and their outflow into the Laptev Sea by the Lena River. *Mar. Chem.* **1993**, *43*, 33–46. [CrossRef]
58. Vdovykin, G.P. Effect of chemical landscape on salt composition of the Lena River water (East Siberia). *Dokl. Acad. Nauk. SSSR* **1991**, *317*, 186–189.
59. Takhteev, V.V.; Lopatovskaya, O.G.; Okuneva, G.L.; Pomazkova, G.I.; Samoilova, E.A.; Rozhkova, N.A. Ecological description of the sodium chloride mineral springs in the Kirenga River Basin and upper reaches of the Lena River: 1. General characteristics of the springs and their hydrofauna. *Inland Water Biol.* **2017**, *10*, 331–341. [CrossRef]
60. Krickov, I.V.; Pokrovsky, O.S.; Manasypov, R.M.; Lim, A.G.; Shirokova, L.S.; Viers, J. Colloidal transport of carbon and metals by western Siberian rivers during different seasons across a permafrost gradient. *Geochim. Cosmochim. Acta* **2019**, *265*, 221–241. [CrossRef]
61. Krickov, I.V.; Lim, A.G.; Manasypov, R.M.; Loiko, S.V.; Vorobyev, S.N.; Shevchenko, V.P.; Dara, O.M.; Gordeev, V.V.; Pokrovsky, O.S. Major and trace elements in suspended matter of western Siberian rivers: First assessment across permafrost zones and landscape parameters of watersheds. *Geochim. Cosmochim. Acta* **2020**, *269*, 429–450. [CrossRef]
62. Guo, L.; Zhang, J.Z.; Guéguen, C. Speciation and fluxes of nutrients (N, P, Si) from the upper Yukon River. *Glob. Biogeochem. Cycles* **2004**, *18*, GB1038. [CrossRef]

63. Cai, Y.; Guo, L.; Douglas, T.A.; Whitledge, T.E. Seasonal variations in nutrient concentrations and speciation in the Chena River, Alaska. *J. Geophys. Res. Biogeosci.* **2008**, *113*, G030035. [CrossRef]
64. Raudina, T.V.; Loiko, S.V.; Kuzmina, D.M.; Shirokova, L.S.; Kulizhsky, S.P.; Golovatskaya, E.A.; Pokrovsky, O.S. Colloidal organic carbon and trace elements in peat porewaters across a permafrost gradient in Western Siberia. *Geoderma* **2021**, *390*, 114971. [CrossRef]
65. Stolpe, B.; Guo, L.; Shiller, A. Binding and transport of rare earth elements by organic and iron-rich nanocolloids in Alaskan rivers, as revealed by field-flow fractionation and ICP-MS. *Geochim. Cosmochim. Acta* **2013**, *106*, 446–462. [CrossRef]
66. Stolpe, B.; Guo, L.; Shiller, A.M.; Aiken, G.R. Abundance, size distributions and trace-element binding of organic and iron-rich nanocolloids in Alaskan rivers, as revealed by field-flow fractionation and ICP-MS. *Geochim. Cosmochim. Acta* **2013**, *105*, 221–239. [CrossRef]
67. Cuss, C.W.; Grant-Weaver, I.; Shotyk, W. AF4-ICPMS with the 300 Da membrane to resolve metal-bearing 'colloids' <1 kDa: Optimization, fractogram deconvolution, and advanced quality control. *Anal. Chem.* **2017**, *89*, 8027–8035.
68. Vasyukova, E.V.; Pokrovsky, O.S.; Viers, J.; Oliva, P.; Dupré, B.; Martin, F.; Candaudap, F. Trace elements in organic- and iron-rich surficial fluids of the boreal zone: Assessing colloidal forms via dialysis and ultrafiltration. *Geochim. Cosmochim. Acta* **2010**, *74*, 449–468. [CrossRef]
69. Pokrovsky, O.S.; Schott, J.; Kudryavtzev, D.I.; Dupré, B. Basalt weathering in Central Siberia under permafrost conditions. *Geochim. Cosmochim. Acta* **2005**, *69*, 5659–5680. [CrossRef]
70. Shogren, A.J.; Zarnetske, J.P.; Abbott, B.W.; Iannucci, F.; Frei, R.J.; Griffin, N.A.; Bowden, W.B. Revealing biogeochemical signatures of Arctic landscapes with river chemistry. *Sci. Rep.* **2019**, *9*, 12894. [CrossRef] [PubMed]
71. Frey, K.E.; McClelland, J.W. Impacts of permafrost degradation on arctic river biogeochemistry. *Hydrol. Process.* **2009**, *23*, 169–182. [CrossRef]
72. Vorobyev, S.N.; Pokrovsky, O.S.; Serikova, S.; Manasypov, R.M.; Krickov, I.V.; Shirokova, L.S.; Lim, A.; Kolesnichenko, L.G.; Kirpotin, S.N.; Karlsson, J. Permafrost boundary shift in Western Siberia may not modify dissolved nutrient concentrations in rivers. *Water* **2017**, *9*, 985. [CrossRef]
73. Mamet, S.D.; Brown, C.D.; Trant, A.J.; Laroque, C.P. Shifting global *Larix* distributions: Northern expansion and southern retraction as species respond to changing climate. *J. Biogeogr.* **2019**, *46*, 30–44. [CrossRef]
74. Rees, W.G.; Hofgaard, A.; Boudreau, S.; Cairns, D.M.; Harper, K.; Mamet, S.; Mathisen, I.; Swirad, Z.; Tutubalina, O. Is subarctic forest advance able to keep pace with climate change? *Glob. Chang. Biol.* **2020**, *26*, 3965–3977. [CrossRef]
75. Liu, H.; Mi, Z.; Lin, L.; Wang, Y.; Zhang, Z.; Zhang, F.; Wang, H.; Liu, L.; Zhu, B.; Cao, G.; et al. Shifting plant species composition in response to climate change stabilizes grassland primary production. *Proc. Natl. Acad. Sci. USA* **2018**, *115*, 4051–4056. [CrossRef] [PubMed]

water

MDPI

Article

Seasonal Variations of Dissolved Iron Concentration in Active Layer and Rivers in Permafrost Areas, Russian Far East

Yuto Tashiro [1,*], Muneoki Yoh [2], Takayuki Shiraiwa [3], Takeo Onishi [4], Vladimir Shesterkin [5] and Vladimir Kim [5]

1. United Graduate School of Agricultural Science, Tokyo University of Agriculture and Technology, Tokyo 183-8509, Japan
2. Institute of Agriculture, Tokyo University of Agriculture and Technology, Tokyo 183-8509, Japan; yoh@cc.tuat.ac.jp
3. Institute of Low Temperature Science, Hokkaido University, Hokkaido 060-0819, Japan; shiraiwa@lowtem.hokudai.ac.jp
4. Faculty of Applied Biological Sciences, Gifu University, Gifu 501-1193, Japan; takeon@gifu-u.ac.jp
5. Separate Subdivision Federal State Budgetary Institution of Science Khabarovsk Federal Research Center of the Far Eastern Branch of the Russian Academy of Sciences, Institute of Water and Ecology Problems of the Far Eastern Branch of the Russian Academy of Sciences, 680000 Khabarovsk, Russia; shesterkin@ivep.as.khb.ru (V.S.); kim@ivep.as.khb.ru (V.K.)

* Correspondence: tassy40y@gmail.com

Received: 20 August 2020; Accepted: 11 September 2020; Published: 15 September 2020

Abstract: Dissolved iron (dFe) in boreal rivers may play an important role in primary production in high-latitude oceans. However, iron behavior in soils and dFe discharge mechanism from soil to the rivers are poorly understood. To better understand iron dynamics on the watershed scale, we observed the seasonal changes in dFe and Dissolved Organic Carbon (DOC) concentrations in the river as well as dFe concentration in soil pore waters in permafrost watershed from May to October. During snowmelt season, high dFe production (1.38–4.70 mg L^{-1}) was observed in surface soil pore waters. Correspondingly, riverine dFe and DOC concentrations increased to 1.10 mg L^{-1} and 32.3 mg L^{-1}, and both were the highest in the year. After spring floods, riverine dFe and DOC concentrations decreased to 0.15 mg L^{-1} and 7.62 mg L^{-1}, and dFe concentration in surface soil pore waters also decreased to 0.20–1.28 mg L^{-1}. In late July, riverine dFe and DOC concentrations increased to 0.33 mg L^{-1} and 23.6 mg L^{-1} in response to heavy rainfall. In August and September, considerable increases in dFe concentrations (2.00–6.90 mg L^{-1}) were observed in subsurface soil pore waters, probably because infiltrated rainwater developed reducing conditions. This dFe production was confirmed widely in permafrost wetlands in valley areas. Overall, permafrost wetlands in valley areas are hotspots of dFe production and greatly contribute to dFe and DOC discharge to rivers, especially during snowmelt and rainy seasons.

Keywords: permafrost; wetland; dissolved iron; Amur river

1. Introduction

Iron is an essential trace element for the growth of all organisms. It plays an important role in in vivo metabolic processes, including photosynthesis, electron transfer, nitrate reduction, and nitrogen fixation [1]. Martin and Fitzwater [2] revealed that iron limits phytoplankton growth in a high-nutrient low-chlorophyll (HNLC) region where the reason for the relatively low phytoplankton productivity despite the abundance of nutrients having long been unclear. Subsequent studies confirmed that dissolved iron (dFe) concentrations in many ocean areas are too low to fully utilize nutrients [3–8].

The main iron source for the ocean was thought to be aeolian dust [2]. However, recent studies suggested that riverine dFe is also an important source to support primary production in the coastal area and the ocean [9–11], stimulating interests in iron dynamics in terrestrial environments.

The characteristic behavior of riverine iron transport is that most of the dFe is complexed with humic substances such as humic acids and fulvic acids [10]. It was reported that most of riverine dFe coagulates and precipitates in brackish zones because of high salinity [12]. However, a part of organically complexed iron can remain dissolved and be utilized by phytoplankton [13–15]. Therefore, there were some attempts to reveal the connection between terrestrial natural environments and marine biological productivity from the perspective of iron cycle [9,16–18].

The Sea of Okhotsk is one of the oceans with the richest marine resources in the world. As regards the reason, some studies indicated that abundant dFe derived from forests and wetlands in the Amur River basin contributes to the creation of high biological productivity in the Sea of Okhotsk [11,19]. Based on these findings, it was suggested that land use change, e.g., reclamation of wetlands for agriculture, influences dFe discharge to the Amur River [20]. In fact, it was reported that the Naoli River, where 87% of wetlands in the basin disappeared from 1949 to 2000 due to large-scale development of farmland, showed a sharp decrease in dFe concentration during this period [21].

On the contrary, a sharp increase in dFe concentration in the Amur River was observed from 1995 to 1998, which could not be explained by the increase in inflow of groundwater for irrigation due to agricultural development [22]. Another reason for this unidentified high dFe concentration in the Amur River, Shamov et al. (2014) [22] pointed out that permafrost thawing due to high summer temperatures in the 1990s has promoted wet and reducing conditions in the active layer, resulting in more dFe production in soils. In order to discuss the possibility that the extent of soil thaw dynamics has an effect on dFe concentration in a river, we need to better understand iron behavior in soils and discharge mechanism from soil to river in the permafrost watershed.

Nowadays, there has been increasing interest in iron behavior in Arctic peat soil because iron potentially controls the degradation of soil organic matter and the emissions of CO_2 and CH_4 [23–25]. In Arctic peat soils that contain large amounts of organic carbon and are generally waterlogged, CH_4 production is suppressed by anaerobic iron respiration because Fe(III) reducing bacteria outcompete methanogens for carbon substrates [26,27]. Recently, studies were conducted to investigate Fe species and their concentrations in soils of the active layer (soil layer which thaws from spring to autumn and freezes during winter) [28–32]. For instance, Herndon et al. [31] showed that the concentrations of dFe(II) were high in deep mineral soil horizon, whereas those of Fe(III) were high in organic soil horizon, suggesting that produced Fe(II) diffuses upward in the soil profile and is oxidized to Fe(III) oxyhydroxides and organically complexed Fe(III). It was also reported that the reduction of iron oxides occurs even in shallow peat soil (0–10 cm) from summer to autumn [30]. However, the overall view of the iron cycle is still poorly understood because iron behavior varies spatially and seasonally in soils of the active layer. Furthermore, little is known about the seasonal change in iron discharge mechanism from the active layer to rivers in permafrost watershed. Bagard et al. [33] reported that seasonal soil thaw dynamics of the active layer in a catchment causes a shift in the source of colloidal trace elements including Fe from organic soil horizon in spring to mineral soil horizon in late summer. It is thus expected that seasonal soil thaw dynamics influences not only iron behavior in soils, but also dFe concentrations and chemical species in permafrost-affected rivers.

The aims of this study were (1) to identify the spatial and seasonal variability of iron behavior in soils of the active layer, with focus on the time and depth of dFe production by iron reduction, and (2) to gain an insight into the influence of seasonal soil thaw dynamics on dFe and DOC discharge mechanism and these concentrations in rivers. We conducted an observation of dFe and DOC concentrations in river and dFe concentration in soil pore waters in a permafrost-affected watershed from May (beginning of soil thawing) to October (beginning of soil freezing). To our knowledge, this study is the first to investigate the watershed-scale iron cycle through the seasons in a permafrost region.

2. Materials and Methods

2.1. Study Site

A field survey was conducted in the Tyrma region, which is approximately 270 km northwest of Khabarovsk in the Russian Far East. Mean annual air temperature is −1.96 °C and annual precipitation is 654.6 mm. The Tyrma region is situated in a sporadic permafrost area [22]. Three large rivers, namely, the Sutyri River, the Yaurin River, and the Gujal River, join the Tyrma River, and eventually join the Bureya River, which is one of the largest tributaries of the Amur River (Figure 1). We fixed the Sofron River watershed (50°5′42.86″ N,132°22′1.20″ E), which is a tributary of the Gujal River, as the intensive experimental site (Figures 1b and 2a). The Sofron River is a second-order stream with a sub-catchment area of approximately 23 km^2 (Figure 2). River width is approximately 10 m and central water depth is usually 10–20 cm at the sampling point without rain.

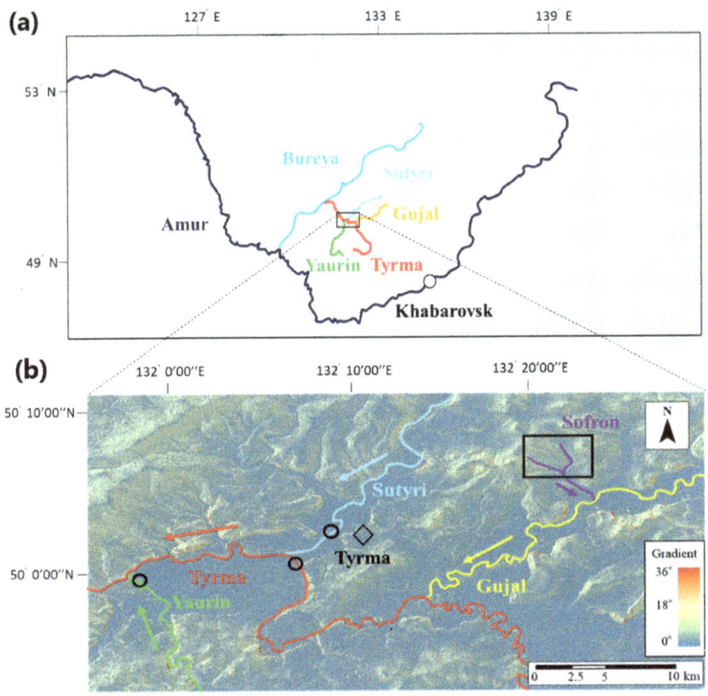

Figure 1. (**a**) map of the Amur River basin; (**b**) map of the Tyrma region with the large rivers where water samples were collected. This map is colored by a gradient (degree) to clearly show terrain characteristics. As for the symbols in map (**b**), the black rectangle denotes the Sofron River watershed where intensive research was conducted; black circles denote the river water sampling sites of the Sturi, Tyrma, and Yaurin Rivers; and black diamond denotes the Tyrma village. This map was created by the authors based on 30 m resolution digital elevation model (DEM) provided by the Japan Aerospace Exploration Agency (JAXA).

Vegetation in the Tyrma region is roughly divided into two types: the forest areas on the ridges and hillslopes are characterized by spruces (*Picea ajanensis*) and white birches (*Betula platyphylla*), and the wetlands in the flat valleys are characterized by shrubs, such as bog blueberry (*Vaccinium uliginosum*), cowberry (*Vaccinium vitis-idaea* L.), and ledum (*Ledum decumbens*). In the wetlands, Sphagnum species widely exist on the ground surface and larches (*Larix gmelinii var. gmelinii*) are scattered. Topsoil layers in the forests and the wetlands are composed of peat soils. In particular, thick peat soil layers are formed in the wetlands due to long-term accumulation of sphagnum biomass. This type of wetland, called *Mari*, is a typical landscape in the flat valleys in the Tyrma region. *Mari* can be seen not only in the flat valleys, but also on the ridges and hillslopes with a gentle slope. In the Tyrma region, permafrost generally exists underneath Mari. The vegetation and permafrost distribution in the Sofron basin are consistent with these typical features as described above.

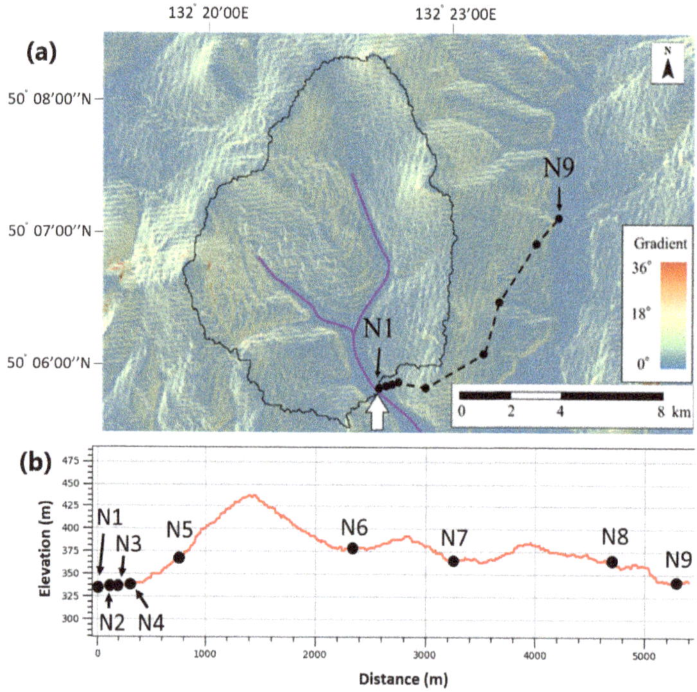

Figure 2. (**a**) nine sampling points for soil pore waters (N1–N9) in the Sofron River and neighboring area. This map is colored by gradient (degree) to clearly show terrain characteristics. Area enclosed with a black line shows the Sofron watershed, and the white arrow denotes the sampling site of the Sofron River. This map was created by the authors based on a 30 m resolution digital elevation model (DEM) provided by the Japan Aerospace Exploration Agency (JAXA); (**b**) topographic profile through the sampling points for soil pore waters.

2.2. Soil Pore Water Sampling and Soil Characteristic Measurement along a Transect

In September 2016, we established a long transect of nine sampling points for pore waters from N1 near the Sofron River to N9 in the adjacent basin (Figure 2). Permafrost wetland Mari is seen not only at N1–N3 and N9 in valley, but also at N7 on ridge and N8 on gentle hillslope. On the other hand, permafrost does not exist at N5 on steep hillslope and N6 on ridge. We dug 1.8 cm diameter holes using a manual drill (DAIKI, DIK-1721, Konosu city, Saitama, Japan) in the ground of each sampling point and installed ceramic cups with a polyvinyl chloride pipe at 20 cm and 40 cm depths for pore water sampling (N4 has different depths as described below and N5 has only 40 cm depth). In April 2017, we installed the ceramic pipes at 5 cm and 10 cm depths at N4 and 10 cm and 25 cm depths at N6 for soil pore water sampling during snowmelt season. Because 10 cm of snow remained at N4 and N6 on this installation, both points were suitable to observe dFe concentration variations in response to snow melting. The frozen ground at that time existed at 10–15 cm depth at N4 and N6; thus, we used an electric drill to install a ceramic cup at 25 cm depth at N6. All the soil pore water sampling depths at each point are compiled in Table 1.

Table 1. Terrain type, sampling depth of soil pore waters, active layer thickness (ALT), and peat layer thickness (PLT) at nine sampling points of the research site in September 2016 (Tyrma, Khabarovsk, Russia). Pore waters at marked (*) depths at N4 and N6 were collected four times in May and twice a month from June to October. Other soil pore waters at nine sampling points were collected twice a month from June to October.

Point Number	Terrain Type	Sampling Depth (cm)	ALT (cm)	PLT (cm)
N1	Valley	20, 40	64	>64
N2	Valley	20, 40	60	>60
N3	Valley	20, 40	82	>82
N4	Boundary between valley and hillslope	5 *, 10 *	102	12
N5	Steep hillslope	40	No permafrost	7
N6	Ridge	10 *, 25 *, 20, 40	No permafrost	42
N7	Ridge	20, 40	75	40
N8	Gentle hillslope	20, 40	127	45
N9	Valley	20, 40	53	>53

At N4 and N6, the soil pore waters (except 20 cm and 40 cm at N6) were collected four times in May. From June, all pore waters at the nine sampling points were collected twice a month until October (see Table 1). The soil pore waters were sucked up overnight by 50 mL disposable syringes (TERUMO, SS-50ESZ, Shibuya ku, Tokyo, Japan) and filtered in situ through 0.45 μm disposable filters made of cellulose acetate (ADVANTEC, DISMIC 25CS045AS, Chiyoda ku, Tokyo, Japan). The filtered waters were preserved in acid-washed 50 mL polypropylene bottles and kept in a refrigerator at 4 °C until analysis.

The peat layer thickness (PLT) and the active layer thickness (ALT) at the nine sampling points were measured in September 2016 by digging a pit down to the permafrost table and examining the soil profile (Table 1). To confirm the ALT data, we checked that the soil temperature of permafrost table was 0 °C. The absence of permafrost at N5 and N6 was judged from the absence of a continuous decline in soil temperatures over 10 cm intervals from the surface. Soils for the analysis of organic carbon content and moisture content were collected in polyethylene freezer bags from 0–10 cm, 10–20 cm, and 20–30 cm depths of the soil profile at the nine sampling points and stored in a freezer until analysis.

When we dug a soil profile at each point in September 2016, we installed geothermal loggers (Onset, U22-001, Bourne, Massachusetts, America) to observe the seasonal thaw dynamics of soil at two depths at N1 (10 cm and 25 cm), three depths at N4 (10 cm, 25 cm, and 50 cm), and three depths at N6 (10 cm, 25 cm, and 50 cm). The soil temperatures were recorded until October 2017.

2.3. River Water Sampling and Hydrological Observation

During snowmelt season from late April to mid May, the water samples were collected from the Sofron River twice a week. From June to October, the water samples were collected twice in a month on the same days as soil pore waters. Two hundred milliliters of water was sampled using a disposable syringe (TERUMO, SS-50ESZ, Shibuya ku, Tokyo, Japan) and immediately filtered through 0.45 μm disposable filters made of cellulose acetate (ADVANTEC, DISMIC 25CS045AS, Chiyoda ku, Tokyo, Japan). One hundred milliliters of the filtered water was preserved in an acid-washed propylene bottle for dFe measurement and the other 100 mL was preserved in a propylene bottle for dissolved organic carbon (DOC) measurement, and both were kept in a refrigerator until analysis.

For the observation of water level of the Sofron River, we installed a water pressure logger (Onset, HOBO U-20-001-04, Bourne, Massachusetts, America) at the river bed of the water sampling point and an air pressure logger (Onset, HOBO U-20-001-04, Bourne, Massachusetts, America) nearby in mid June 2017 when the river bed adequately thawed. In addition, daily precipitation data in the

Tyrma region from June to October were obtained from the weather site in Russia (http://ru8.rp5.ru/Weather_archive_in_Tyrma_(Sutyr)).

Water samples were also collected from the Sutyri River (catchment area: 2129 km^2), the Tyrma River (6168 km^2), and the Yaurin River (3175 km^2) (Figure 1b). The frequency, date, and way of sampling were same as those for the Sofron River.

2.4. Chemical Analyses

All the analyses were done in the Institute of Mining in Khabarovsk. The dFe concentrations were measured for soil pore waters and river waters. Prior to Fe analysis, the samples were acidified (~pH 2) with HNO_3, and Fe concentration was determined with Agilent 7500cx Inductively Coupled Plasma-Mass Spectrometer (ICP-MS) using helium modes to diminish the interferences. The detection limit for Fe by ICP-MS was 10 $ng\,L^{-1}$, and an uncertainty was 5–10%. Standard solutions for Fe analysis were prepared by using Environmental Calibration Standard (Agilent Technologies). In this paper, we define dFe as Fe that passed through 0.45 μm disposable filter in samples. Moreover, for the river water samples, DOC concentrations were also determined with a TOC analyzer (SHIMADZU TOC-LCSH) using the catalytic combustion oxidation method. The detection limit for TOC by TOC analyzer was 0.1 $mg\,L^{-1}$, and an uncertainty was less than 1.5%. Standard solutions for DOC analysis were prepared by using Potassium Hydrogen Phthalate ($C_6H_4(COOK)(COOH)$) (Nacalai tesque). For the data of DOC and dFe concentrations in the Sofron River, simple liner regression analysis and Student's *t*-test were performed.

Soil organic carbon content was measured by Tyurin's method (wet combustion) based on Bel'chikova (1975) [34]. Weight moisture content was determined from the difference in mass before and after drying soil.

3. Results

3.1. Seasonal Variations in dFe and DOC Concentrations in Sofron River and Large Rivers

The dFe concentration in the Sofron River showed substantial seasonal variation, ranging from 0.10 $mg\,L^{-1}$ to 1.10 $mg\,L^{-1}$ (Figure 3a). During snowmelt season, dFe concentration increased rapidly within two weeks from 0.27 $mg\,L^{-1}$ to the annual highest value of 1.10 $mg\,L^{-1}$. Although it once dropped through June, a relatively high dFe concentration was recorded again in late July and late September. The annual lowest level of 0.10 $mg\,L^{-1}$ was noted in October, immediately before soil freezing. DOC concentration in the Sofron River showed a similar seasonal variation to dFe concentration. DOC concentration rapidly increased during snowmelt season from 17.0 $mg\,L^{-1}$ to the annual highest value of 32.3 $mg\,L^{-1}$. After that, it increased again in late July and late September, when dFe concentration also increased.

The water level of the Sofron River showed frequent increases accompanied with heavy rainfall of over 10 mm from July to mid August (Figure 3b). The highest water level of 36 cm was recorded in early August. The increases in dFe and DOC concentrations in late July and late September were associated with the water level risings, that is, the seasonal variations in dFe and DOC concentrations almost synchronized with the change in water level throughout the year. Whereas dFe concentration greatly increased from 0.15 $mg\,L^{-1}$ to 1.07 $mg\,L^{-1}$ during the period affected by snow melting, the degrees of increases associated with heavy rainfall in late July and late September were relatively small. In the separated periods before and after July, two distinct positive correlations were found between dFe concentration and DOC concentration ($r^2 = 0.51$, $p = 0.056$ from April to June, and $r^2 = 0.75$, $p < 0.01$ from July to October) (Figure 3c), suggesting a difference in the discharge mechanisms of these two substances among seasons.

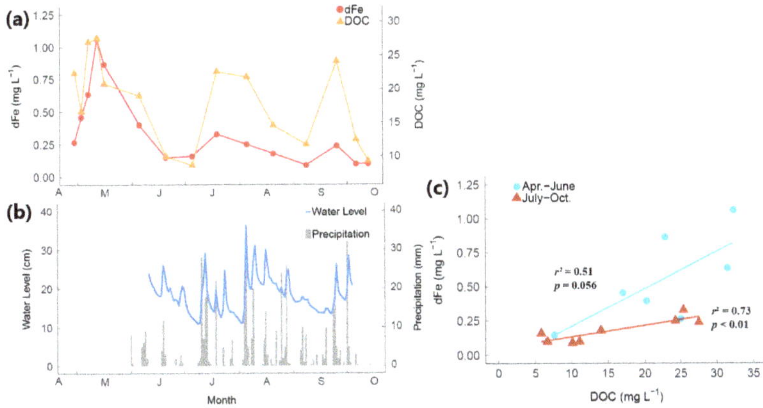

Figure 3. (**a**) seasonal changes in dFe and DOC concentrations in the Sofron River; (**b**) daily change in water level of the Sofron River and daily amount of rainfall in the Tyrma region; (**c**) liner regression plot of DOC and dFe concentrations in the Sofron River. Blue circles show the data from April to June (first seven samples in (**a**)), and red triangles show the data from July to October (late nine samples in (**a**)). The regression lines were calculated by the least squares method.

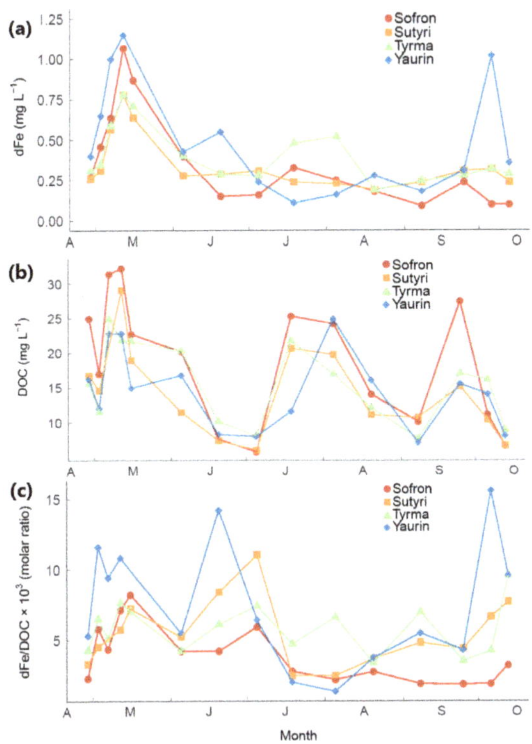

Figure 4. Seasonal changes in dFe and DOC concentrations in the Sofron River and the large rivers (Sturi, Tyrma, and Yaurin) [(**a**) dFe concentration; (**b**) DOC concentration; (**c**) dFe/DOC molar ratio].

The seasonal variations in dFe and DOC concentrations in the large rivers (Sutyri, Tyrma, and Yaurin) were almost consistent with those in the Sofron River throughout the year (Figure 4a,b). During snowmelt season, similar large increases in dFe and DOC concentrations were also observed. The increases in DOC concentrations corresponding to rainfall as seen in the Sofron River were also observed in the large rivers in late July and late September. Although the increases in dFe concentrations during these storm events were not clearly observed in the large rivers, these concentration levels matched those of the Sofron River. In a rainless season (from June to early July and from September to October), however, it appeared that dFe concentrations in the large rivers were generally higher than those in the Sofron River. This difference between the Sofron River and the large rivers was more clearly seen in the seasonal change in dFe/DOC ratio (Figure 4c).

3.2. Organic Carbon Contents and Weight Moisture Contents of Soils along a Transect

Organic carbon contents and weight moisture contents of soils at the nine sampling points in the research site are shown in Table 2. The organic carbon contents at the uppermost layer at points located in valley (N1–3 and N9) were 144–478 gC kg^{-1}, whereas those at points located on hillslope and ridge (N4–8) were 48–358 gC kg^{-1}, slightly lower than those at valley points. Interestingly, permafrost wetlands Mari in valley had high organic carbon contents of 100–474 gC kg^{-1} even at 20–30 cm depths, whereas Mari on ridge (N7) and gentle hillslope (N8) had low organic carbon contents of 52 and 23 gC kg^{-1}, respectively. Although N5 on steep hillslope and N6 on ridge are both no-permafrost points, the organic carbon contents were relatively high at N6 and low at N5.

Table 2. Organic carbon contents and weight moisture contents at each point of the research site in September 2016 (Tyrma, Khabarovsk, Russia).

Point Number	Terrain Type	Organic Carbon Content (gC kg^{-1})			Weight Moisture Content (%)		
		0–10 cm	10–20 cm	20–30 cm	0–10 cm	10–20 cm	20–30 cm
N1	Valley	405	451	331	92.7	93.0	89.8
N2	Valley	478	403	474	99.7	99.0	98.4
N3	Valley	411	311	100	99.5	92.1	90.3
N4	Boundary between valley and hillslope	358	143	29	92.4	45.8	32.4
N5	Steep hillslope	142	62	61	90.0	64.4	58.3
N6	Ridge	355	262	197	74.3	75.9	69.3
N7	Ridge	331	162	52	49.4	42.6	40.5
N8	Gentle hillslope	48	248	23	68.1	48.7	38.7
N9	Valley	144	293	297	99.8	99.5	78.0

The weight moisture contents showed similar spatial variations to the organic carbon contents. Weight moisture contents were remarkably high, i.e., 92.7–99.8%, at the uppermost layer at points located in valley (N1–3 and N9), but were relatively low, i.e., 49.4–92.4% on hillslope and ridge (N4–8). In short, wetlands Mari in valley had the highest organic carbon and moisture contents within the watershed. However, *Mari* on ridge (N7) and gentle hillslope (N8) had relatively low weight moisture contents of 49.4% and 68.1% even at the uppermost layer relative to *Mari* in valley points. At hillslope and ridge points, weight moisture contents decreased significantly with depths. At N6 that had contained relatively high organic carbon contents among hillslope and ridge points, weight moisture contents were also high.

3.3. Seasonal Soil Thaw Dynamics in Different Terrain Types

The seasonal variations in soil temperature at several depths at N1, N4, and N6 are shown in Figure 5. In addition, the thawing date for each depth and the thawing rate, which is calculated simply by dividing the difference in depth by the number of required days for thawing, are shown in Table 3. The thawing of 10 cm depth at N1, which is the closest point to the river in valley, was completed on 24 May, approximately one month later than that at N4 on the boundary between valley and hillslope and that at N6 on ridge. The 25 cm depth at N1 thawed on 15 June, which was also the latest among the same depths.

The 10 cm and 25 cm depths at N4 and N6 showed remarkable rises of soil temperatures after thawing. Soil temperatures as of July 1 for 10 cm and 25 cm were 9.1 °C and 4.2 °C at N4 and 8.5 °C and 3.3 °C at N6, which were considerably higher than those at N1 of 2.4 °C and 0.6 °C. At N6, where permafrost does not exist, soil temperatures always tended to be higher than those at N1 and N4. In fact, the annual highest soil temperatures at N6 (10 cm: 13.1°C, 25 cm: 9.6 °C, 50 cm: 7.9 °C) were the highest compared with those at N1 (10 cm: 10.9 °C, 25 cm: 6.9 °C) and N4 (10 cm: 11.5 °C, 25 cm: 8.2 °C, 50 cm: 5.2 °C). Moreover, the thawing rate from 25 cm to 50 cm depth at N6 was 1.6 times faster than that for the same depth at N4. Therefore, the seasonal downward thawing of soil from spring to summer was fastest at N6 on ridge, followed by N4 on boundary between valley and hillslope, and the slowest at N1 in valley. This order agreed with the difference in active layer thickness at those points (N1: 64 cm, N4: 102 cm, N6: No permafrost) (Table 1). Combined with organic carbon content data and weight moisture content data (Table 2), permafrost wetland *Mari* in valley area is characterized by an environment with the highest organic matter content and the highest moisture content, and the lowest soil temperature within the watershed.

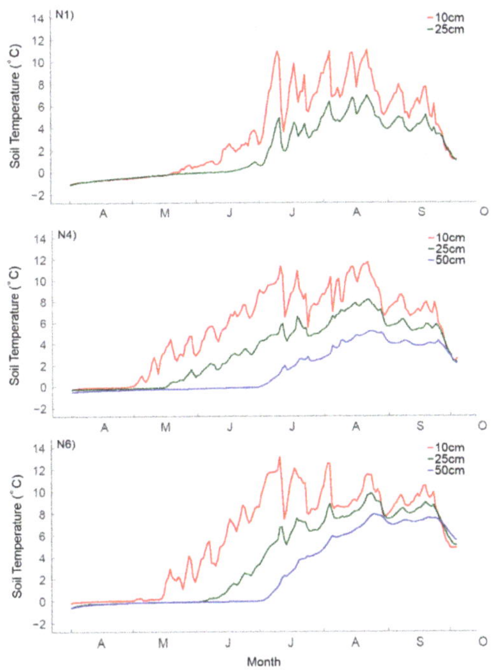

Figure 5. Seasonal changes in daily mean soil temperatures at N1, N4, and N6.

Table 3. Thawing dates for each depth at N1, N4, and N6, and thawing rates between depths. Thawing dates were defined as the day when the average daily soil temperature exceeded 0 degree. Thawing rate was calculated by dividing the difference in depth by the number of required days for thawing.

	Depth (cm)	N1	N4	N6
Thawing Date (mm/dd)	10	5/24	4/29	4/18
	25	6/15	5/18	6/5
	50		7/1	7/2
Thawing Rate ($cm\ day^{-1}$)	10→25	0.68	0.79	0.31
	25→50		0.57	0.93

3.4. Seasonal Changes in dFe Concentrations in Soil Pore Waters along a Transect

Seasonal changes in dFe concentration in soil pore waters at N4 and N6, the observations of which only started from snowmelt season, are shown in Figure 6. In early May, when snowmelt was in progress, dFe concentrations in the surface soil pore waters were 1.38–4.70 mg L^{-1}, clearly higher than those in summer and autumn. These remarkably high dFe concentrations in the soil pore waters were consistent with the increases in dFe concentration in the Sofron River and the large rivers during snowmelt (Figure 4a). From spring to summer, dFe concentrations at N4 (5 cm and 10 cm depths) and N6 (10 cm and 25 cm depths) decreased.

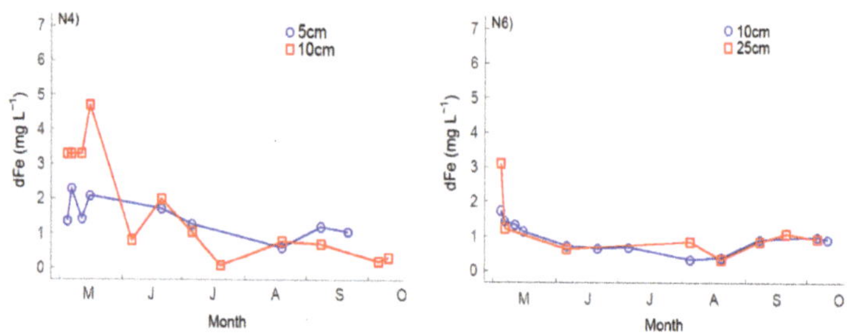

Figure 6. Seasonal changes in dFe concentrations in surface soil pore waters at N4 and N6, observed from the snowmelt season.

Seasonal changes in dFe concentrations in soil pore waters at the transect points where the observation started after snowmelt season are shown in Figure 7. The sampling depths at these points were 20 cm and 40 cm, which were larger than the depths at N4 and N6 observed in snowmelt season. dFe concentrations for both depths at each point except N3, were less than 1.00 mg L^{-1} for two months from the start of observation. However, from August to September, dFe concentrations at 20 cm and 40 cm depths in permafrost wetland *Mari* (N1–3 and N7–9) increased rapidly to 2.00–6.90 mg L^{-1}. In contrast, these increases in dFe concentrations were not observed at N6, where permafrost does not exist. According to the changes in soil temperature (Figure 5), it appears that the seasonal downward thawing of soil from 20 cm to 40 cm took place in May on ridge and hillslope and in June in valley. Thus, dFe production in these depths occurred two or three months after thawing. After that, these dFe concentrations in the soil pore waters decreased to less than 1.00 mg L^{-1} in October, when soil started to freeze from the top layer.

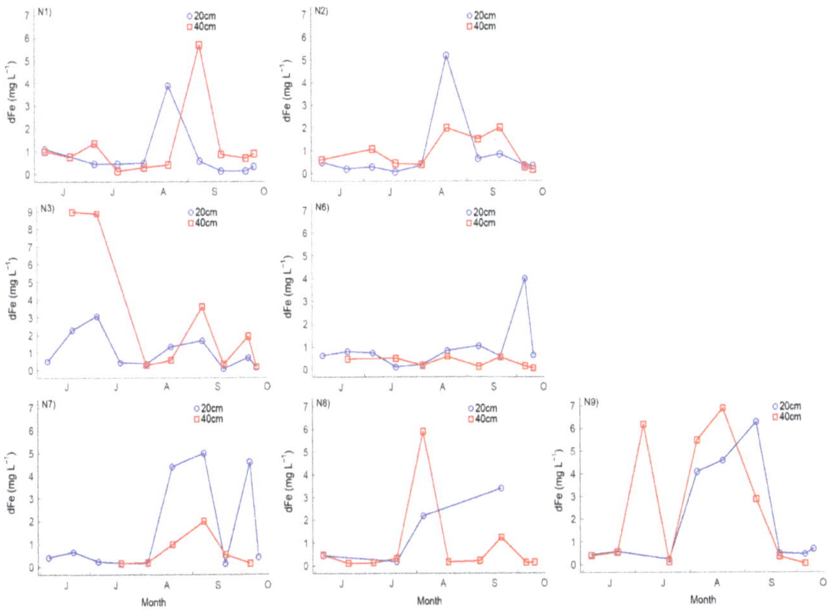

Figure 7. Seasonal changes in dFe concentrations in soil pore waters of 20 cm and 40 cm depths at the points along a transect. Note that the observation at these points started after snowmelt season. There is no result for N5, which is located on steep hillslope because the soil pore water could not be collected throughout the year.

4. Discussion

4.1. dFe Production in Uppermost Soils and Its Discharge into Rivers during Snowmelt Season

From late April to mid May, when snow was melting, dFe and DOC concentrations in the Sofron River rapidly increased, reaching the highest concentration in the year (Figure 3a). In accordance with this, surface soil pore waters also showed the highest dFe concentrations in this period (Figure 6). These elevated dFe concentrations in the Sofron River and the soil pore waters can be explained by snow melting and the existence of frozen ground at shallow depths. According to the seasonal changes in soil temperature (Figure 5), frozen ground existed at approximately 0–10 cm depth in valley and 10–25 cm depth in hillslope and ridge. It is known that frozen ground impedes downward infiltration of snowmelt water [35,36], resulting in a waterlogged condition near the soil surface. This was actually the case at N4 and N6 where soil pore waters were collected during this period. In soil saturated with snowmelt water, anoxic conditions could develop through dissolved oxygen consumption by microbial respiration [37,38]. Under such conditions, Fe(III) can be utilized as an electron acceptor by anaerobic microorganisms, and produced Fe(II) is probably complexed with DOC that seeps from ambient organic-rich soils (Table 2). Although we did not observe DOC concentrations in surface soil pore waters, the positive correlation between DOC and dFe concentrations in the Sofron River during snowmelt season (Figure 3c) suggests that exported dFe is mainly associated with organic compounds, as reported in other boreal regions [39–42]. The large amount of organically complexed iron produced in this way should be responsible for the high dFe concentrations in the uppermost soil pore waters (Figure 6) and the Sofron River (Figure 3a).

During snowmelt season, elevated dFe concentrations in surface soil pore waters were observed not only at N4 located on the boundary between valley and hillslope but also at N6 located on the ridge (Figure 6). In hillslope and ridge without permafrost, soil water usually infiltrates downward easily, and thus the uppermost soil is unlikely to be saturated. However, the seasonally frozen ground at a shallow depth would have led to waterlogging and dFe production in soil pore waters even in these terrains. It is thus expected that dFe production in waterlogged surface soils in this period may occur over a wide range of areas in these terrains where soil freezes and snow falls in winter.

However, in case frozen ground is formed with low water content in winter, snowmelt water may be able to infiltrate it because of discontinuous frozen soil [43,44]. It is therefore likely that, on ridge and hillslope, where the water content before freezing was relatively low (Table 2), snowmelt water probably flowed out while infiltrating frozen ground. Thus, even if snowmelt water on ridge and hillslope once had high dFe concentration, its contribution to the increase in dFe concentration in rivers would be minor. In contrast, in the valley, where the surface soils retained high water content before freezing (Table 2), snowmelt water would have flowed into the rivers over continuous frozen ground without infiltration. Consequently, the valley area most likely contributed to the rapid increase in dFe concentration in the rivers during the snowmelt season.

The rapid increases in dFe and DOC concentrations were also observed in the large rivers during snowmelt season (Figure 4a,b). It is thus suggested that the small watershed dominated by permafrost wetlands *Mari* like the Sofron river greatly contributes to the terrestrial DOC and dFe transport in this period. Moreover, both DOC and dFe concentrations in the large rivers during snowmelt season had the highest values in the year, similarly to those in the Sofron River, strongly indicating that snowmelt season is the most important period for transporting large amounts of DOC and dFe. Elevated dFe concentrations during snowmelt season were also observed in large tributaries of the Amur River [45,46], and other boreal rivers [39–41,47]. To our knowledge, the present study is the first to corroborate this large-scale dFe transport that resulted from the high dFe concentrations in the uppermost soil pore waters.

4.2. Seasonal Changes in Iron Behavior in Thawed Active Layer after Snowmelt Season

The dFe concentrations in surface soil pore waters at N4 and N6 declined after snowmelt season (Figure 6). As soil thawed downward towards summer, the surface soils most likely changed gradually to oxygenated conditions because of the decline of groundwater level. Such a situation that produces Fe(II) in pore waters should be oxidized chemically and/or biologically to insoluble Fe oxides [48], possibly resulting in the declines in dFe concentrations. The early declines in dFe concentrations of soil pore waters at N6 relative to that at N4 are explainable by the faster downward thawing of soil at N6 than N4 from spring to summer (Figure 5 and Table 2).

The dFe concentrations at 20 cm and 40 cm depths at the transect points, where observation was started in June, were less than 1 mg L^{-1} until August except N3, but increased markedly to 2–7 mg L^{-1} from August to September (Figure 7). This abrupt increase in dFe concentration would be associated with heavy rainfall from July to August (Figure 3b). The total amount of rainfall in July and August was 298.4 mm, which is equivalent to 45.6% of the annual average precipitation of 654.6 mm. Abundant rainfall would make subsurface (20–40 cm) soils saturated as permafrost would prevent infiltration of soil water. Under this situation, such electron acceptors as NO_3^-, Mn^{4+}, and humic substances would be utilized by being utilized for microbial anaerobic respiration [30,49–52]. As a result, microbial iron reduction was likely activated from August to September and large amounts of Fe(II) were produced in soil pore waters at 20 cm and 40 cm depths. Street et al. [53] observed that the redox potential (Eh) of soil in the active layer did not change markedly for one month after thawing but started to decline after the rainy season in summer, in agreement with our interpretation. Moreover, it was reported that iron reduction was facilitated in peat soils with artificial flooding relative to soils without this treatment [29]. It is thus likely that dFe production in soil prevails widely in permafrost wetland *Mari* after the rainy season. Only N6 did not show the increase in dFe concentrations in late summer, probably because the thawing of seasonally frozen soil at N6 was faster than at the other points in permafrost wetland *Mari*, leading to infiltration of soil water and an unsaturated condition even after abundant rainfall.

The elevated dFe concentrations in late summer subsequently decreased to approximately 1 mg L^{-1} towards October. The possible reason for this would be the decline of groundwater level in soils of the active layer in September when there was little heavy rainfall. Soil conditions probably changed to oxidative during this rainless season, and most of the produced Fe(II) in soil pore waters should have been oxidized to insoluble Fe oxyhydroxides by incoming oxygen and/or iron-oxidizing bacteria [54,55]. It was reported that reducible Fe oxyhydrox-ides exist in abundance in soils where oxidizing and reducing conditions are repeated [28,56], and the newly formed Fe oxyhydroxides will be reduced again by iron-reducing bacteria when reducing conditions are developed.

4.3. Seasonal Changes in dFe Discharge Mechanism Associated with Downward Shift of Flow Path after Snowmelt Season

After snowmelt season, groundwater table most likely declined as frozen ground thawed downward with rising air temperature. With this hydrological change from spring to summer, both dFe and DOC concentrations in the Sofron River and the large rivers decreased (Figure 4a,b). This agrees with the decline in dFe concentration in surface soil pore waters after snowmelt season (Figure 6). In addition, because the riverine dFe/DOC ratio also decreased in this period (Figure 4c), the decreases in dFe concentrations in the rivers after snowmelt season probably implies the decrease in discharge of organically complexed iron from the peat soil layer with the downward shift of flow path. This is supported by previous studies reporting that the discharge of DOC derived from humic substances decreased from spring to summer in the boreal region [57–59].

In the Sofron River, increases in dFe and DOC concentrations were observed in response to rising water levels due to heavy rainfall in late July and late September (Figure 3a,b). The elevated DOC concentrations were as high as those during snowmelt season, but the degree of these increases in dFe concentrations was less. Although storm events often increase dFe discharge from the upper organic-rich soil in boreal region [60], this result emphasizes the importance of snowmelt flooding to terrestrial dFe transport. The most likely explanation for the smaller increases in riverine dFe concentration due to heavy rainfall compared to snowmelt season is short period of waterlogged conditions in surface soils. It was reported that a high water level after heavy rainfall (~10 mm) lasts for only 1–2 days in permafrost peatland [61], whereas the waterlogged conditions due to snowmelt lasted for at least two weeks in this study. It is therefore suggested that summer rainfall causes the flushing of dFe into river, but, unlike the snowmelt season, it does not contribute to the develop of reducing conditions in surface peat soils. As mentioned in Section 4.2, the summer heavy rainfall probably contributed to the large amount of dFe production in subsurface soil pore waters in late summer (Figure 7), but, nevertheless, dFe concentration in the Sofron River hardly increased (Figure 3a). This finding is consistent with that of Street et al. [53], i.e., deep soil methane concentration in the active layer increased over time through the season but did not appear to influence stream methane concentration. Possible interpretation for the luck of increase in dFe concentration in the Sofron river in late summer would be a low discharge rate of subsurface soils. According to Quinton et al. [62], the hydraulic conductivity of peat soils below 20 cm depth is approximately less than one tenth of surface peat soils; therefore, even though a large amount of dFe is produced in subsurface soil pore waters, there might be little immediate influence on riverine dFe concentration.

Seasonal changes in dFe and DOC concentrations in the large rivers showed similar patterns to those in the Sofron River throughout the year; DOC variations in particular were exact matches (Figure 4a,b). In addition to snowmelt season, DOC and dFe concentrations in the large rivers during rainfall were also consistent with the elevated concentrations of both in the Sofron River. It is therefore suggested that small watershed with permafrost wetlands *Mari* like the Sofron River greatly contribute to terrestrial transport of DOC and dFe to the large rivers throughout the year, especially during snowmelt season as well as during the rainy season.

Interestingly, however, dFe concentration and dFe/DOC ratio in the large rivers after September, when there was less rainfall, were higher than those in the Sofron River (Figure 1a,c). The reason for this remains unclear, but likely explanation is the inflow of mineral-rich deep groundwater into overlying river through unfrozen ground (taliks) that commonly lies underneath the large rivers [33,63]. As river discharge decreases in rainless September, the contribution of such groundwater to river water might increase, which could explain the high dFe concentrations and the increases in dFe/DOC ratio in the large rivers in this period. dFe flowing into river in this way is believed to be Fe-bearing organic colloids produced in the hyporheic zone, where Fe(II)-rich groundwater and organic-rich river water are mixed [64,65]. It is therefore suggested that, for small rivers like the Sofron River, such deep ground water may not be important to chemical composition because of no talik. More studies are needed about why dFe concentrations in the large rivers are higher than those in the small river after September because it may be an important mechanism to support terrestrial dFe transport even in rainless periods when dFe concentrations in small rivers are low.

5. Conclusions

In this study, we observed dFe and DOC concentrations in rivers as well as dFe concentration in the soil pore waters during the water active period (from May to October) to understand the spatial and seasonal iron behavior on the watershed scale. To our knowledge, this study is the first to investigate the seasonal changes in dFe concentration in both river and soil pore water. The results highlight the following conclusions:

- In snowmelt season, high dFe production occur in the waterlogged surface soils, which leads to the largest terrestrial dFe transport in the year.
- Summer rainfall not only increases in dFe and DOC concentrations in river but probably has the effect of promoting dFe production in subsurface soils of permafrost wetlands in valley area.
- Overall, permafrost wetlands in valley areas are important environment in which dFe production occurs in response to seasonal hydrological events (spring snowmelt and summer rainfall) and soil thaw depth, and play a significant role in supplying dFe and DOC to rivers.

Author Contributions: Conceptualization, M.Y., T.S., T.O., V.S., and Y.Y.; methodology, M.Y., T.S., and T.O.; formal analysis, V.S. and Y.Y.; investigation, Y.T., M.Y., and T.O.; data curation, Y.T.; writing—original draft preparation, Y.T.; writing—review and editing, M.Y., T.S., and T.O.; visualization, Y.T.; supervision, T.S., V.S., and V.K.; project administration, T.S., V.S., and V.K.; funding acquisition, T.S. All authors have read and agreed to the published version of the manuscript.

Funding: This research was funded by the Japan Society for the Promotion of Science (JSPS) KAKENHI Grant No. JP15H05208.

Acknowledgments: We would like to thank S. I. Levshina, A. L. Antonov, D. V. Matveenko, and R. S. Kozyrev for supporting field research and sample analyses in the laboratory. We are also grateful to D. N. Markina for mediating this work as an interpreter. We thank N. P. Mihailov and V. G. Stebunov for their help in field research as a guide and for the regular sampling of soil pore waters and river waters in the Tyrma region.

Conflicts of Interest: The authors declare no conflict of interest.

References

1. Sunda, W.G. Feedback interactions between trace metal nutrients and phytoplankton in the ocean. *Front. Microbiol.* **2012**, *3*, 204. [CrossRef] [PubMed] [CrossRef]
2. Martin, J.H.; Fitzwater, S.E. Iron deficiency limits phytoplankton growth in the North-east Pacific subarctic. *Nature* **1988**, *336*, 403–405. [CrossRef] [CrossRef]
3. Martin, J.H.; Gordon, R.M.; Fitzwater, S.E.; Broenkow, W.W. Vertex: Phytoplankton/iron studies in the Gulf of Alaska. *Deep Sea Res. Part A* **1989**, *36*, 649–680. [CrossRef] [CrossRef]
4. Martin, J.H.; Gordon, R.; Fitzwater, S.E. Iron in Antarctic waters. *Nature* **1990**, *345*, 156–158. [CrossRef] [CrossRef]
5. Martin, J.H.; Coale, K.H.; Johnson, K.S.; Fitzwater, S.E.; Gordon, R.M.; Tanner, S.J.; Hunter, C.N.; Elrod, V.A.; Nowick, J.L.; Coley, T.L.; et al. Testing the iron hypothesis in ecosystems of the equatorial Pacific Ocean. *Nature* **1994**, *371*, 123–129. [CrossRef] [CrossRef]

6. Price, N.M.; Ahner, B.A.; Morel, F.M.M. The equatorial Pacific Ocean: Grazer-controlled phytoplankton populations in an iron-limited ecosystem. *Limnol. Oceanogr.* **1994**, *39*, 520–534. [CrossRef] [CrossRef]
7. Takeda, S.; Obata, H. Response of equatorial Pacific phytoplankton to subnanomolar Fe enrichment. *Mar. Chem.* **1995**, *50*, 219–227. [CrossRef] [CrossRef]
8. Bruland, K.W.; Lohan, M.C. Controls of Trace Metals in Seawater. *Treatise Geochem.* **2003**, *6*, 23–47.
9. Matsunaga, K.; Nishioka, J.; Kuma, K.; Toya, K.; Suzuki, Y. Riverine input of bioavailable iron supporting phytoplankton growth in Kesennuma Bay (Japan). *Water Res.* **1998**, *32*, 3436–3442. [CrossRef] [CrossRef]
10. Laglera, L.M.; Vandenberg, C.M.G. Evidence for geochemical control of iron by humic substances in seawater. *Limnol. Oceanogr.* **2009**, *54*, 610–619. [CrossRef] [CrossRef]
11. Nishioka, J.; Nakatsuka, T.; Ono, K.; Volkov, Y.N.; Scherbinin, A.; Shiraiwa, T. Quantitative evaluation of iron transport processes in the Sea of Okhotsk. *Prog. Oceanogr.* **2014** *126*, 180–193. [CrossRef]
12. Boyle, E.A.; Edmond, J.M.; Sholkovitz, E.R. The mechanism of iron removal in estuaries. *Geochim. Cosmochim. Acta* **1977**, *41*, 1313–1324. [CrossRef] [CrossRef]
13. Morel, F.M.M.; Kustka, A.B.; Shaked, Y. The role of unchelated Fe in the iron nutrition of phytoplankton. *Limnol. Oceano.* **2008**, *53*, 400–404. [CrossRef] [CrossRef]
14. Kranzler, C.; Lis, H.; Shaked, Y.; Keren, N. The role of reduction in iron uptake processes in a unicellular, planktonic cyanobacterium. *Environ. Microbiol.* **2011**, *13*, 2990–2999. [CrossRef] [CrossRef] [PubMed]
15. Morrissey, J.; Bowler, C. Iron utilization in marine cyanobacteria and eukaryotic algae. *Front. Microbiol.* **2012**, *3*, 1–13. [CrossRef] [PubMed] [CrossRef] [PubMed]
16. Krachler, R.; Jirsa, F.; Ayromlou, S. Factors influencing the dissolved iron input by river water to the open ocean. *Biogeosciences* **2005**, *2*, 311–315. [CrossRef] [CrossRef]
17. Krachler, R.; Krachler, R.F.; Von Der Kammer, F.; Süphandag, A.; Jirsa, F.; Ayromlou, S.; Hofmann, T.; Keppler, B.K. Science of the Total Environment Relevance of peat-draining rivers for the riverine input of dissolved iron into the ocean. *Sci. Total Environ.* **2010**, *408*, 2402–2408. [CrossRef] [PubMed] [CrossRef]
18. Klunder, M.B.; Bauch, D.; Laan, P.; De Baar, H.J.W.; Van Heuven, S.; Ober, S. Dissolved iron in the Arctic shelf seas and surface waters of the central Arctic Ocean: Impact of Arctic river water and ice-melt. *J. Geophys. Res. Ocean.* **2012**, *117*, 1–18. [CrossRef] [CrossRef]
19. Suzuki, K.; Hattori-Saito, A.; Sekiguchi, Y.; Nishioka, J.; Shigemitsu, M.; Isada, T.; Liu, H.; McKay, R.M.L. Spatial variability in iron nutritional status of large diatoms in the Sea of Okhotsk with special reference to the Amur River discharge. *Biogeosciences* **2014**, *11*, 2503–2517. [CrossRef] [CrossRef]
20. Shiraiwa, T. "Giant Fish-Breeding Forest": A New Environmental System Linking Continental Watershed with Open Water. In *The Dilemma of Boundaries*; Taniguchi, M., Shiraiwa, T., Ed.; Springer: Tokyo, Japan, 2012; pp. 73–85.
21. Pan, X.F.; Yan, B.X.; Yoh, M. Effects of land use and changes in cover on the transformation and transportation of iron: A case study of the Sanjiang Plain, Northeast China. *Sci. China Earth Sci.* **2011**, *54*, 686–693. [CrossRef] [CrossRef]
22. Shamov, V.V.; Onishi, T.; Kulakov, V.V. Dissolved Iron Runoff in Amur Basin Rivers in the Late XX Century. *Water Resour.* **2014**, *41*, 201–209. [CrossRef]
23. Gu, B.; Schmitt, J.; Chen, Z.; Liang, L.; McCarthy, J.F. Adsorption and desorption of natural organic matter on iron oxide: Mechanisms and models. *Environ. Sci. Technol.* **1994**, *28*, 38–46. [PubMed] [CrossRef]
24. Gu, B.; Schmitt, J.; Chen, Z.; Liang L.; McCarthy, J.F. Adsorption and desorption of different organic matter fractions on iron oxide. *Geochim. Cosmochim. Acta* **1995**, *59*, 219–229. [CrossRef]
25. Baldock, J.A.; Skjemstad, J.O. Role of the soil matrix and minerals in protecting natural organic materials against biological attack. *Org. Geochem.* **2000**, *31*, 697–710. [CrossRef]
26. Roden, E.E.; Wetzel, R.G. Kinetics of microbial Fe (III) oxide reduction in freshwater wetland sediments. *Limnol. Oceanogr.* **2002**, *47*, 198–211. [CrossRef]
27. Miller, K.E.; Lai, C.T.; Friedman, E.S.; Angenent, L.T.; Lipson, D.A. Methane suppression by iron and humic acids in soils of the Arctic Coastal Plain. *Soil Biol. Biochem.* **2015**, *83*, 176–183. [CrossRef]
28. Lipson, D.A.; Jha, M.; Raab, T.K.; Oechel, W.C. Reduction of iron (III) and humic substances plays a major role in anaerobic respiration in an Arctic peat soil. *J. Geophys. Res. Biogeosci.* **2010**, *115*, 1–13. [CrossRef]
29. Lipson, D.A.; Zona, D.; Raab, T.K.; Bozzolo, F.; Mauritz, M.; Oechel, W.C. Water-table height and microtopography control biogeochemical cycling in an Arctic coastal tundra ecosystem. *Biogeosciences* **2012**, *9*, 577–591. [CrossRef]

30. Lipson, D.A.; Raab, T.K.; Goria, D.; Zlamal, J. The contribution of Fe(III) and humic acid reduction to ecosystem respiration in drained thaw lake basins of the Arctic Coastal Plain. *Glob. Biogeochem. Cycles* **2013**, *27*, 399–409. [CrossRef]
31. Herndon, E.M.; Yang, Z.; Bargar, J.; Janot, N.; Regier, T.Z.; Graham, D.E.; Wullschleger, S.D.; Gu, B.; Liang, L. Geochemical drivers of organic matter decomposition in arctic tundra soils. *Biogeochemistry* **2015**, *126*, 397–414. [CrossRef]
32. Herndon, E.; Albashaireh, A.; Singer, D.; Roy, T.; Gu, B.; Graham, D. Influence of iron redox cycling on organo-mineral associations in Arctic tundra soil. *Geochim. Cosmochim. Acta* **2017**, *207*, 210–231. [CrossRef] [CrossRef]
33. Bagard, M.L.; Chabaux, F.; Pokrovsky, O.S.; Viers, J.; Prokushkin, A.S.; Stille, P.; Rihs, S.; Schmitt, A.; Dupré, B. Seasonal variability of element fluxes in two Central Siberian rivers draining high latitude permafrost dominated areas. *Geochim. Cosmochim. Acta* **2011**, *75*, 3335–3357. [CrossRef] [CrossRef]
34. Bel'chikova, N.P. Determination of humus in soil by I.V. Tyurin method. In *Agrochemical Methods of Soil Studies*; Sokolov, A.V., Ed.; Nauka: Moscow, Russia, 1975; pp. 56–62. (In Russian)
35. Thunholm, B.; Lundln, L.; Lindell, S.; Meteorological, S. Infiltation into a Frozen Heavy Clay Soil. *Hydrol. Res.* **1989**, *20*, 153–166. [CrossRef] [CrossRef]
36. Stähli, M.; Jansson, P.E.; Lundin, L.C. Preferential water flow in a frozen soil—A two-domain model approach. *Hydrol. Process.* **1996**, *10*, 1305–1316. [CrossRef] [CrossRef]
37. Seto, M.; Akagi, T. Influence of snow on iron release from soil. *Geochem. J.* **2005**, *39*, 173–183. [CrossRef] [CrossRef]
38. Morse, J.L.; Durán, J.; Groffman, P.M. Soil Denitrification Fluxes in a Northern Hardwood Forest: The Importance of Snowmelt and Implications for Ecosystem N Budgets. *Ecosystems* **2015**, *18*, 520–532. [CrossRef] [CrossRef]
39. Rember, R.D.; Trefry, J.H. Increased concentrations of dissolved trace metals and organic carbon during snowmelt in rivers of the alaskan arctic. *Geochim. Cosmochim. Acta* **2004**, *68*, 477–489. [CrossRef] [CrossRef]
40. Andersson, K.; Dahlqvist, R.; Turner, D.; Stolpe, B.; Larsson, T.; Ingri, J.; Andersson, P. Colloidal rare earth elements in a boreal river: Changing sources and distributions during the spring flood. *Geochim. Cosmochim. Acta* **2006**, *70*, 3261–3274. [CrossRef] [CrossRef]
41. Ingri, J.; Malinovsky, D.; Rodushkin, I.; Baxter, D.C. Iron isotope fractionation in river colloidal matter. *Earth Planet. Sci. Lett.* **2006**, *245*, 792–798. [CrossRef] [CrossRef]
42. Björkvald, L.; Buffam, I.; Laudon, H.; Mörth, C.M. Hydrogeochemistry of Fe and Mn in small boreal streams: The role of seasonality, landscape type and scale. *Geochim. Cosmochim. Acta* **2008**, *72*, 2789–2804. [CrossRef] [CrossRef]
43. Iwata, Y.; Hasegawa, S.; Suzuki, S.; Nemoto, M.; Hirota, T. Effects of soil frost depth and soil temperature on downward soil water movement during snowmelt period. *J. Jpn. Soc. Soil Phys.* **2011**, *117*, 11–21. (In Japanese)
44. Ågren, A.; Buffam, I.; Berggren, M.; Bishop, K.; Jansson, M.; Laudon, H. Dissolved organic carbon characteristics in boreal streams in a forest-wetland gradient during the transition between winter and summer. *J. Geophys. Res. Biogeosci.* **2008**, *113*, 1–11. [CrossRef] [CrossRef]
45. Nagao, S.; Motoki, T.; Kodama H.; Kim V.I.; Shesterkin P.V.; Makhinov A.N. Migration Behavior of Fe in the Amur River Basin. *Rep. Amur Okhotsk Proj.* **2007**, *4*, 37–48.
46. Guan, J.; Yan, B.; Yuan, X. Variations of total dissolved iron and its impacts during an extreme spring flooding event in the Songhua River. *J. Geochem. Explor.* **2016**, *166*, 27–32. [CrossRef] [CrossRef]
47. Ingri, J.; Conrad, S.; Lidman, F.; Nordblad, F.; Engström, E.; Rodushkin, I.; Porcelli, D. Iron isotope pathways in the boreal landscape: Role of the riparian zone. *Geochim. Cosmochim. Acta* **2018**, *239*, 49–60. [CrossRef]
48. Weber, K.A.; Achenbach, L.A.; Coates, J.D. Microorganisms pumping iron: Anaerobic microbial iron oxidation and reduction. *Nat. Rev. Microbiol.* **2006**, *4*, 752–764. [CrossRef] [CrossRef]
49. Lovley, D.R.; Coates, J.D.; Blunt-Harris, E.L.; Phillips, E.J.P.; Woodward, J.C. Humic substances as electron acceptors for microbial respiration. *Nature* **1996**, *382*, 445–448. [CrossRef] [CrossRef]
50. Benz, M.; Schink, B.; Brune, A. Humic acid reduction by Propionibacterium freudenreichii and other fermenting bacteria. *Appl. Environ. Microbiol.* **1998**, *64*, 4507–4512. [CrossRef] [CrossRef]

51. Coates, J.D.; Ellis, D.J.; Roden, E.; Gaw, K.; Blunt-Harris, E.L.; Lovley, D.R. Recovery of humics-reducing bacteria from a diversity of sedimentary environments. *Appl. Environ. Microbiol.* **1998**, *64*, 1504–1509. [CrossRef] [CrossRef]
52. Kappler, A.; Benz, M.; Schink, B.; Brune, A. Electron shuttling via humic acids in microbial iron(III) reduction in a freshwater sediment. *FEMS Microbiol. Ecol.* **2004**, *47*, 85–92. [CrossRef] [CrossRef]
53. Street, L.E.; Lessels, J.S.; Subke, J.; Tetzlaff, D.; Wookey, P.A. Redox dynamics in the active layer of an Arctic headwater catchment; examining the potential for transfer of dissolved methane from soils to stream water. *J. Geophys. Res. Biogeosci.* **2016**, *121*, 2776–2792. [CrossRef]
54. Fuss, C.B.; Driscoll, C.T.; Johnson, C.E.; Petras, R.J.; Fahey, T.J. Dynamics of oxidized and reduced iron in a northern hardwood forest. *Biogeochemistry* **2010**, *104*, 103–119. [CrossRef]
55. Lüdecke, C.; Reiche, M.; Eusterhues, K.; Nietzsche, S.; Küsel, K.; Schiller, F. Acid-tolerant microaerophilic Fe(II)-oxidizing bacteria promote Fe(III)-accumulation in a fen. *Environ. Microbiol.* **2010**, *12*, 2814–2825. [CrossRef]
56. Poulton, S.W.; Canfield, D.E. Development of a sequential extraction procedure for iron: Implications for iron partitioning in continentally derived particulates. *Chem. Geol.* **2005**, *214*, 209–221. [CrossRef]
57. Ågren, A.; Buffam, I.; Jansson, M.; Laudon, H. Importance of seasonality and small streams for the landscape regulation of dissolved organic carbon export. *J. Geophys. Res. Biogeosci.* **2007**, *112*, 1–11. [CrossRef]
58. Prokushkin, A.S.; Pokrovsky, O.S.; Shirokova, L.S.; Korets, M.A.; Viers, J.; Prokushkin, S.G.; Amon, R.M.W.; Guggenberger, G.; McDowell, W.H. Sources and the flux pattern of dissolved carbon in rivers of the Yenisey basin draining the Central Siberian Plateau. *Environ. Res. Lett.* **2011**, *6*,1–14. [CrossRef]
59. Koch, J.C.; Runkel, R.L.; Striegl, R.; McKnight, D.M. Hydrologic controls on the transport and cycling of carbon and nitrogen in a boreal catchment underlain by continuous permafrost. *J. Geophys. Res. Biogeosci.* **2013**, *118*, 698–712. [CrossRef]
60. Abesser, C.; Robinson, R.; Soulsby, C. Iron and manganese cycling in the storm runoff of a Scottish upland catchment. *J. Hydrol.* **2006**, *326*, 59–78. [CrossRef]
61. Quinton, W.L.; Baltzer, J.L. The active-layer hydrology of a peat plateau with thawing permafrost (Scotty Creek , Canada). *Hydrogeol. J.* **2013**, *21*, 201–220. [CrossRef]
62. Quinton, W.L.; Hayashi, M.; Carey, S.K. Peat hydraulic conductivity in cold regions and its relation to pore size and geometry. *Hydrol. Process.* **2008**, *22*, 2829–2837. [CrossRef]
63. Watson, V.; Kooi, H.; Bense, V. Potential controls on cold-season river flow behavior in subarctic river basins of Siberia. *J. Hydrol.* **2013**, *489*, 214–226. [CrossRef]
64. Ilina, S.M.; Poitrasson, F.; Lapitskiy, S.A.; Alekhin, Y.V. Extreme iron isotope fractionation between colloids and particles of boreal and temperate organic-rich waters. *Geochim. Cosmochim. Acta* **2013**, *101*, 96–111. [CrossRef]
65. Pokrovsky, O.S.; Manasypov, R.M.; Loiko, S.V.; Krickov, I.A.; Kopysov, S.G.; Kolesnichenko, L.G.; Vorobyev, S.V.; Kirpotin, S.N. Trace element transport in western Siberian rivers across a permafrost gradient. *Biogeosciences* **2016**, *13*, 1877–1900. [CrossRef]

water

MDPI

Article

Spatial and Seasonal Variations of C, Nutrient, and Metal Concentration in Thermokarst Lakes of Western Siberia Across a Permafrost Gradient

Rinat M. Manasypov [1,*], Artem G. Lim [1], Ivan V. Krickov [1], Liudmila S. Shirokova [2,3], Sergey N. Vorobyev [1], Sergey N. Kirpotin [1] and Oleg S. Pokrovsky [1,2,3]

1 BIO-GEO-CLIM Laboratory, Tomsk State University, 36 Lenina av., 634004 Tomsk, Russia; lim_artyom@mail.ru (A.G.L.); krickov_ivan@mail.ru (I.V.K.); soil@green.tsu.ru (S.N.V.); kirp@mail.tsu.ru (S.N.K.); oleg.pokrovski@get.omp.eu (O.S.P.)

2 Federal Center for Integrated Arctic research, Institute of Ecological Problem of the North, 23 Nab Severnoi Dviny, 163000 Arkhangelsk, Russia; lshirocova@ya.ru

3 GET UMR 5563 CNRS University of Toulouse, 14 Avenue Edouard Belin, 31400 Toulouse, France

* Correspondence: rmmanassypov@gmail.com

Received: 26 May 2020; Accepted: 22 June 2020; Published: 26 June 2020

Abstract: Thermokarst lakes and ponds formed due to thawing of frozen peat in high-latitude lowlands are very dynamic and environmentally important aquatic systems that play a key role in controlling C emission to atmosphere and organic carbon (OC), nutrient, and metal lateral export to rivers and streams. However, despite the importance of thermokarst lakes in assessing biogeochemical functioning of permafrost peatlands in response to climate warming and permafrost thaw, spatial (lake size, permafrost zone) and temporal (seasonal) variations in thermokarst lake hydrochemistry remain very poorly studied. Here, we used unprecedented spatial coverage (isolated, sporadic, discontinuous, and continuous permafrost zone of the western Siberia Lowland) of 67 lakes ranging in size from 10^2 to 10^5 m^2 for sampling during three main hydrological periods of the year: spring flood, summer baseflow, and autumn time before ice-on. We demonstrate a systematic, all-season decrease in the concentration of dissolved OC (DOC) and an increase in SO_4, N-NO_3, and some metal (Mn, Co, Cu, Mo, Sr, U, Sb) concentration with an increase in lake surface area, depending on the type of the permafrost zone. These features are interpreted as a combination of (i) OC and organically bound metal leaching from peat at the lake shore, via abrasion and delivery of these compounds by suprapermafrost flow, and (ii) deep groundwater feeding of large lakes (especially visible in the continuous permafrost zone). Analyses of lake water chemical composition across the permafrost gradient allowed a first-order empirical prediction of lake hydrochemical changes in the case of climate warming and permafrost thaw, employing a substituting space for time scenario. The permafrost boundary shift northward may decrease the concentrations and pools of dissolved inorganic carbon (DIC), Li, B, Mg, K, Ca, Sr, Ba, Ni, Cu, As, Rb, Mo, Sr, Y, Zr, rare Earth elements (REEs), Th, and U by a factor of 2–5 in the continuous permafrost zone, but increase the concentrations of CH_4, DOC, NH_4, Cd, Sb, and Pb by a factor of 2–3. In contrast, the shift of the sporadic to isolated zone may produce a 2–5-fold decrease in CH_4, DOC, NH_4, Al, P, Ti, Cr, Ni, Ga, Zr, Nb, Cs, REEs, Hf, Th, and U. The exact magnitude of this response will, however, be strongly seasonally dependent, with the largest effects observable during baseflow seasons.

Keywords: permafrost; lake; trace metals; carbon; season; thermokarst; Western Siberia

1. Introduction

The impact of ongoing climate change on the functioning of aquatic ecosystems and biogeochemical cycles of chemical elements poses the main environmental threat in the arctic and subarctic regions [1].

Western Siberia is a key region and the most convenient platform to study the fundamental issues of climate–permafrost interaction, examine the applied aspects of these changes, and assess their possible social impacts [2–4]. Thawing of frozen peatlands under increasing air and water temperature and greenhouse gas (GHG) concentrations are the most crucial issues related to behavior of aquatic ecosystems in the north of Western Siberia, the largest permafrost peatland in the world.

The permafrost-affected part of the West Siberian Lowland (WSL) covers an area of 1.05 million km^2 [5]. The majority of lakes here are of thermokarst origin, formed during thawing of frozen peat bogs, similar to other arctic and subarctic regions of the northern hemisphere. The lake coverage of the lowland can attain 40–60% in some areas. Recent estimates obtained using high-resolution satellite imagery show that the total area of the lakes amounts to 61,900 km^2 [5], while the total number of thermokarst lakes of the WSL permafrost zone is 727,700 [6]. Thermokarst lakes in Western Siberia are the key components of GHG exchange between surface waters and the atmosphere [7]. Such lakes represent highly dynamic water systems [8–11], which are substantially fed by atmospheric precipitation [12,13]. Thermokarst lakes, as the most common water bodies in the arctic and subarctic regions, are also the main pools [14,15] and sources [16] of dissolved metals and dissolved organic carbon (DOC), and they play a crucial role for the adjacent hydrological network during flooding or drainage [17,18]. The surface of frozen peat bogs in the north of Western Siberia is smooth and flat; therefore, the water surface area of lakes is almost equal to the catchment area [13]. Most of the chemical elements and DOC enter the lakes during coastal abrasion of the reactive frozen peat [13,18,19], via surface runoff in the snowmelt period [12], via atmospheric transport [20], and after ground fires [21].

In high-latitude peatlands, thermokarst (thaw) lakes are formed due to ice melting and soil subsidence in frozen wetlands as a result of thermokarst processes [22–31]. Former studies of the history of thermokarst lakes revealed a cyclic nature of their formation and development [3,18,32–35], although these cycles are not observed in all circumpolar regions (i.e., References [25,36–39]). In the WSL, the cycle starts when frozen peat subsides due to warming, and a depression filled by thaw water is formed in the moss/lichen cover [3,18]. After some time, the depression attains the size of a lake due to ongoing subsidence of peat soil and degradation of its unstable edges. A mature thermokarst lake can eventually drain into another lake or into a hydrological system (local streams and rivers). The bottom of the drained lake is covered with gramineae-like plant communities alternated with mosses and lichens. The soil gradually heaves due to renewed permafrost, and small isolated mounds form and merge into a flat uniform palsa plateau. Another wet depression at the former basin of the thermokarst lake can start a new cycle. In this regard, the size of the water body can be indicative of the lake maturity, with small actively growing thaw ponds being a few years to a few decades old, and large, mature lakes with stable borders persisting for centuries (i.e., Reference [28]).

Thermokarst lakes in Western Siberia feature shallow depth (0.5–1.5 m), complete freezing in winter [40], and strong contact of thick sediments with overlaying water column via exchange of gases and solutes between porewaters and the water column [7,32]. Due to their shallow depth and unshaded (from the wind) position, the lakes lack vertical stratification in the water temperature, O_2, and dissolved elements. Moreover, thermokarst lakes are typically isolated and not connected to hydrological networks. The watershed area of thermokarst lakes is typically very small, and it rarely exceeds the open water surface area. As a result, two main sources of solutes to the lake are peat abrasion/vegetation litter degradation on the lake shore, and subsoil inflow of the suprapermafrost ground waters above the frozen peat at the depth of 100 to 40 cm, depending on the active layer thickness (ALT). During the open water period from May/June to October, this suprapermafrost water originates from the rain/snowmelt water penetrating through moss and lichen layer and thawed peat ice, following progressively increasing ALT.

Thermokarst lakes in Western Siberia were studied primarily in summer [13,18,19,32,41–44], and the seasonal dynamics of the elemental composition was studied fragmentarily in one selected site of discontinuous permafrost zone [40]. As such, the seasonal dynamics of the elemental composition of lake water should be studied over a more extensive latitudinal transect of Western Siberia, encompassing several permafrost zones. By analogy with lakes of permafrost peatlands in Canada [12,26] or northern Sweden [45,46], one may expect that the hydrochemical pattern of lake water in WSL will follow the main hydrological events during the "active" (unfrozen) period of the year [47]: dilution by snowmelt in spring, evaporation and autochthonous biological processes in summer when the lake is fed by suprapermafrost waters, and beginning of freezing and sediment–bottom water exchange in autumn. However, it is not clear how and to which extent these processes will affect lake water chemical composition across small depressions, thaw ponds, and large thermokarst lakes abundant in different permafrost zones.

Ongoing climate warming in the Arctic can affect natural zones and cause spread of permafrost northward [48–50]. To predict changes in the elemental composition of surface waters [13,16,51,52], it is necessary to study the latitudinal gradient of the concentration of C, nutrients, and trace elements in water bodies. Generally, a latitudinal gradient allows using a "substituting space for time" approach as tested in Western Siberia for rivers [53–55], when contemporary spatial phenomena can be used to model future events (i.e., Reference [56]).

This work is aimed at getting new insights into the processes of formation of the lake water elemental composition across four permafrost zones of Western Siberia (from isolated to continuous permafrost zone), assessing a within-year variability of lake hydrochemistry, and predicting the evolution of elemental composition under ongoing climate change. For this, we addressed the following questions: (1) What are the relationships between element concentration and water surface area in different permafrost zones?; (2) How do the concentration and stock of DOC, DIC, nutrients, and trace elements in lake waters respond to the change in the main hydrological seasons?; (3) How do the concentrations and stocks of DOC, nutrients, and trace elements in thermokarst lakes change throughout the latitudinal profile of Western Siberia, and how may the elemental composition of the lake water respond to ALT increase and the permafrost boundary shift?

To shed light on these issues, we performed thorough seasonal measurements across all permafrost zones of the WSL in 2016. We studied major components of the C biogeochemical cycle in high-latitude water bodies—dissolved organic and inorganic carbon. Furthermore, we measured the concentrations and discussed the behavior of major anions (Cl, SO_4), cations (Na, Mg, Ca, K), macro- (N, P, Si) and micro-nutrients (B, Mn, Fe, Co, Ni Cu, Zn, Rb, Mo, Ba), toxicants (As, Cd, Sb, Pb), and geochemical tracers (Li, Al, Ti, Cr, Ga, Sr, Y, rare Earth elements (REEs), Zr, Cs, Nb, Hf, Th, and U). As such, we covered all individual solutes of both ecological and general geochemical significance in the dissolved (<0.45 μm) fraction of the lake water. This set of data is novel compared to previous assessment of WSL thermokarst lake chemical composition performed in summer 2016 in the northern part of the lowland [13] and over four hydrological seasons in 2014 on a limited number of lakes in the sporadic permafrost zone [40].

2. Study Site Description, Sampling, Analytical and Statistical Methods

2.1. Study Site Description

The West Siberian lowland is covered by homogeneous taiga, forest–tundra, and tundra landscapes. The thermokarst lakes are developed in sporadic, discontinuous, and continuous permafrost zones, whereas raised bogs and ridge–pool complexes occur in isolated and sporadic permafrost zones [57]. The thermokarst lakes of Western Siberia are formed due to thawing and subsidences of frozen peatlands. The general cycles of their development are described elsewhere [2,3,32]. In the studied territory, the thermokarst lakes are shallow, isometric water bodies with sharp, rather abrupt shorelines. The mature thermokarst lake has stable frozen shores and it occupies clearly distinguishable depression in frozen peatlands. The height of the lake border can achieve 1.5 m which is comparable with the height of frozen mounds in the surrounding palsa peatbogs. The change of hydrological seasons leads to the change in lake depth due to evaporation and atmospheric input.

Lake water was sampled along the permafrost transect of Western Siberia, which includes four permafrost zones with four key sites in these zones: the southern site in the environs of the Kogalym town, taiga biome of the isolated permafrost zone; the environs of the Khanymey village, northern taiga biome of the sporadic permafrost zone; the environs of the Urengoy town, forest–tundra biome of the discontinuous permafrost zone; and most northern key site in the environs of the Tazovsky town, tundra biome of the continuous permafrost zone (Figure 1). In the description of the results obtained, the key study sites are referred to in accordance with the permafrost zones. The four key study sites are located essentially on watershed divides within the catchments of the Ob, Nadym, Pur, and Taz rivers. The average annual temperatures in these four territories are −4.0, −5.6, −6.4, and −9.1 °C when going from south to north [58]. During summer sampling, the average thickness of the active layer ranged from 200–300 cm in the south to 65 cm in the north [50]. A more detailed description of the study sites is provided elsewhere [7,13,49,50,59,60].

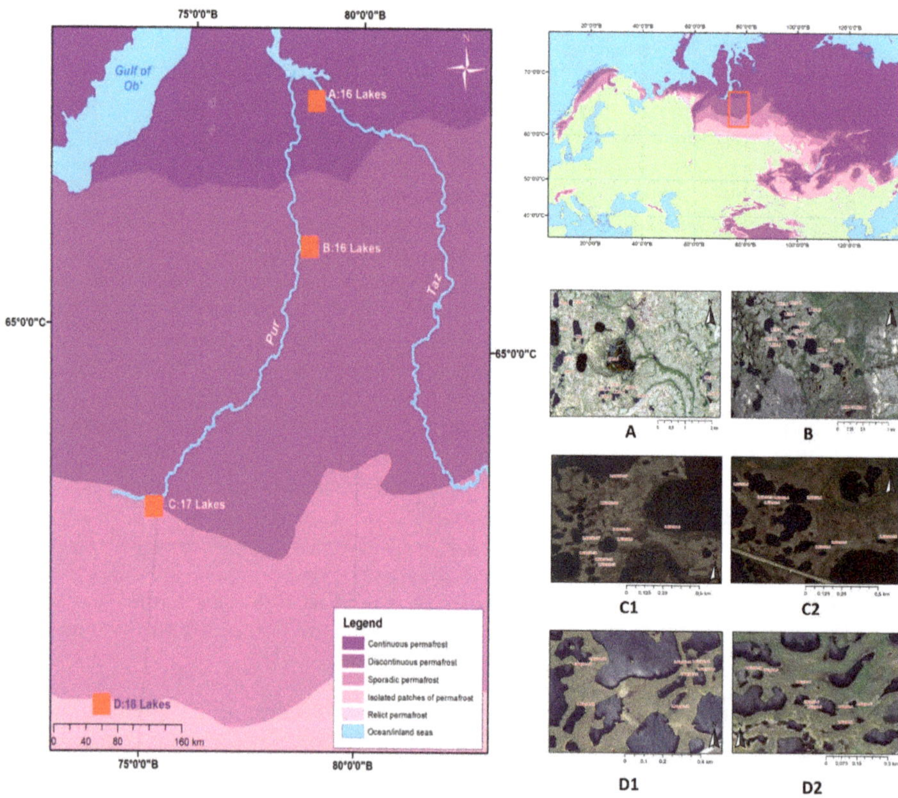

Figure 1. Map of the study area in the Western Siberia Lowland (WSL), Russia. Color shading represents different permafrost zones of the WSL according to ref. [61]. Red squares indicate the studied sites. Panels (**A**), (**B**), (**C1**,**C2**), and (**D1**,**D2**) refer to the sites in the continuous, discontinuous, sporadic, and isolated permafrost zones, respectively. The inserts (**A**,**B**,**C1**,**C2**,**D1**,**D2**) are from Google Maps®.

In the latitudinal transect of Western Siberia, we studied 67 lakes during three main hydrological seasons—melting of lake ice cover in spring, base flow in summer, and at the beginning of lake freeze-up in autumn. Specifically, we worked in 2016 at the ice-off event (20 May–13 June 2016), in the middle of the summer (8–24 August 2016), and just before lakes were ice-covered (25 September–8

October 2016). The sampling was performed in all four permafrost zones and included 18, 17, 16, and 16 lakes in the isolated, sporadic, discontinuous, and continuous permafrost zones, respectively.

The studied lakes are mostly rounded flat-concave shallow basins [7,13,62]. All the lakes have small catchments, with a watershed area almost similar to the area of the lake. The size of the lakes varies from several hundred meters to 1 km^2 across all the permafrost zones. The studied lakes are characterized by shallow depth that varies significantly in different seasons since the main source of water supply for lakes is atmospheric precipitation and spring snowmelt [12,19,40], and parts of the small water bodies dry out in summer. The depth of the studied lakes in spring, during active snowmelt and abundant precipitation, was 122 ± 35 cm for the isolated zone, 50 ± 20 cm for the sporadic zone, 81 ± 37 cm for the discontinuous zone, and 140 ± 87 cm for the continuous permafrost zone. In summer, the depth of the studied lakes was observed to significantly decrease from the isolated to continuous zone and amounted to 90 ± 35, 36 ± 25, 45 ± 27, and 125 ± 77 cm, respectively. During autumn ice-on period and heavy rains, the depth of the lakes insignificantly increased, which was especially characteristic of sporadic and discontinuous zones (Figure S1A, Supplementary Materials).

The average temperature of lake water in the four key study sites ranged from 9.3 to 19.3 °C in the spring, 17 to 21.4 °C in the summer, and 3 to 8.1 °C in the autumn (Figure S1B, Supplementary Materials). There was no sizable temperature stratification between the most surface (0–20 cm) and most deep (0–10 m above sediments) water layers. All the lakes studied are characterized by high oxygen content, especially in the autumn (Figure S1C, Supplementary Materials). In order to obtain comparable results for the same season across full permafrost gradient, in spring, we moved from the south to the north, progressively sampling the lakes, following the northward advancement of the snowmelt. In summer and autumn, we moved from the north to the south. Note that, for seasonal observations, ultra-small water bodies (<100 m^2), which often dry out in the summer, were excluded from sampling.

2.2. Sampling and Analyses

Water samples were collected from the lake surface (15–20 cm) using sterile plastic syringes and vinyl gloves. The lake water was filtered in situ immediately after sampling through disposable MILLEX Filter units (0.45 µm). The first 20–50 mL of the filtrate was not analyzed. To determine nutrients, samples (50 mL) of water filtered through pre-combusted acid-washed 0.45-µm Whatman GF/F glass fiber filters at 550 °C for 4 h (Thermo Fisher Scientific, Waltham, MA, USA) were immediately frozen for subsequent analysis of N-NH_4^+, N-NO_3^-, and P-PO_4^{3-}.

The filtered samples of lake water were divided into two parts, each of which was placed in a polypropylene vial (12 and 30 mL) pre-washed in a clean room (class A 10,000). One vial was acidified with bidistilled nitric acid to pH = 2 for cation analysis, and the other one was not acidified to analyze DOC, DIC, major anions (Cl, SO_4), and ultraviolet (UV) absorbance. Prior to the analysis, the samples were stored in a refrigerator at 4 °C. At each sampling site, the temperature, pH, specific conductivity (SC), and dissolved oxygen (O_2) were measured using a WTW Multi 3430 meter. The lake depth was measured using an echo-sonder (Cole-Parmer) at five points and represented as the median of a set of five numbers.

The nutrient concentration was analyzed using an automated flow injection analyzer (FIA star 5000, FOSS, Denmark) with detection limits of 1 $\mu M \cdot L^{-1}$ for N-NH_4, 0.5 $\mu M \cdot L^{-1}$ for N-NO_3, and 0.5 $\mu M \cdot L^{-1}$ for P-PO_4. The CO_2 and CH_4 concentrations were measured by gas chromatography (see References [44,50] for details). The concentration of DOC, DIC, Cl, SO_4, Si, major cations, and trace elements was measured by standard methods used for boreal organic-rich and low-mineralized water samples [18,43,63] as briefly described below. Trace elements were measured by inductively coupled plasma mass spectrometry using an ICP-MS quadrupole (Agilent Technologies 7500 ce, Santa Clara, CA, USA), with an In + Re internal standard and an uncertainty of ±5%. International geo-standard SLRS-5 (river water reference material for trace element analysis certified by the National Research Council of Canada) was used to validate the analyses [64]. The agreement of the results with certified SLRS-5 values ranged from 10% to 15% for 40 elements, except for B and P (≤30%). Non acidified samples were used for the following types of analysis: (1) measurement of DOC concentration by complete combustion at 800 °C using a platinum catalyst with subsequent determination of CO_2 by infrared spectroscopy (TOC-VCSN, Shimadzu, Kyoto, Japan), with an accuracy of 5%, and a detection limit of 0.1 $mg \cdot L^{-1}$; (2) determination of chlorides and sulfates by high-performance liquid chromatography in the range of 0.05–10 $mg \cdot L^{-1}$ (Dionex ICS-2000, San Jose, CA, USA), with an accuracy of 2%, and a detection limit of 0.02 $mg \cdot L^{-1}$; (3) determination of ultraviolet radiation absorption at 254 nm using a Varian CARY-50 UV–visible light (Vis) spectrophotometer (Agilent Technologies, Santa Clara, CA, USA) and 10-mm quartz cuvette.

2.3. *Statistical Treatment*

The data on element concentrations were, firstly, checked for normal distribution using the Shapiro–Wilk test. Nonparametric statistics methods were employed even if a part of the data used in statistical processing was normally distributed and another part was distributed abnormally. The values of the median and interquartile range (median ± IQR) were used to describe the uncertainty due to a large number of extrema and outliers in the datasets and abnormally distributed data. For comparison, we also present the average values and standard deviation (mean ± SD) (Table A1, Appendix A). A correlation analysis with the calculated Spearman's correlation coefficient (*Rs*) ($p < 0.05$) was used to identify the relationship between the elemental composition of lake waters and the water surface area, and to identify the shape of latitudinal patterns of chemical elements. The non-parametric Kruskal–Wallis test (H-test), which determines whether three or more independent samples originate from the same median population, was used to estimate the total difference in the concentration of major and trace elements between four sampling sites during three hydrological seasons. In this regard, this method is similar to the parametric one-way analysis of variance (ANOVA). In addition, a paired comparison analysis was performed using the non-parametric Mann–Whitney test (U-test) to identify statistically significant differences between two independent datasets based on one given parameter. Further statistical treatment of a complete set of element concentration in thermokarst lake waters included hierarchical cluster analysis (HCA) and principal components analysis (PCA) using a variance estimation method [65]. The PCA allowed testing the effect of various parameters, latitude and hydrological seasons in particular, on the behavior of DOC and the concentration of chemical elements. All graphs and drawings were plotted using Microsoft Excel 2016 and STATISTICA version 8 software package (StatSoft Inc., Tulsa, OK, USA) (http://www.statsoft.com).

3. Results and Discussion

3.1. *Dependence of the Concentration of DOC, DIC, Anions, and Trace Elements on the Water Body Size*

It is known that the main sources of chemical elements in thermokarst lakes in the north of Western Siberia are coastal abrasion of peat strata of frozen peat bogs (via leaching processes) and lateral soil (suprapermafrost) flow from the lake watershed [18,42,49,50]. As a result, the concentration of dissolved chemical elements, DOC, CO_2, and CH_4 strongly depends on lake surface area [13,19,40,42], as it is also known in various regions of the northern hemisphere [66,67].

The Spearman's rank correlation coefficient ($p < 0.05$) was used to reveal the relationship between the concentration of chemical elements in different seasons and the water surface area for each study area. Correlation analysis data are presented in Table S1A (Supplementary Materials), which lists the elements that exhibit statistically significant values of the correlation coefficient during three hydrological seasons (spring–summer–autumn).

3.1.1. Isolated Permafrost Zone

In the isolated permafrost zone, the DOC concentration decreased ($Rs_{\text{spring}} = -0.72$, $Rs_{\text{summer}} = -0.79$, and $Rs_{\text{autumn}} = -0.76$, $p < 0.05$) in all three hydrological seasons as the water surface area increased (Figure 2A), which can be due to progressive decrease of allochthonous DOC supply from the surrounding peat [18] upon lake surface area (S_{area}) increase due to lake maturation. It is known that, when the lake surface area increases, the impact of peat (coastal abrasion) decreases [19]. This is accompanied by a replacement of allochthonous by autochthonous DOC [68,69]. A statistically significant increase in the SO_4 concentration in lake water with the increase in S_{area} (Figure 2B) during all three seasons ($0.76 \leq Rs \leq 0.83$, $p < 0.05$) can be attributed to an increased effect of the groundwater and to diffusion of sulfate from sediment porewater, where it accumulates, due to autochthonous organic matter (OM) [32], to the water column. These effects become more pronounced in the course of lake maturation, leading to elevated SO_4 concentrations in the largest lakes.

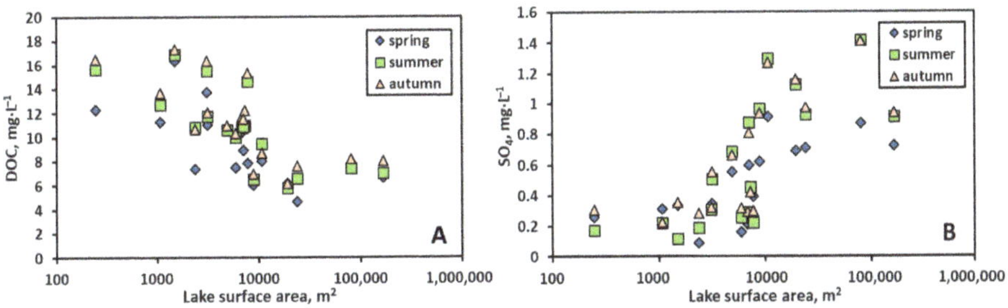

Figure 2. Dependence of the concentration of dissolved organic carbon (DOC) (**A**) and SO_4 (**B**) in lake waters of the isolated permafrost zone on the water surface area (Kogalym).

The Mann–Whitney U-test ($p < 0.05$) did not show a statistically significant difference between two size classes (<10,000 m^2 and >10,000 m^2) for lakes in the isolated permafrost zone. In the spring, when the S_{area} increased due to lateral input of snowmelt water, a statistically significant increase in concentration with S_{area} could be observed for Cl, N-NO_3, B, K, Co, Rb, and Cs ($0.51 \leq Rs \leq 0.66$). In the summer base flow period, the concentrations of Li ($Rs = -0.63$), Cr ($Rs = -0.5$), Zn ($Rs = -0.55$), and Ga ($Rs = -0.47$) decreased with S_{area} increase and the concentrations of Sb ($Rs = 0.49$) and Cs ($Rs = 0.62$) increased. In autumn, at the beginning of freeze up, the SC ($Rs = -0.54$), Li ($Rs = -0.78$), B ($Rs = -0.49$), and Cr ($Rs = -0.54$) showed a negative correlation with S_{area}, whereas a positive correlation was observed for Rb ($Rs = 0.53$), Cd ($Rs = 0.61$), and Sb ($Rs = 0.62$). The other solutes did not demonstrate any sizable effect of lake area on their concentrations.

3.1.2. Sporadic Permafrost Zone

As described previously, the lakes of the sporadic permafrost zone exhibit strong dynamics of changes in the elemental composition of water depending on the lake surface area [19,40,42,60]. However, in this study, we could not track strong dynamics between these indicators during all hydrological seasons, probably due to a small number of studied lakes and the lack of ultra-small water bodies. Only lakes with S_{area} > 150–200 m^2, which do not dry out in summer, were used to study spatial and seasonal dynamics.

Statistically significant decreases in concentrations with S_{area} increase in all three seasons were found only for SC ($Rs_{\ spring} = -0.54$, $Rs_{\ summer} = -0.65$, and $Rs_{\ autumn} = -0.71$, $p < 0.05$) and DOC ($Rs_{\ spring} = -0.79$, $Rs_{\ summer} = -0.51$, and $Rs_{\ autumn} = -0.51$, $p < 0.05$) (Figure 3A). The mechanisms responsible for elevated DOC values in small size lakes are peat abrasion and enhanced suprapermafrost water input, with limited processing of allochthonous DOC due to its stability with respect to bio- and photo-degradation (i.e., Reference [70]). Typically, young thermokarst thaw ponds are marked by the presence of organic matter from the degrading permafrost [69]. Another element exhibiting systematic change of concentration over all seasons is nitrite. The N-NO_3 concentration increased with S_{area} ($Rs_{\ spring} = 0.9$, $Rs_{\ summer} = 0.73$, and $Rs_{\ autumn} = 0.9$, $p < 0.05$, Figure 3B), which can be indicative of nitrate diffusion from lake sediments, where it is produced due to enhanced input of allochthonous OM from decaying phytoplankton, which is typical for large lakes.

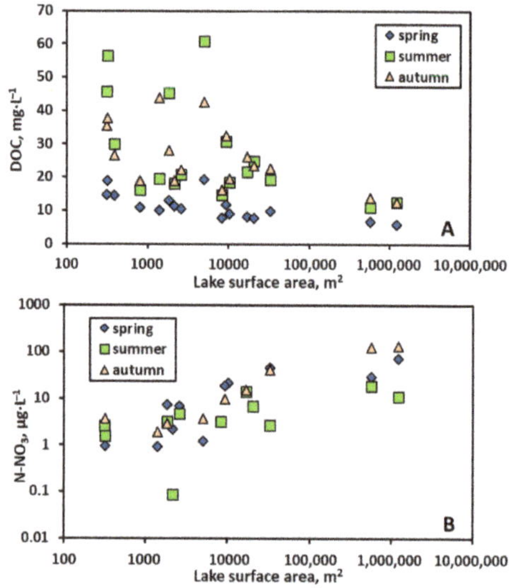

Figure 3. Dependence of the concentration of DOC (**A**) and N-NO_3 (**B**) in lake waters of the sporadic permafrost zone on the water surface area (Khanymey).

In the spring period of increased water volumes in lakes and breakup of ice, when the lake water surface area increases, an increase in the concentration of Cl, SO_4, specific ultraviolet absorbance ($SUVA_{245}$), N-NH_4, K, V, Rb, and Mo was found to be statistically significant ($0.50 \leq Rs \leq 0.84$, $p < 0.05$). Only Cd exhibited a decrease in concentration with an increase in S_{area} ($Rs = 0.52$, $p < 0.05$). The autumn was characterized by an increase in SO_4, $SUVA_{245}$, P-PO_4, P_{tot}, Ti, V, Mo, Cs, rare earth elements (REEs), and Th concentration with an increase in S_{area} ($0.51 \leq Rs \leq 0.71$, $p < 0.05$). In the summer low water period, the concentration of nutrients (K, Rb), Y, Zr, Hf, and REEs increased significantly (Rs = from 0.51 to 0.71). The behavior of these lithogenic elements, likely originated from alumosilicate dissolution in mineral soil and subsoil horizons, may mark significant influence of groundwater on the formation of the elemental composition of lake water. This can be highly sensitive to a change in geocryological conditions due climate warming and the shift of permafrost zones to the north (see Section 3.6 below).

3.1.3. Discontinuous Permafrost Zone

In the discontinuous permafrost zone in spring, most chemical elements exhibited statistically significant differences between the two distinguished lake size classes (<10,000 m^2 and >10,000 m^2). The concentration of both lithogenic (Al, Si, Ti, trivalent and tetravalent hydrolysates) and some labile elements (Li, Na, Mg, Ca, Sr, As, Mo, U) increased with an increase in S_{area}, which can be explained by active lateral input during snowmelt and ice abrasion of the peat shoreline. Therefore, for these elements, the ratio of the lake circumference to lake volume is not a driver of their concentrations, similar to the case for sporadic/discontinuous permafrost zone [40,42]. Presumably, the main factor controlling the increase in concentration of these elements is snowmelt and leaching of trace elements (TE) from the vegetation of frozen peat bogs followed by their lateral transfer to lakes.

At the same time, the pH and concentration of Li, Mn, and Co significantly increased with S_{area} during all three seasons ($p < 0.05$, Figure 4). Here, both mobilization from groundwater (Li) or lake sediments, due to redox reactions (Mn, Co; see Reference [32]) in large lakes, may be the governing factor of element increase during the baseflow period.

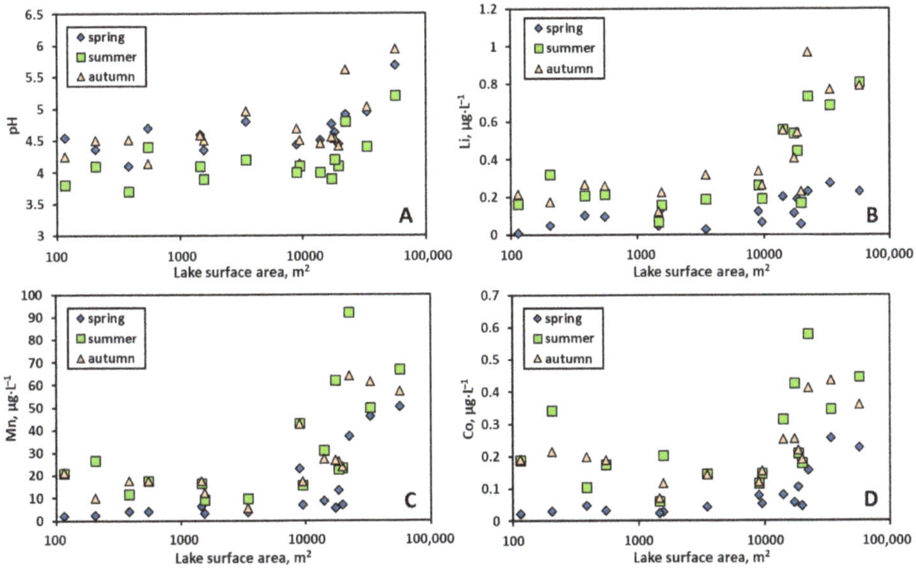

Figure 4. Dependence of the concentration of pH (**A**), Li (**B**), Mn (**C**), and Co (**D**) in lake waters of the discontinuous permafrost zone on the water surface area (Urengoy).

3.1.4. Continuous Permafrost Zone

The lakes in the continuous permafrost zone neither show a statistically significant difference between the size classes nor a distinct dependence of the concentration of chemical elements on the water surface area. Presumably, the presence of thawed mineral horizon [71] interferes with peat as the main source of chemical elements entering the thermokarst lake water, and this distinguishes the continuous permafrost zone from southern regions. In summer, a number of labile, presumably seawater-originated elements exhibited an increase in concentrations with increased S_{area}: pH, Ca, Sr, Cu, Sb, Mo, U (Figure 5). The mineral horizons here are formerly marine clays represented by diatomites with illite and montmorillonite with some chlorite and allophanes (see Reference [51] for soil profile and references therein). These clays may contain a sizable amount of labile elements originated from seawater via adsorption during diagenesis (Ca, Sr, Mo, Sb, U). The high affinity of these elements to clay minerals such as smectite, chlorite and montmorillonite is fairly well known [72,73]. Therefore, we hypothesize that large thermokarst lakes of continuous permafrost zone are capable of eroding the peat underlying mineral horizons, and, after some threshold S_{area} (around 50,000 km^2, see Figure 5), the exposed mineral layers on the lake bottom start to release a sizable amount of labile elements to the water column.

Considering all permafrost zones together, it was found that there was an increase in pH, DIC, alkali and alkaline-earth metals (Li, Mg, K, Ca, and Sr), divalent heavy metals (Co, Ni and Cu), trivalent and tetravalent hydrolysates (Y, Zr, REEs, and Th), Mo, As, and U with lake surface area. The increase in labile soluble elements (alkali and alkaline-earth metals, Mo, As, and U) in the southern, sporadic to discontinuous permafrost zone can be interpreted as due to the input of deep underground waters which contact with carbonate rocks and penetrate into large lakes via taliks or subsurface flow. In the continuous permafrost zone, this increase may be due to a desorption of soluble elements from marine clay mineral surfaces (illite, montmorillonite) exposed at the bottom of large lakes. The increase in concentration of insoluble trivalent and tetravalent hydrolysates with S_{area} is pronounced mostly during baseflow period and likely stems from silicate-bearing rocks such as clays, whose dissolution in deeper soil horizons provide tri- and tetravalent hydrolysates (TE^{3+} and TE^{4+}) to the lake water.

These elements are then stabilized in solution via organic and organo-ferric colloids. In contrast, concentrations of DOC and Pb decreased with S_{area} (Table S1B, Supplementary Materials), which reflects their dominant origin from peat, abrading at the lake shores. Subsequent delivery of these elements to the lake water column occurs via suprapermafrost flow over frozen organic horizons. These factors are mostly pronounced in small thaw ponds, having a high ratio of lake circumference (and lake watershed area) to the water volume.

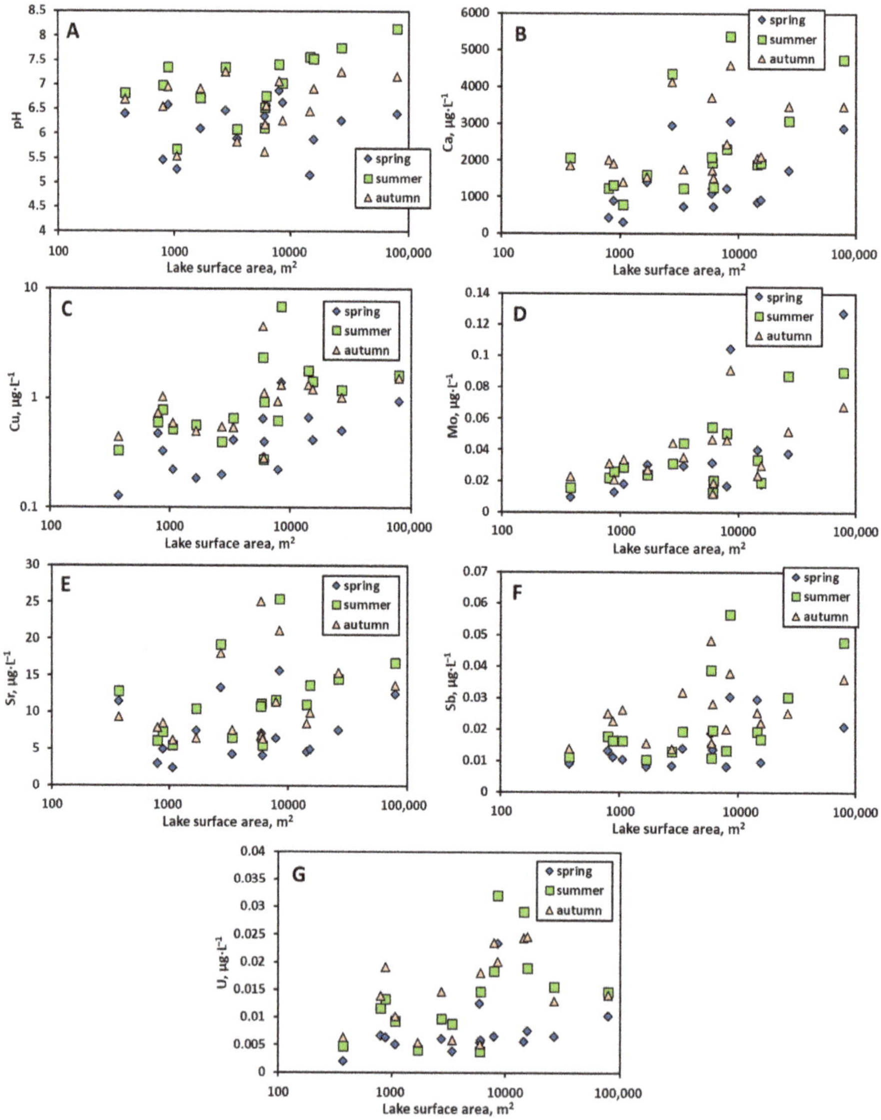

Figure 5. Dependence of pH and element concentration in lake waters of continuous permafrost zone on the water surface area (Tazovsky): pH (**A**), Ca (**B**), Cu (**C**), Mo (**D**), Sr (**E**), Sb (**F**), and U (**G**).

3.2. Impact of Hydrological Season on the Concentration of DOC, DIC, Anions, and Trace Elements in Thermokarst Lake Water

Mean and median concentrations of lake water components measured in the WSL territory for each season are listed in Table A1 (Appendix A). Statistical analysis of the difference in concentrations of chemical elements among different hydrological season was performed for each of the four permafrost zones. For comparison, we used the non-parametric Mann–Whitney test (U-test) for unrelated samples. The correlation was based on a statistically significant difference between the seasons ($p < 0.05$). Table S2 (Supplementary Materials) shows the relationships where at least one value (compared pair) was statistically significant. During the analysis, the size classes of lakes were not divided due to the absence of very small and very large water bodies in the sample set. Following a previously developed approach, we considered a more than two-fold change in the concentration between the dataset to be significant [40].

There was a progressive increase in the concentration of Mg, Li, B, Al, and Ca from spring to autumn (an increase of 1.5–2-fold or more) in lakes located in the isolated permafrost zone (Figure 6A). Compared to spring season, significantly higher (50–120%) concentrations were observed in summer and slightly higher (10–40%) concentrations were observed in the autumn. The concentrations of DIC, Cl, Si, Cr, V, and Hf were the highest in the spring and autumn periods of heavy rains, and they decreased sharply in the summer during low water period. A particularly noticeable increase in their concentration could be observed in the autumn. The concentration of DOC, SO_4, Fe, Mn, Pb, Na, Cd, Cu, Rb, K, Ti, Mo, Ba, Ni, Zr, Cs, Sr, Co, Sb, Nb, Th, U, and most REEs did not show a statistically significant ($p < 0.05$) difference between seasons.

Thermokarst lakes located in the sporadic permafrost zone showed a 0.5–3-fold increase ($p < 0.05$) in the concentration of $SUVA_{245}$, DOC, P_{tot}, Fe, Sb, As, U, Ba, Zr, Sr, Cu, and Th and a 3–5-fold increase in the concentration of Mn, Al, Cd, Co, Ni, Ti, Cr, Li, Nb, and REEs in the summer relative to the spring (Figure 6B). The most pronounced increase was observed for Ca (6.3) and B (6.8). No statistically significant difference in concentrations between summer and autumn was observed.

Seasonal variations in the elemental composition of lake waters in the discontinuous permafrost zone are similar to those in the sporadic zone (Figure 6C). A sharp increase in the concentration of most of the studied elements was observed from spring to summer. Furthermore, a progressive increase in the concentration from spring to autumn (2–5-fold increase) was characteristic of DOC, Al, Ca, B, Cu, Li, Cr, Pb, Nb, Sr, Hf, Th, and some REEs (Y, Ce, Pr, Sm, Eu, Gd, Tb, Dy, Ho, Er, Yb, and Lu). The SUVA and concentration of P_{tot}, Mn, Cd, As, Ni, and Co increased sharply in summer and slightly decreased in autumn. Significant accumulation of Si (five-fold), Zn, Rb, and SO_4 (1.5–2-fold) in thermokarst lake waters occurred in spring and autumn. This may indicate the inflow of these elements from plant litter which is flooded during snowmelt and heavy rains. Accumulation of SO_4, especially in spring (a two-fold decrease in summer concentration), may indicate diffusive transport of sulfates from bottom sediments. The concentration of DIC, Cl, Fe, Na, K, Mg, Sb, Ti, Cs, Zr, V, U, Ba, and Mo remained stable over all three hydrological seasons.

A statistical analysis of the elemental composition of lakes located in the continuous permafrost zone, where most of chemical elements originate from thawed mineral horizon, showed a 1.5–3-fold gradual increase in the concentration of DOC, Cl, Al, Cu, Ti, Cr, Th, U, Li, Ni, Na, Si, Sr, Hf, Zr, Y, and REEs from spring to summer and autumn (Figure 6D). Lower concentrations of Fe, Mn, Cd, V, K, Rb, and Nb in summer relative to spring, when the water level in thermokarst lakes in this zone is minimal, may indicate a preferential lateral inflow of these elements into lake waters with surface runoff, which is mostly pronounced in the spring. This surface inflow is replaced by a subsurface (supra-permafrost) flow in the summer. The decomposition of aquatic vegetation and precipitation can represent additional sources of Zn and B, which show a eight- and 12-fold increase in summer and a four- and two-fold decrease in the autumn relative to the spring, respectively. The elevated concentrations of K, Rb, and Si during autumn compared to summer may indicate (i) ceasing of these nutrient uptake by macrophytes, planktonic, and periphytic diatoms, and (ii) degradation and release of autochthonous freshly produced organic matter. These effects (at least for Si) are also visible in the discontinuous and isolated permafrost zones.

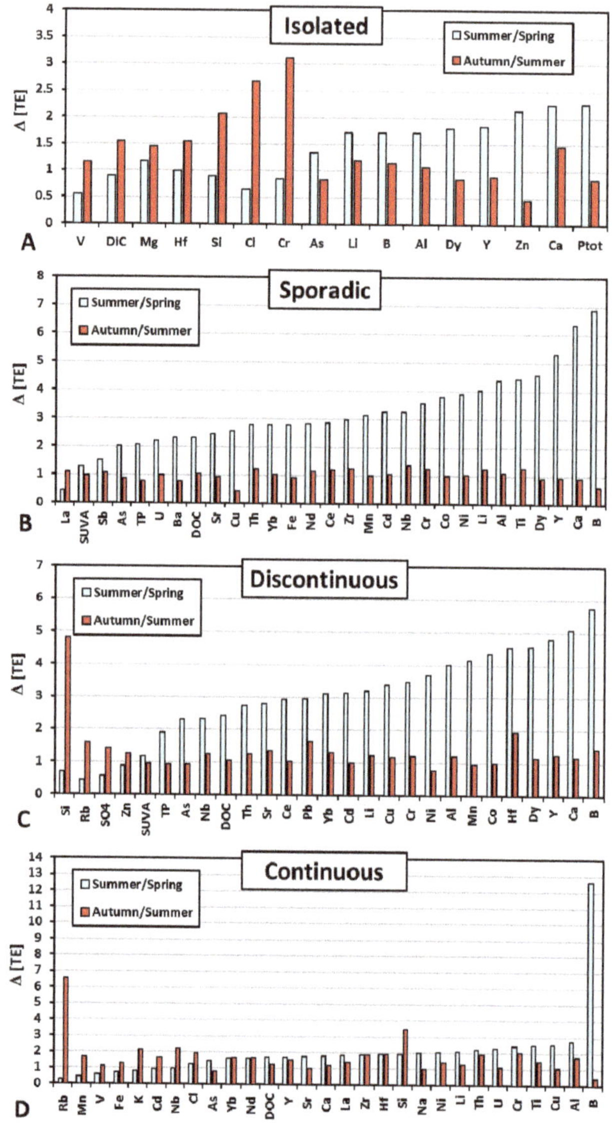

Figure 6. The ratio of concentrations of chemical elements between different hydrological seasons (summer/spring and autumn/summer): isolated (**A**), sporadic (**B**), discontinuous (**C**), and continuous (**D**) permafrost zones. The diagrams show elements with statistically significant differences recorded for at least one compared pair ($p < 0.05$).

3.3. *Effect of the Permafrost Gradient on the Concentration of Elements in Thermokarst Lakes*

Prior to testing the concentration dependence of chemical elements on the geographical latitude, we conducted the Mann–Whitney U-test to identify variations in concentration in the four areas over different seasons (spring, summer, and autumn) (Table S3, Supplementary Materials). The revealed variation shows that the concentrations of most elements change significantly across the areas in all three seasons. Sporadic and discontinuous permafrost zones show the greatest similarity in elemental composition, which can be due to similar conditions for lateral element delivery to the lakes via water leaching of the peat deposits and surrounding vegetation under condition of discontinuous/sporadic permafrost, drastically different from those in most southern (isolated) and most northern (continuous) permafrost zones.

The impact of permafrost gradients on thermokarst lake hydrochemistry was determined via analysis of correlation between element concentration in the lake water and geographical latitude. Spearman's nonparametric correlation coefficient (*Rs*, $p < 0.05$) was used to analyze the data with abnormal distribution (Table S4, Supplementary Materials). Note that, for this treatment, we considered only trends of element concentration with latitude that were significant ($p < 0.05$) over all three hydrological seasons. Based on latitudinal dynamics of lake water chemical composition, the following group of elements were distinguished:

(1) The value of pH and concentrations of CO_2, DIC, Mg, Ca, Mo, Sr, U, P_{tot}, Li, Na, Ti, V, Fe, Co, Ni, Cu, Y, Zr, REEs, Hf, and Th increased northward (Figure 7). Specifically, the pH and concentration of CO_2, DIC, Mg, Ca, Mo, Sr, and U changed insignificantly from south to north (from the isolated to discontinuous permafrost zone) and increased sharply (2–3-fold) in the continuous permafrost zone. A possible factor affecting this pattern is shallower peat deposits in the north [3], such that the thawed layer reaches the mineral horizon (marine clays containing illite, chlorite, and montmorillonite) in the tundra zone. This promotes accumulation of low-mobile lithogenic elements and rapidly increases the concentration of dissolved inorganic carbon in lake waters of the continuous permafrost zone. Note that this behavior is somewhat different from that observed for rivers of this territory [16,51]. The elevated pH in the north may explain the enrichment in labile elements; unlike rivers, thermokarst lakes are essentially disconnected from groundwaters and, thus, former marine clays in Tazovsky are most likely the cause of elevated pH and concentrations of DIC, Mg, Ca, Sr, Mo, U, and partially K.

(2) Elements decreased their concentrations northward in all three seasons ($p < 0.05$): SO_4, Cd, Pb, Sb, and Cs (Figure 8). The decrease in SO_4 may be due to isolation of sulfate-bearing groundwater reservoirs from surface waters occurring in the continuous permafrost zone. Trace elements Cd, Pb, Sb, and Cs are known to enrich the peat and peat porewaters [49,50], and their decreased concentration in the north may reflect a decreased release of these elements to adjacent surface waters due to decreased peak thickness in the continuous permafrost zone [3]. Furthermore, a decreased input of these elements to the lake water in the north may be due to a decrease in the impact of atmospheric aerosol deposition [20].

(3) The elements which did not show statistically significant changes with latitude in all three seasons ($p < 0.05$) were SC, Cl, DOC, P-PO_4, N-NO_3, N-NH_4, B, Al, Si, K, Cr, Mn, Zn, As, and Rb. However, some elements such as DOC, Al, Cr, and Mn demonstrated a clear maximum in the sporadic or discontinuous permafrost zone (Figure 9). Presumably, this permafrost belt with its highest ALT during summer and autumn baseflow and at the same time sizable amount of frozen peat provided the most optimal redox conditions for Mn (and Cr) mobilization from the lake sediments to the water column and DOC and Al leaching via peat abrasion and transport in the form of colloids [60].

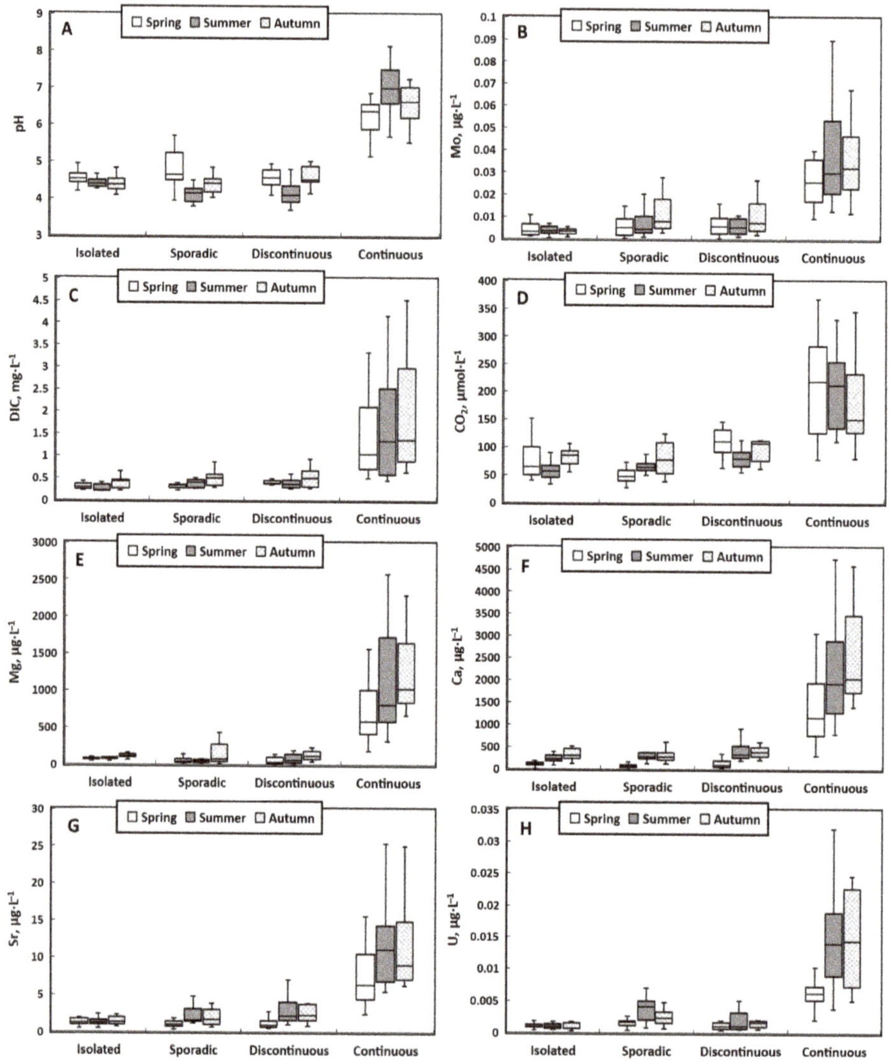

Figure 7. Increase in pH and concentrations of some chemical elements in different hydrological seasons in the permafrost profile of Western Siberia: pH (**A**), Mo (**B**), DIC (**C**), CO_2 (**D**), Mg (**E**), Ca (**F**), Sr (**G**), and U (**H**).

It is noteworthy that, considering all lakes and all seasons simultaneously, there was no impact of latitude (or permafrost coverage) on major nutrients (P-PO_4, N-NO_3, N-NH_4) (Table S4, Supplementary Materials). The only exceptions were a weak increase in PO_4 concentration ($Rs = 0.31$) and a decrease of NH_4 with latitude ($Rs = -0.38$), detectable in spring. In summer, there was also an increase in PO_4 concentration with latitude, as illustrated in Figure S2 (Supplementary Materials). Other nutrients such as Si demonstrated an increase in concentration northward, which was, however, visible only in autumn (Figure 10A), presumably due to progressive involvement of soil mineral horizons, capable of supplying Si due to the shallow peat layer in the north. Alternatively, the autochthon uptake of Si by aquatic plants and diatoms could be much less pronounced in the north due to lower biomass and productivity. This phenomenon was described in WSL rivers [51]. A sharp increase in Si concentration in the discontinuous permafrost zone in spring may indicate active inflow during snowmelt due to Si leaching from forest–tundra litter. In the continuous permafrost zone, K concentration increased by a factor of 2–3 relative to southern regions (Figure 10B), which can be attributed to its atmospheric transport as part of the marine aerosols and inflow with spring and autumn rains. Finally, a micronutrient (Zn) could be strongly affected by a release from tundra vegetation, which intercepts atmospheric deposits and contributes to this element transport from the watershed to the lake. This may produce a sharp increase in Zn concentration in the continuous permafrost zone (Figure 10C).

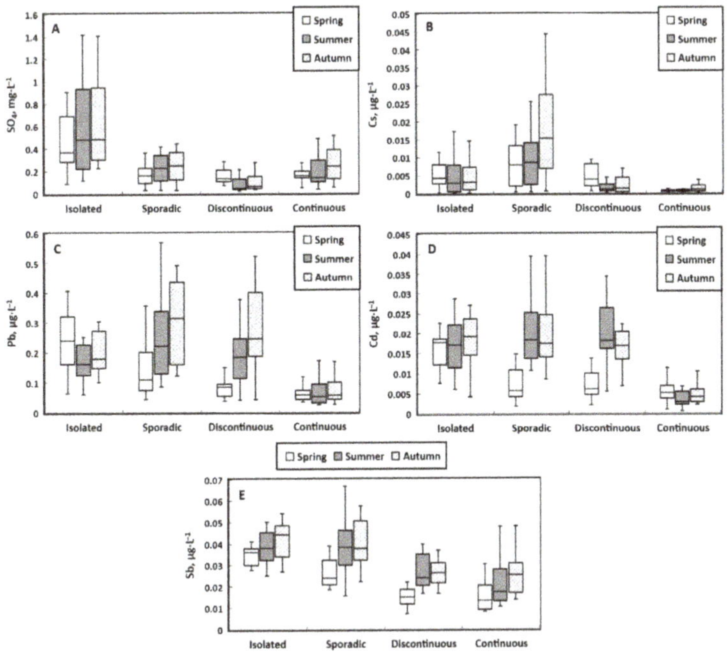

Figure 8. Decrease in concentrations of chemical elements during all three hydrological seasons across the permafrost profile of Western Siberia: SO_4 (**A**), Cs (**B**), Pb (**C**), Cd (**D**), and Sb (**E**).

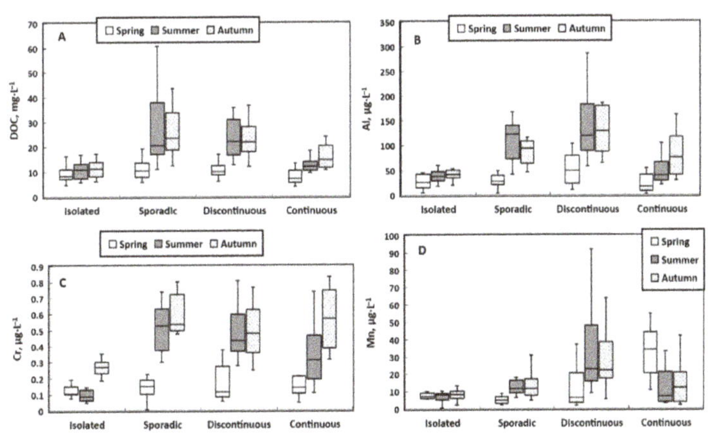

Figure 9. Elements whose concentrations exhibited a maximum in the sporadic and discontinuous permafrost zones: DOC (**A**), Al (**B**), Cr (**C**), and Mn (**D**).

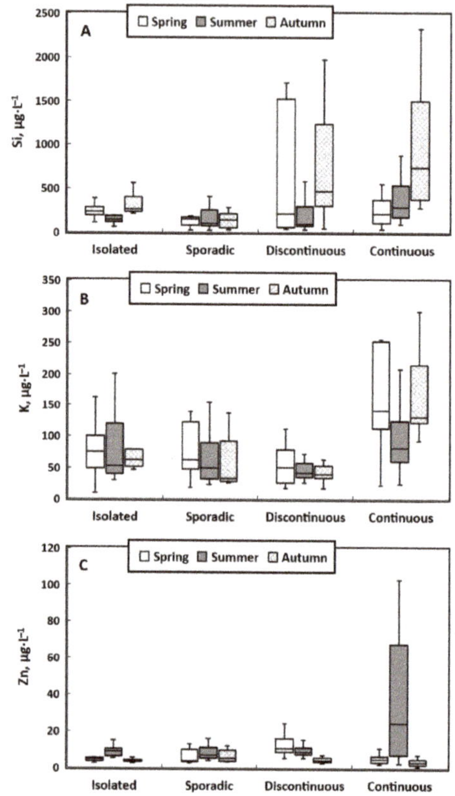

Figure 10. Elements that showed a sharp increase in concentrations in the discontinuous and continuous permafrost zones during individual seasons: Si (**A**), K (**B**), and Zn (**C**).

3.4. Element Pools in Thermokarst Lakes of Western Siberia: Seasonal Variations and Dependence on the Geographic Latitude

Because the depth of lakes varies across permafrost zones and seasons, we estimated the seasonal dynamics of element pools in each permafrost zone based on evolution of concentration and depth, thus quantifying the role of lakes in accumulation of solutes in the water column. Note that, recently, we calculated the summer stock of carbon and chemical elements for the lakes located in the entire permafrost-affected WSL territory [6]. However, the seasonal dynamics of elementary pools is not yet studied. Here, we calculated changes in the pool of DOC, DIC, and trace elements in the studied lakes as a ratio of two most contrasting periods, summer and spring ($R_{summer/spring}$). The changes in elemental pools across the seasons are provided in Table S5 (Supplementary Materials).

The degree of excess of summer over spring pool of DOC, DIC, and many trace elements (Li, B, Na, Mg, Al, Si, Ti, Cr, Ni, Sr, Ba, Se, Zr, Sb, REEs, Th, and U) in lakes increased northward. The overall increase ranged from 200% to 400% (Figure 11). Note the quite low $R_{summer/spring}$ of most elements in the isolated permafrost zone; this indicates that the main source of these elements can be lateral transfer to the lakes as a result of decomposition of plant litter during active snowmelt and rather minor inflow in summer. In this permafrost zone, the most significant decrease (20% to 80%) in $R_{summer/spring}$ value occurred for DOC, DIC, Se, Na, Mg, Al, Ti, V, Cr, Mn, Fe, Co, Ni, Cu, Rb, Sr, Zr, Cd, Sb, Ba, light REEs, Pb, Th, and U.

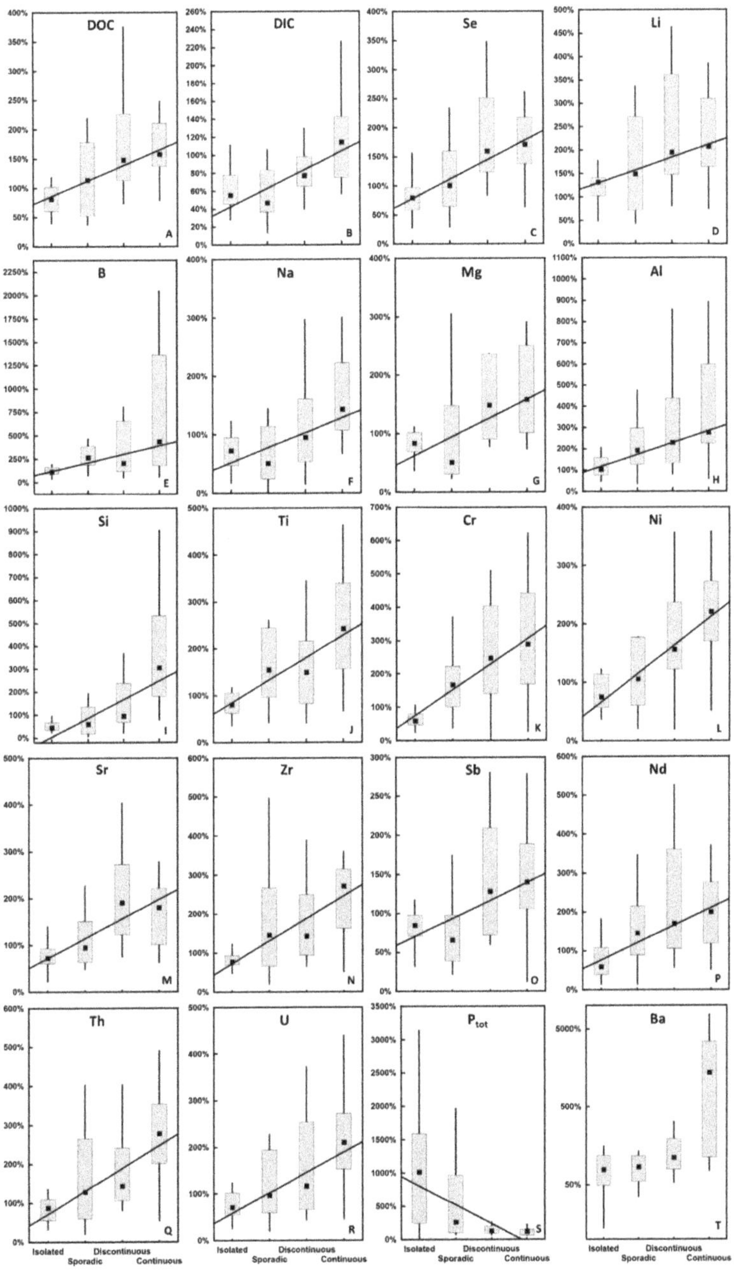

Figure 11. Seasonal increase in pools of chemical elements in thermokarst lakes in summer relative to spring, across the permafrost gradient of Western Siberia: DOC (**A**), DIC (**B**), Se (**C**), Li (**D**), B (**E**), Na (**F**), Mg (**G**), Al (**H**), Si (**I**), Ti (**J**), Cr (**K**), Ni (**L**), Sr (**M**), Zr (**N**), Sb (**O**), Nd (**P**), Th (**Q**), U (**R**), P_{tot} (**S**), and Ba (**T**). A value of 100% ($R_{summer/spring}$ = 1) means that the summer and spring periods exhibit the same pools of elements.

Summer pools of K and Rb were lower than in spring in all four permafrost zones (Figure S3A,B, Supplementary Materials), whereas Mo exhibited high seasonal stability of its pool in the four permafrost zones (Figure S3C, Supplementary Materials). Summer pools of Ca, Co, Cd, Mn, Fe, and Pb were 150–300% higher than in spring in the sporadic and discontinuous permafrost zones (Figure S3D–I, respectively, Supplementary Materials), while the pools of these elements in the isolated and continuous zones were comparable and even lower in summer relative to spring. In the continuous permafrost zone, summer pools of Zn and Ba exhibited a three- and 14-fold increase, respectively, relative to spring (Figure S3J,K, Supplementary Materials). The pool of total dissolved P (P_{tot}) was sizably higher in summer relative to spring (a factor of 2–10 for various permafrost zones), but the $R_{summer/spring}$ strongly decreased northward (Figure 11). Thus, the summertime inflow of P from watershed to lakes decreased northward, suggesting that the main source of P is thawed rather than frozen peat and mineral horizons, as also confirmed by experimental modeling of peat core interaction with aqueous solutions [74].

Overall, the evolution of $R_{summer/spring}$ for DOC, DIC, Li, B, Na, Mg, Al, Si, P (P_{tot}), Ti, Cr, Ni, Cu, As, Se, Sr, Zr, Sb, Ba, REEs, Th, and U reflects the impact of ALT, climate, and biota on the combination of processes responsible for element delivery from the watershed to the lakes. The increase in $R_{summer/spring}$ northward for DIC, Li, B, Na, Si, Mg, Sr, Ba, Sb, and U may reflect progressive thawing of rather shallow depth of peat with involvement of underlying mineral horizons. Marine clays composed of illite, chlorite, and montmorillonite are, thus, capable of supplying these elements in summer, when the ALT exceeds the peat layer. The increase in summertime DOC storage northward may be due to enhanced delivery of DOC from thawing peat yet weak biodegradability of dissolved organic matter (DOM) in the north. The excess of DOM stabilizes low-soluble elements such as Al, Ti, Cr, Ni, Se, Zr, REEs, Th, and U in the form of organic and organo-mineral colloids; thus, the $R_{summer/spring}$ increases northward. Finally, more pronounced biotic input of K, Rb, and Zn in the north during summer may be linked to intensive cycling of these elements by lake macrophytes. The other elements exhibited stable pools over season and latitude, or they demonstrated a local maximum in $R_{summer/spring}$ in the sporadic to discontinuous permafrost zone, thus reflecting a combination of biotic uptake/release processes, deep groundwater and shallow suprapermafrost water feeding, lake sediment, and underlying rock/peat interactions with the water column.

3.5. Multiparametrical Statistical Analysis (PCA and HCA)

Further distinction of different groups of elements was achieved via a PCA treatment. It revealed two main factors controlling major and TE concentrations in WSL lakes across a latitudinal profile, although the explicatory power of these factors was rather low (Figure 12A). The first factor, providing 25% of overall variation, included (Al), Ti, (Fe), Cu, Y, Zr, REEs, Hf, Th, and U, i.e., elements of lithogenic (mineral) origin, controlled by organic and organo-mineral colloids. The second factor (7.7% variation) acted on mobile, soluble elements affected by groundwater feeding in the south and desorption from marine clays in the north (DIC, pH, Mg, (Ca, K, and Mo)).

The HCA treatment of all dataset on WSL lakes also demonstrated a distinct group of soluble highly mobile elements (Ca, Mg, DIC, Mo, Sr) linked to latitude and low-soluble, trivalent and tetravalent hydrolysates, Se and U, linked to DOC (Figure 12B). Note that a similar distinction of elements between two main factors was revealed for WSL rivers [16] and, as such, it can be considered as a general (universal) feature of surface waters in the WSL territory. These two main group of elements are (*i*) labile, highly soluble alkalis, alkaline-earths, and oxyanions, linked to groundwater reservoirs, and (*ii*) low-soluble lithogenic trivalent and tetravalent hydrolysates, controlled by organic and organo-mineral colloids. These colloids are formed via mixing of Fe(II) and Al-rich groundwater and DOM-rich surface waters due to adsorption onto and coprecipitation with Fe and Al hydroxides stabilized by OM. The two main groups of elements are also encountered in large rivers of the boreal zone, based on the element concentration–river discharge relationship and seasonal behavior of river water solutes [75,76]; thus, they reflect general principles of geochemistry of surface waters in high-latitude regions.

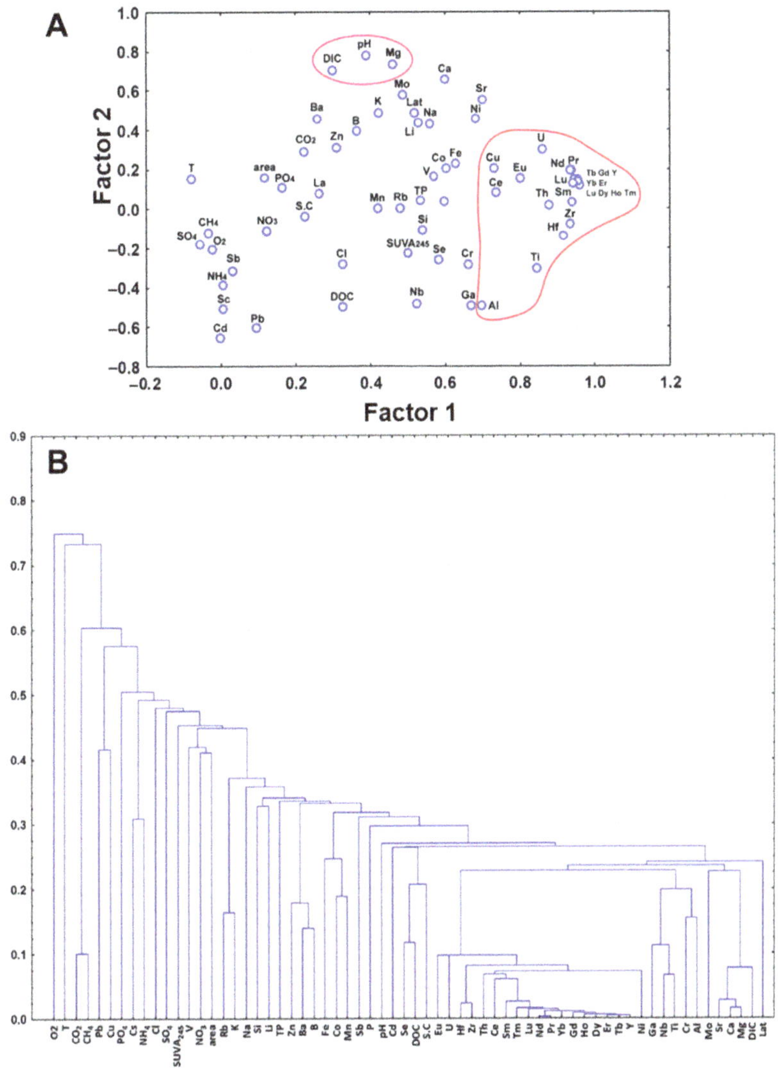

Figure 12. Principal component analysis (PCA) (**A**) and HCA (**B**) treatment of dissolved elements and studied parameters of lake waters during all seasons.

3.6. How Northward Shift in Permafrost Zones May Affect the Chemical Composition of Lakes

In the Arctic regions, temperature increases faster than in other areas [77,78], which contributes to permafrost thawing and the release of labile organic carbon into surface waters [11]. An estimated carbon loss of up to 15% is expected in the frozen landscapes of the Northern Hemisphere in the next 300 years [79].

In Western Siberia, common climate change scenarios predict a northward shift of permafrost boundaries and an increase in the depth of the seasonally thawed layer [80–86]. Within a space for time substituting approach [49,50,53,54], the northward shift of permafrost boundaries will transform the continuous permafrost zones into a discontinuous one, and the discontinuous zone into sporadic/isolated zones. As a result, contemporary permafrost zones located along the climate gradient in the WSL and measured lake ecosystem parameters can be used to assess the future status of the territory provided that there will be a lateral shift northward of aquatic and terrestrial landscapes. Such a highly empirical yet straightforward prediction takes into account the integral change of biomes, at the level of both lake watershed (vegetation, soil) and lake ecosystem (water temperature and aquatic biota). Below, we consider the effects of climate changes on the elemental composition of lake waters and pools of DOC, DIC, and trace elements in thermokarst lakes across the permafrost gradient.

The increase in temperature and ALT will alter the proportion of the main sources supplying the components to the lakes, especially in the sporadic and discontinuous permafrost zones; therefore, chemical changes in lake waters will be most pronounced in these areas. The number of small and ultra-small lakes in these permafrost zones is expected to increase [2,3], and thawing of frozen permafrost and discharge of the water into the river network is likely to be enhanced in this transitional permafrost zone [13,18,87]. The main sources of elements entering thermokarst lakes in the continuous permafrost zone are atmospheric depositions and leaching from thawed mineral horizon. Thus, an increased depth of the active layer can cause a rapid influx of dissolved elements from the mineral horizons and deep underground waters to the lake.

For a long-term forecast of changes in the concentrations of chemical elements in the WSL lake waters (due to lateral shift of the permafrost boundaries northward), from isolated to continuous zones, we used previously published data for the Arctic coastal zone of Western Siberia [13] (Figure S4, Supplementary Materials). We also calculated the ratio of mean element concentrations in lakes located in the continuous permafrost zone to those encountered in the discontinuous permafrost zone and similar ratios for the adjacent discontinuous/sporadic and sporadic/isolated permafrost zones (Table 1). As can be seen from Figure S4 (Supplementary Materials) and Table 1, a shift of permafrost zones by 2° to 4° northward will lead, in the region which is now the continuous permafrost zone, to a 2–5-fold decrease in the concentration of most labile components (DIC, Li, B, Mg, K, Ca, Ni, Cu, As, Rb, Sr, Ba, Mo, U), as well as of insoluble low-mobile elements present in colloidal forms (Y, Zr, REEs, Th). The pH will decrease by ca. two units and the concentrations of CH_4, DOC, NH_4, Cd, Sb, and Pb may increase by a factor of 2–3, while the other elements in northern thermokarst lakes will remain unaffected by this permafrost boundary shift. The changes in concentrations of other elements might not exceed a factor of 1.5 to 2.0, which, given the large lateral and temporal variability of lake hydrochemical composition, can be considered as of low significance. The changes in lake water chemical composition (virtually all components) due to transformation of discontinuous to sporadic zone will also be quite low, within a factor of 1.5 to 2.0. In contrast, the transformation of sporadic to isolated permafrost will lead to a sizable (a factor of 2–5) decrease in the southern part of the permafrost-affected WSL territory of CH_4, DOC, NH_4, Al, P_{tot}, Ti, Cr, Ni, Ga Zr, Nb, Cs, REEs, Hf, Th, and U concentrations. Of special interest is a two-fold decrease in DOC concentration in thermokarst lakes of the southern, currently sporadic zone and its comparable increase in the northern, currently continuous permafrost zone.

As previously described [6], the stocks of carbon and most of the elements in the water column of thermokarst lakes account for 10–20% of the riverine export in the area; however, at the same time, lakes may play an important role in the storage of a number of toxic elements in bottom sediments, thus preventing element transfer to the river [32]. The shift in the permafrost zones and change in the elementary supply may cause a 2–5-fold change in the overall dissolved (<0.45 μm) pools of a large number of elements in both the most northern and the most southern permafrost zones of the WSL. Note, however, that these predictions do not take into account the changes in the amount of precipitation and the duration of warm/cold seasons, as well as development of lake macrophytes (i.e., Reference [88]). As such, further integrative ecosystem and climate modeling should allow more precise predictions on changes in element concentrations and pools in lentic waters of the WSL.

Table 1. Prediction of change in concentrations (factor of increase or decrease relative to the current state) of DOC, DIC, and trace elements in lake waters on climate change and 2° to 4° northward (N) shift of the permafrost zones. SC—specific conductivity; $SUVA_{245}$—specific ultraviolet absorbance.

Element	Continuous into Discontinuous	Discontinuous into Sporadic	Sporadic into Isolated	Element	Continuous into Discontinuous	Discontinuous into Sporadic	Sporadic into Isolated
SC	1.05	0.92	1.34	Ga	0.57	1.11	2.45
pH	1.46	1	0.99	As	2.17	0.71	0.97
CH_4	0.21	1.96	4.21	Se	1.13	0.74	1.63
CO_2	1.64	1.3	1.24	Rb	3.2	0.51	1.03
Cl^-	0.74	1.32	0.9	Sr	3.3	1.73	1.05
SO_4^{2-}	1.76	0.49	0.43	Y	2.78	1.81	1.97
$SUVA_{245}$	0.98	0.99	1.24	Zr	2.51	0.84	4.22
DOC	0.63	0.93	2.05	Nb	0.81	0.67	5.64
DIC	3.55	1.15	1.23	Mo	4.84	1.08	1.99
$P\text{-}PO_4$	0.94	1.25	1.59	Cd	0.35	0.87	1.04
$N\text{-}NO_3$	3.21	0.16	1.43	Sb	1.02	0.69	0.92
$N\text{-}NH_4$	0.14	0.53	3.48	Cs	0.4	0.25	2.46
Li	3.1	1.03	0.98	Ba	6.84	1.86	1.66
B	3.23	0.6	1.56	La	2.25	0.52	4.48
Na	1.71	1.29	1.27	Ce	1.72	3.18	1.51
Mg	9.86	0.97	0.99	Pr	3.07	2.13	1.97
Al	0.54	1.37	2.4	Nd	3.14	2.05	1.98
Si	0.9	2.63	0.77	Sm	1.76	2.3	1.23
P_{tot}	0.96	1.31	4.93	Eu	4.93	1.59	1.56
K	3.52	0.75	0.81	Gd	2.69	2.13	1.67
Ca	5.28	1.53	1	Tb	2.77	1.76	1.76
Ti	1.18	0.81	4.04	Dy	2.67	1.7	2.05
V	1.81	0.84	1.65	Ho	2.57	1.7	2.2
Cr	0.91	1.07	2.45	Er	2.6	1.82	2.11
Mn	1.01	2.2	1.35	Tm	2.45	1.83	2.1
Fe	1.84	1.75	1.19	Yb	2.55	1.89	1.92
Co	1.87	2.29	1.87	Lu	2.61	1.98	1.81
Ni	3.59	1.13	4.5	Hf	1.44	1.33	4.18
Cu	2.17	1.59	1.34	Pb	0.39	0.81	1.02
Zn	1.48	1.29	1.12	Th	2.02	1.11	3.45
				U	5.87	0.9	2.6

4. Conclusions

The chemical composition (DOC, nutrients, major and trace elements) of thermokarst lakes in the largest permafrost peatland in the world, the Western Siberia Lowland, was studied across four permafrost zones (isolated, sporadic, discontinuous, and continuous) and three main hydrological open-water periods (spring, summer, and autumn). Using this unprecedented geographical and seasonal resolution, we identified and quantified the primary environmental factors controlling thermokarst lake water chemistry such as lake size and type of permafrost distribution. These factors determine the link of the lake with groundwaters, soil suprapermafrost waters, and underlying mineral sediments, as well as and the intensity of peat leaching at the lake shore. Using thorough statistical treatment, we revealed two distinct group of elements—soluble highly mobile DIC, Ca, Mg, Sr, Mo, U, linked to groundwater reservoirs and marine clay minerals (chlorite, montmorillonite) and low-soluble lithogenic trivalent and tetravalent hydrolysates, controlled by organic and organo-mineral colloids. Such a distinction can be considered as a general feature of all surface waters in the WSL territory. In order to predict possible changes in thermokarst lake water concentration of C, nutrient, major and trace elements in response to climate warming and permafrost boundary shift northward, we used a "substitution space for time" approach. A shift of permafrost zones by 2° to 4° northward may lead to a 2–5-fold decrease in the concentration of most of the dissolved components (CO_2, DIC, Ca, Mg, Sr, Al, Fe, Ti, Mn, Ni, Co, V, Mo, Sr, Zr, Hf, Th, REEs, and U) in lakes located in the continuous permafrost zones. However, the elements supplied from leaching of the peat mass and top humus (DOC, Si, K, Cr, Zn, As, and Rb) will exhibit rather small changes in concentration (i.e., 30–50% decrease). In the southern part of the WSL (sporadic permafrost zone), a sizable decrease in concentrations of CH_4, DOC, NH_4, P_{tot}, Al, Ti, Cr, Ni, Ga, Zr, Nb, REEs, Hf, Th, and U can be anticipated. These predictions can be considered of first order only and require further verifications via coupled ecosystem, vegetation, groundwater, and climate modeling.

Supplementary Materials: The following are available online at http://www.mdpi.com/2073-4441/12/6/1830/s1: Figure S1. Changes in lake depth (A), water temperature (B), and oxygen concentration (C) across different permafrost zones of Western Siberia over three hydrological seasons; Figure S2. Dependence of nutrient concentrations on the permafrost gradient (latitude): N-NO_3 (A), N-NH_4 (B), P-PO_4 (C), and Ptot (D); Figure S3. Elements that do not show a significant latitudinal change in stocks in lake waters between spring and summer: K (A), Rb (B), and Mo (C); elements that increase their concentration in summer relative to spring in the sporadic and discontinuous permafrost zones: Ca (D), Co (E), Cd (F), Mn (G), Fe (H), and Pb (I); Figure S4. Summer element concentrations dependence of the permafrost gradient (latitude): DOC (A), Al (B), Fe (C), Mn (D), Si (E), Pb (F), Ca (G), Mg (H), Na (I), K (J), La (K), Ce (L). The data for the Arctic coastal zone of Western Siberia are from ref. [13]; Table S1. Dependence of physicochemical parameters and element concentrations on the water surface area for each permafrost zone during different hydrological seasons; Table S2. Mann–Whitney U test of the difference in element concentration between different seasons; Table S3. Mann–Whitney U test of the difference in element concentration between different permafrost zones; Table S4. Dependence of changes in the concentrations of the studied elements on the geographical latitude in different hydrological seasons; Table S5. Change in pools of DOC, DIC, and trace elements in lake waters in summer relative to spring.

Author Contributions: R.M.M. and O.S.P. conceptualized and designed the measurements of trace elements in lakes; R.M.M., A.G.L., and I.V.K. performed the field sampling; R.M.M., O.S.P., and L.S.S. analyzed the data; R.M.M., A.G.L., and L.S.S. contributed the data on lake water chemistry; S.N.V. provided the statistical treatment of the data; S.N.K. supervised the work and contributed to interpretation and permafrost transect design; R.M.M. and O.S.P. wrote the paper. All authors have read and agreed to the published version of the manuscript.

Funding: This work was supported by RSF grant No. 19-77-00073 (50%, sampling and interpretation), as well as RFBR grants No 19-05-50096, 19-55-15002, 20-05-00729.

Conflicts of Interest: The authors declare no conflicts of interest.

Appendix A

Table A1. Seasonal elemental composition of lake waters in different permafrost zones (mean ± SD is the numerator, median ± interquartile range (IQR) is the denominator). The last column represents global element concentrations for the whole WSL permafrost-affected territory, averaged over three seasons.

Element	Units	Isolated			Sporadic			Discontinuous			Continuous			Average
		Spring	Summer	Autumn	Spring	Summer	Autumn	Spring	Summer	Autumn	Spring	Summer	Autumn	
SC	μS·cm^{-1}	14 ± 5.2 12.5 ± 5	15 ± 2.4 14 ± 4	18 ± 3.6 17.5 ± 4	13 ± 5.2 11 ± 4	24 ± 9.9 21.5 ± 13	26 ± 7.9 27 ± 8	13 ± 4.6 11.5 ± 7	22 ± 7.8 19 ± 12	23 ± 8.2 22 ± 12	16 ± 7.2 14.5 ± 9	21 ± 11 16.5 ± 14	24 ± 8.9 20 ± 14	18 ± 8.4 17 ± 9.4
pH		4.57 ± 0.2 4.53 ± 0.2	4.51 ± 0.4 4.4 ± 0.2	4.51 ± 0.5 4.39 ± 0.3	4.9 ± 0.8 4.65 ± 0.6	4.14 ± 0.2 4.15 ± 0.3	4.4 ± 0.2 4.42 ± 0.3	4.62 ± 0.4 4.57 ± 0.4	4.18 ± 0.4 4.1 ± 0.4	4.7 ± 0.5 4.53 ± 0.3	6.18 ± 0.5 6.38 ± 0.7	6.99 ± 0.7 7.01 ± 0.8	6.58 ± 0.6 6.63 ± 0.8	5 ± 1 4.52 ± 1.3
CH_4	μmol·L^{-1}	1.12 ± 0.9 0.73 ± 1.1	1.35 ± 1.49 1.02 ± 1.3	1.72 ± 2.3 0.9 ± 1.5	0.661 ± 0.89 0.33 ± 0.37	1.08 ± 2.4 0.34 ± 0.58	15.9 ± 47.8 0.49 ± 1.2	16.3 ± 23 13.1 ± 13	4.45 ± 7.8 2.2 ± 3.7	13.7 ± 27 5 ± 9.9	5.26 ± 5.9 2.2 ± 9.3	1.18 ± 0.7 1.2 ± 1	0.639 ± 0.7 0.34 ± 0.54	5.17 ± 18 0.97 ± 2.3
CO_2	μmol·L^{-1}	75.6 ± 30 65 ± 48	61.6 ± 27.6 57 ± 18	86.2 ± 24 86 ± 19.4	49.7 ± 12 47.7 ± 18	71.4 ± 26 65 ± 7.7	156 ± 251 78.2 ± 49	120 ± 50 112 ± 37	100 ± 80 80 ± 24	139 ± 128 108 ± 34	211 ± 85 218 ± 142	205 ± 73 212 ± 114	175 ± 75 151 ± 92	119 ± 106 88 ± 69
Cl^-	mg·L^{-1}	0.188 ± 0.09 0.171 ± 0.1	0.118 ± 0.09 0.09 ± 1.7	0.188 ± 0.08 0.189 ± 0.07	0.1 ± 0.08 0.08 ± 0.06	0.148 ± 0.1 0.119 ± 0.18	0.199 ± 0.2 0.18 ± 0.11	0.106 ± 0.08 0.08 ± 0.03	0.223 ± 0.19 0.171 ± 0.2	0.261 ± 0.17 0.187 ± 0.21	0.123 ± 0.04 0.12 ± 0.04	0.136 ± 0.1 0.11 ± 0.08	0.178 ± 0.04 0.17 ± 0.06	0.163 ± 0.12 0.14 ± 0.13
SO_4^{2-}	mg·L^{-1}	0.466 ± 0.25 0.368 ± 0.4	0.607 ± 0.42 0.479 ± 0.7	0.64 ± 0.39 0.49 ± 0.6	0.192 ± 0.1 0.16 ± 0.09	0.243 ± 0.2 0.226 ± 0.18	0.302 ± 0.3 0.247 ± 0.22	0.154 ± 0.6 0.135 ± 0.1	0.09 ± 0.09 0.045 ± 0.08	0.117 ± 0.11 0.063 ± 0.09	0.159 ± 0.1 0.15 ± 0.06	0.192 ± 0.1 0.14 ± 0.19	0.281 ± 0.2 0.24 ± 0.23	0.295 ± 0.29 0.2 ± 0.23
$SUVA_{245}$	L·mg-C^{-1}·m^{-1}	3.8 ± 0.45 3.8 ± 0.7	3.4 ± 0.46 3.3 ± 0.6	3.2 ± 0.36 3.2 ± 0.5	3.6 ± 0.5 3.6 ± 0.6	4.7 ± 0.7 4.5 ± 0.8	4.5 ± 0.7 4.4 ± 1	3.9 ± 0.74 3.8 ± 1.1	4.5 ± 0.66 4.2 ± 1	4.2 ± 0.61 4.1 ± 0.8	4.4 ± 0.4 4.3 ± 0.6	4.1 ± 1 3.9 ± 0.7	3.9 ± 0.7 3.8 ± 1.1	4 ± 0.7 3.9 ± 1
DOC	mg·L^{-1}	9.3 ± 3.02 8.5 ± 3.8	10.8 ± 3.4 10.8 ± 5.3	11.3 ± 3.43 11.2 ± 5.4	11.2 ± 3.9 10.5 ± 4.8	27.4 ± 15 20.7 ± 13	25.9 ± 9.6 23.5 ± 14	10.8 ± 2.9 10.2 ± 3.4	24.9 ± 12 22.3 ± 12	24.1 ± 9.2 22 ± 9.3	8.46 ± 3.9 7.27 ± 4.6	13.5 ± 5.2 12.2 ± 3.3	15.9 ± 4.4 14.9 ± 8.6	16 ± 10 12 ± 9
DIC	mg·L^{-1}	0.307 ± 0.06 0.289 ± 0.11	0.274 ± 0.11 0.221 ± 0.14	0.395 ± 0.12 0.434 ± 0.19	0.313 ± 0.05 0.327 ± 0.05	0.395 ± 0.1 0.401 ± 0.17	0.496 ± 0.2 0.51 ± 0.25	0.408 ± 0.06 0.388 ± 0.06	0.424 ± 0.19 0.369 ± 0.15	0.55 ± 0.28 0.508 ± 0.34	1.41 ± 0.9 1.04 ± 1.3	1.67 ± 1.2 1.34 ± 1.9	1.82 ± 1.2 1.36 ± 1.9	0.689 ± 0.76 0.4 ± 0.3
P-PO_4	μg·L^{-1}	1.42 ± 0.86 1.46 ± 1.3	1.76 ± 2.14 1.06 ± 1.1	1.66 ± 0.99 1.63 ± 1.3	1.87 ± 1.3 1.44 ± 0.84	2.52 ± 1.7 2.24 ± 2.4	3.31 ± 3 2.24 ± 4	2.02 ± 1.4 1.62 ± 0.8	5.51 ± 8.2 1.96 ± 3.8	2.08 ± 1.6 1.72 ± 2.4	2.27 ± 0.9 2.07 ± 1.4	4.11 ± 2.2 3.48 ± 2.8	2.6 ± 1.2 2.44 ± 1.5	2.61 ± 3 1.9 ± 1.8
N-NO_3	μg·L^{-1}	11.9 ± 17 7 ± 9.2	7.03 ± 9.3 1.6 ± 10	24.4 ± 42 4.1 ± 14	18.6 ± 22 7.5 ± 28	6.2 ± 5.9 3.2 ± 8.5	37.2 ± 53 9.8 ± 37	3.33 ± 3.8 2.39 ± 1.7	2.83 ± 1.9 2.19 ± 2.7	3.48 ± 2.07 2.57 ± 3.6	4.1 ± 2.1 3.72 ± 2.8	1.5 ± 1.3 1.44 ± 2.8	25.4 ± 3.5 12.4 ± 29	12.2 ± 25 3.2 ± 8
N-NH_4	μg·L^{-1}	108 ± 102 120 ± 135	20.6 ± 44 5.13 ± 4.8	38.9 ± 87 7.6 ± 12	130 ± 98 137 ± 193	105 ± 151 55 ± 118	348 ± 277 361 ± 564	88.8 ± 124 38 ± 105	54.2 ± 76 20 ± 80	164 ± 302 7.6 ± 95	11.8 ± 12 8.3 ± 12	8.74 ± 9.1 5.36 ± 5.2	23 ± 46 6.62 ± 17	89 ± 161 14 ± 86
Li	μg·L^{-1}	0.167 ± 0.15 0.124 ± 0.02	0.331 ± 0.49 0.21 ± 0.04	0.381 ± 0.52 0.244 ± 0.08	0.108 ± 0.1 0.103 ± 0.07	0.336 ± 0.2 0.294 ± 0.21	0.422 ± 0.1 0.4 ± 0.17	0.128 ± 0.08 0.112 ± 0.15	0.357 ± 0.24 0.24 ± 0.37	0.403 ± 0.25 0.24 ± 0.32	0.511 ± 0.3 0.43 ± 0.39	1.01 ± 0.4 0.97 ± 0.63	1.23 ± 0.5 1.19 ± 0.7	0.443 ± 0.44 0.26 ± 0.36
B	μg·L^{-1}	2.02 ± 1.12 1.8 ± 1.2	2.76 ± 0.71 2.7 ± 1.2	2.91 ± 0.77 2.7 ± 0.49	1.77 ± 1.3 1.38 ± 1.6	6.38 ± 2.3 6.2 ± 3.3	3.88 ± 1.8 3.8 ± 2.3	1.25 ± 1.3 0.916 ± 1.1	2.63 ± 0.97 2.17 ± 1.2	3.4 ± 1.7 3.13 ± 2	2.24 ± 1.8 1.49 ± 2.4	17.6 ± 12 14.9 ± 20	3.61 ± 1.5 3.57 ± 2.1	4.26 ± 5.7 2.7 ± 2.6
Na	μg·L^{-1}	201 ± 61 183 ± 62	200 ± 124 181 ± 54	169 ± 170 136 ± 43	232 ± 114 211 ± 102	297 ± 184 302 ± 159	193 ± 120 179 ± 129	225 ± 167 182 ± 101	363 ± 303 242 ± 415	341 ± 345 227 ± 279	362 ± 157 320 ± 138	686 ± 440 675 ± 388	535 ± 257 470 ± 316	314 ± 265 221 ± 228
Mg	μg·L^{-1}	81.8 ± 49 74 ± 13	97.6 ± 83 79 ± 18	144 ± 129 115 ± 33	48.6 ± 36 43 ± 36	109 ± 149 52 ± 37	161 ± 133 74 ± 213	54.8 ± 47 34 ± 64	105 ± 117 62 ± 71	148 ± 120 113 ± 107	709 ± 392 583 ± 539	1098 ± 746 811 ± 1006	1225 ± 537 1029 ± 744	331 ± 501 92 ± 239

Table A1. *Cont.*

Element	Units	Isolated			Sporadic			Discontinuous			Continuous			Average
		Spring	Summer	Autumn	Spring	Summer	Autumn	Spring	Summer	Autumn	Spring	Summer	Autumn	
Al	μg·L^{-1}	27.4 ± 14 26 ± 27	38.7 ± 12 38 ± 17	41 ± 9.7 42 ± 15	51.3 ± 94 28 ± 14	111 ± 40 123 ± 63	94.8 ± 36 94 ± 37	52.1 ± 33 50 ± 56	137 ± 65 120 ± 88	163 ± 111 129 ± 87	31.2 ± 32 18 ± 32	61.9 ± 67 40 ± 37	97.6 ± 87 76 ± 70	74.6 ± 72 47 ± 64
Si	μg·L^{-1}	239 ± 69 235 ± 85	236 ± 338 147 ± 62	425 ± 451 266 ± 144	134 ± 49 151 ± 78	273 ± 446 107 ± 143	282 ± 539 145 ± 161	660 ± 725 217 ± 1391	185 ± 174 98 ± 201	970 ± 1167 471 ± 750	239 ± 156 217 ± 269	379 ± 253 294 ± 361	1021 ± 702 744 ± 1067	415 ± 581 219 ± 262
P_{tot}	μg·L^{-1}	0.518 ± 0.66 0.16 ± 0.31	3.53 ± 2.5 3.14 ± 2.1	3.66 ± 2.04 3.3 ± 2.6	4.25 ± 2.7 4.6 ± 4	24.8 ± 7.7 22 ± 15	8.97 ± 3.8 9.4 ± 5.6	10.3 ± 6 7.6 ± 9.7	26.4 ± 13 21 ± 11	13.2 ± 6.3 11 ± 8.9	14.2 ± 8.9 13 ± 8.9	20.9 ± 23 12 ± 10	13 ± 6 12 ± 5	11.6 ± 11 9.1 ± 13
K	μg·L^{-1}	77.7 ± 44 75 ± 44	85.1 ± 68 52 ± 61	85.5 ± 63 62 ± 27	77.2 ± 39 63 ± 71	68.4 ± 44 49 ± 47	56.4 ± 38 33 ± 61	54.9 ± 31 50 ± 51	52.6 ± 33 42 ± 22	43.7 ± 20 40 ± 18	198 ± 153 141 ± 138	141 ± 166 82 ± 58	194 ± 142 132 ± 93	95 ± 97 67 ± 75
Ca	μg·L^{-1}	132 ± 137 107 ± 58	273 ± 232 227 ± 120	363 ± 237 301 ± 202	73 ± 45 68 ± 61	378 ± 284 273 ± 113	321 ± 169 283 ± 143	124 ± 108 92 ± 57	462 ± 297 329 ± 242	593 ± 599 387 ± 201	1408 ± 888 1165 ± 1076	2329 ± 1364 1937 ± 1418	2488 ± 1031 2042 ± 1734	746 ± 1009 302 ± 755
Ti	μg·L^{-1}	1.41 ± 0.84 1.24 ± 1.4	1.71 ± 1.5 1.37 ± 0.8	1.9 ± 1.2 1.68 ± 0.9	2.08 ± 0.9 2.07 ± 0.9	8.94 ± 5 8.67 ± 6.6	9.24 ± 5.4 9 ± 4.8	3.45 ± 2.3 2.85 ± 3.9	5.62 ± 4.5 4.1 ± 2.6	7.31 ± 4.5 6.1 ± 4.6	3.17 ± 2.1 2.48 ± 2.4	7.6 ± 9 3.63 ± 5.4	8.5 ± 6.1 6.9 ± 4.5	5.04 ± 5.2 3.2 ± 4.9
V	μg·L^{-1}	0.594 ± 0.28 0.736 ± 0.57	0.201 ± 0.05 0.193 ± 0.04	0.232 ± 0.05 0.23 ± 0.06	0.491 ± 0.4 0.295 ± 0.43	0.606 ± 0.3 0.542 ± 0.1	0.596 ± 0.3 0.575 ± 0.2	0.515 ± 0.41 0.289 ± 0.83	0.45 ± 0.13 0.447 ± 0.17	0.453 ± 0.16 0.428 ± 0.27	1.21 ± 0.3 1.08 ± 0.23	0.731 ± 0.7 0.47 ± 0.38	0.628 ± 0.4 0.56 ± 0.19	0.552 ± 0.41 0.46 ± 0.46
Cr	μg·L^{-1}	0.121 ± 0.03 0.106 ± 0.04	0.107 ± 0.07 0.088 ± 0.06	0.287 ± 0.09 0.269 ± 0.07	0.143 ± 0.1 0.15 ± 0.08	0.523 ± 0.1 0.529 ± 0.25	0.595 ± 0.1 0.539 ± 0.18	0.312 ± 0.61 0.115 ± 0.17	0.479 ± 0.17 0.436 ± 0.22	0.56 ± 0.3 0.481 ± 0.24	0.224 ± 0.2 0.14 ± 0.1	0.394 ± 0.3 0.31 ± 0.27	0.61 ± 0.3 0.57 ± 0.33	0.352 ± 0.3 0.28 ± 0.39
Mn	μg·L^{-1}	8.93 ± 5.2 7.28 ± 2.7	7.54 ± 3.8 8.1 ± 3.5	8.53 ± 3.8 8.33 ± 4.1	5.14 ± 2 5 ± 3.7	15.4 ± 9.5 11.8 ± 5.1	13.2 ± 7.1 11.6 ± 9	13.9 ± 16 6.5 ± 14	32.4 ± 24 23 ± 30	28 ± 18 22.6 ± 17	41.8 ± 34 33.4 ± 22	13.9 ± 14 7.33 ± 15	19.2 ± 21 12 ± 16	17 ± 19 9.9 ± 13
Fe	μg·L^{-1}	122 ± 68 101 ± 101	107 ± 98 79 ± 73	97.6 ± 92 75 ± 56	62.7 ± 26 55.6 ± 32	171 ± 70 175 ± 139	154 ± 53 148 ± 44	160 ± 161 95 ± 138	215 ± 207 163 ± 90	303 ± 346 185 ± 177	520 ± 345 379 ± 468	361 ± 304 191 ± 432	366 ± 217 306 ± 309	215 ± 231 134 ± 161
Co	μg·L^{-1}	0.042 ± 0.01 0.044 ± 0.02	0.039 ± 0.02 0.035 ± 0.02	0.047 ± 0.01 0.046 ± 0.02	0.03 ± 0.01 0.028 ± 0.02	0.11 ± 0.1 0.104 ± 0.08	0.1 ± 0.03 0.1 ± 0.04	0.08 ± 0.07 0.051 ± 0.06	0.249 ± 0.14 0.196 ± 0.2	0.22 ± 0.1 0.195 ± 0.1	0.389 ± 0.3 0.31 ± 0.31	0.297 ± 0.2 0.23 ± 0.27	0.342 ± 0.3 0.28 ± 0.22	0.158 ± 0.18 0.1 ± 0.15
Ni	μg·L^{-1}	0.079 ± 0.04 0.061 ± 0.03	0.095 ± 0.06 0.083 ± 0.04	0.065 ± 0.02 0.066 ± 0.03	0.166 ± 0.1 0.132 ± 0.13	0.39 ± 0.2 0.374 ± 0.38	0.523 ± 0.8 0.337 ± 0.15	0.219 ± 0.11 0.203 ± 0.13	0.558 ± 0.3 0.551 ± 0.33	0.438 ± 0.22 0.465 ± 0.32	0.858 ± 0.5 0.74 ± 0.66	1.6 ± 0.7 1.48 ± 0.69	1.91 ± 0.5 2.08 ± 0.51	0.56 ± 0.69 0.27 ± 0.57
Cu	μg·L^{-1}	0.402 ± 1.08 0.116 ± 0.05	0.118 ± 0.05 0.107 ± 0.02	0.101 ± 0.04 0.09 ± 0.06	0.211 ± 0.1 0.161 ± 0.13	0.459 ± 0.2 0.436 ± 0.2	0.16 ± 0.1 0.134 ± 0.1	0.178 ± 0.11 0.16 ± 0.1	0.453 ± 0.24 0.387 ± 0.34	0.69 ± 1.1 0.212 ± 0.23	0.465 ± 0.3 0.41 ± 0.36	1.31 ± 1.6 0.72 ± 0.97	1.1 ± 1 0.97 ± 0.72	0.47 ± 0.79 0.23 ± 0.38
Zn	μg·L^{-1}	5.48 ± 3 4.67 ± 1.5	11.3 ± 11 8.95 ± 3.5	3.96 ± 1.2 3.94 ± 1	8.32 ± 10 3.8 ± 6.4	8.37 ± 3.7 7.13 ± 4.5	6.55 ± 3.2 5.37 ± 5.4	12.4 ± 5.6 10.7 ± 6.6	9.29 ± 2.9 8.67 ± 3.9	8.35 ± 14 4.45 ± 2	4.79 ± 2.4 4.59 ± 3.1	36.4 ± 31 24.5 ± 60	3.31 ± 1.9 3.06 ± 3	9.56 ± 13 5.75 ± 5.8
Ga	μg·L^{-1}	0.003 ± 0.003 0.002 ± 0.005	0.01 ± 0.002 0.008 ± 0.002	0.01 ± 0.002 0.008 ± 0.003	0.008 ± 0.01 0.006 ± 0.003	0.021 ± 0.01 0.022 ± 0.01	0.019 ± 0.01 0.018 ± 0.01	0.007 ± 0.003 0.006 ± 0.005	0.016 ± 0.008 0.015 ± 0.01	0.031 ± 0.03 0.019 ± 0.01	0.008 ± 0.004 0.01 ± 0.007	0.009 ± 0.007 0.01 ± 0.006	0.014 ± 0.02 0.01 ± 0.006	0.012 ± 0.01 0.009 ± 0.009
As	μg·L^{-1}	0.386 ± 0.08 0.37 ± 0.08	0.51 ± 0.09 0.508 ± 0.15	0.424 ± 0.08 0.437 ± 0.14	0.267 ± 0.1 0.266 ± 0.09	0.552 ± 0.2 0.501 ± 0.15	0.465 ± 0.11 0.459 ± 0.13	0.184 ± 0.09 0.164 ± 0.08	0.386 ± 0.15 0.342 ± 0.21	0.347 ± 0.12 0.332 ± 0.16	0.585 ± 0.25 0.48 ± 0.4	0.801 ± 0.37 0.72 ± 0.24	0.601 ± 0.14 0.56 ± 0.2	0.458 ± 0.23 0.42 ± 0.2
Se	μg·L^{-1}	25.8 ± 5.1 25.5 ± 6.1	28.8 ± 8.4 30.4 ± 11	28.7 ± 5.3 29 ± 9.6	25 ± 5.9 25.7 ± 7.6	57 ± 21 45.8 ± 26	53.1 ± 12 55 ± 17	16.8 ± 3.4 17 ± 4.6	45.7 ± 17 42.2 ± 19	38.2 ± 9.8 38 ± 17	24.5 ± 11 20 ± 10	46.6 ± 22 38.4 ± 15	42.6 ± 10 42 ± 9.3	36 ± 17 33 ± 19

Table A1. *Cont.*

Element	Units	Isolated			Sporadic			Discontinuous			Continuous			Average
		Spring	Summer	Autumn	Spring	Summer	Autumn	Spring	Summer	Autumn	Spring	Summer	Autumn	
Rb	μg·L^{-1}	0.194 ± 0.12 0.212 ± 0.15	0.149 ± 0.19 0.065 ± 0.18	0.131 ± 0.17 0.054 ± 0.11	0.178 ± 0.12 0.178 ± 0.21	0.163 ± 0.16 0.091 ± 0.2	0.145 ± 0.15 0.065 ± 0.24	0.145 ± 0.09 0.122 ± 0.15	0.048 ± 0.04 0.039 ± 0.02	0.057 ± 0.07 0.044 ± 0.04	0.35 ± 0.25 0.32 ± 0.26	0.09 ± 0.11 0.03 ± 0.13	0.359 ± 0.44 0.21 ± 0.17	0.169 ± 0.2 0.11 ± 0.2
Sr	μg·L^{-1}	1.52 ± 1.1 1.22 ± 0.73	1.73 ± 1.8 1.27 ± 0.6	1.84 ± 1.9 1.36 ± 0.98	1.07 ± 0.45 1.02 ± 0.59	2.39 ± 1.6 1.6 ± 1.3	1.87 ± 1.1 1.7 ± 1.6	1.3 ± 0.91 0.947 ± 0.76	3.02 ± 2 2.14 ± 2.4	4.91 ± 7 2.3 ± 2.1	7.28 ± 3.9 6.37 ± 5.1	11.8 ± 5.4 11.1 ± 7.1	11.4 ± 5.7 8.98 ± 7.3	4.12 ± 5 1.7 ± 4.3
Y	μg·L^{-1}	0.012 ± 0.01 0.01 ± 0.007	0.019 ± 0.01 0.015 ± 0.008	0.018 ± 0.01 0.015 ± 0.008	0.01 ± 0.006 0.009 ± 0.01	0.05 ± 0.03 0.056 ± 0.04	0.035 ± 0.03 0.027 ± 0.04	0.022 ± 0.02 0.015 ± 0.03	0.063 ± 0.04 0.052 ± 0.04	0.088 ± 0.1 0.042 ± 0.04	0.09 ± 0.05 0.07 ± 0.03	0.16 ± 0.14 0.12 ± 0.09	0.231 ± 0.17 0.18 ± 0.15	0.065 ± 0.09 0.03 ± 0.06
Zr	μg·L^{-1}	0.014 ± 0.02 0.012 ± 0.006	0.019 ± 0.03 0.013 ± 0.008	0.023 ± 0.03 0.014 ± 0.005	0.032 ± 0.02 0.025 ± 0.03	0.106 ± 0.06 0.096 ± 0.08	0.1 ± 0.08 0.068 ± 0.07	0.042 ± 0.03 0.029 ± 0.06	0.075 ± 0.05 0.06 ± 0.08	0.084 ± 0.04 0.068 ± 0.06	0.087 ± 0.07 0.06 ± 0.04	0.149 ± 0.15 0.11 ± 0.11	0.269 ± 0.27 0.23 ± 0.18	0.081 ± 0.12 0.05 ± 0.09
Nb	μg·L^{-1}	0.0031 ± 0.002 0.003 ± 0.003	0.002 ± 0.001 0.002 ± 0.002	0.0035 ± 0.003 0.003 ± 0.003	0.006 ± 0.004 0.004 ± 0.003	0.023 ± 0.012 0.024 ± 0.02	0.019 ± 0.012 0.017 ± 0.01	0.007 ± 0.004 0.008 ± 0.007	0.012 ± 0.006 0.011 ± 0.007	0.014 ± 0.005 0.013 ± 0.01	0.008 ± 0.007 0.005 ± 0.006	0.006 ± 0.006 0.004 ± 0.003	0.012 ± 0.006 0.011 ± 0.006	0.01 ± 0.009 0.007 ± 0.009
Mo	μg·L^{-1}	0.005 ± 0.004 0.003 ± 0.005	0.004 ± 0.003 0.004 ± 0.003	0.003 ± 0.001 0.004 ± 0.002	0.006 ± 0.005 0.005 ± 0.007	0.007 ± 0.005 0.004 ± 0.006	0.011 ± 0.008 0.008 ± 0.011	0.007 ± 0.005 0.006 ± 0.007	0.009 ± 0.01 0.005 ± 0.007	0.01 ± 0.008 0.008 ± 0.011	0.035 ± 0.033 0.026 ± 0.017	0.052 ± 0.062 0.03 ± 0.06	0.038 ± 0.02 0.032 ± 0.02	0.016 ± 0.03 0.007 ± 0.01
Cd	μg·L^{-1}	0.016 ± 0.004 0.018 ± 0.006	0.017 ± 0.006 0.017 ± 0.01	0.018 ± 0.006 0.019 ± 0.009	0.008 ± 0.005 0.006 ± 0.005	0.025 ± 0.019 0.018 ± 0.008	0.021 ± 0.011 0.018 ± 0.009	0.007 ± 0.003 0.006 ± 0.005	0.022 ± 0.011 0.018 ± 0.01	0.018 ± 0.008 0.017 ± 0.007	0.006 ± 0.004 0.005 ± 0.003	0.004 ± 0.003 0.003 ± 0.003	0.006 ± 0.005 0.004 ± 0.003	0.014 ± 0.01 0.01 ± 0.01
Sb	μg·L^{-1}	0.035 ± 0.004 0.036 ± 0.007	0.038 ± 0.008 0.038 ± 0.013	0.041 ± 0.008 0.044 ± 0.014	0.026 ± 0.006 0.024 ± 0.011	0.04 ± 0.014 0.038 ± 0.016	0.04 ± 0.011 0.038 ± 0.018	0.015 ± 0.005 0.015 ± 0.007	0.027 ± 0.008 0.024 ± 0.014	0.031 ± 0.023 0.026 ± 0.009	0.026 ± 0.047 0.013 ± 0.01	0.022 ± 0.014 0.017 ± 0.012	0.026 ± 0.009 0.025 ± 0.012	0.03 ± 0.02 0.03 ± 0.02
Cs	μg·L^{-1}	0.005 ± 0.004 0.004 ± 0.005	0.005 ± 0.006 0.003 ± 0.007	0.005 ± 0.004 0.003 ± 0.006	0.008 ± 0.006 0.008 ± 0.011	0.011 ± 0.011 0.009 ± 0.01	0.018 ± 0.014 0.015 ± 0.017	0.005 ± 0.003 0.004 ± 0.006	0.0021 ± 0.003 0.001 ± 0.002	0.002 ± 0.002 0.001 ± 0.004	0.001 ± 0.0003 0.0005 ± 0.0004	0.001 ± 0.001 0.001 ± 0.0006	0.002 ± 0.004 0.001 ± 0.001	0.006 ± 0.008 0.003 ± 0.007
Ba	μg·L^{-1}	1.63 ± 1.3 1 ± 1.7	1.4 ± 0.99 1.1 ± 0.57	1.34 ± 1.2 0.984 ± 0.81	1.65 ± 0.84 1.49 ± 1.2	3.35 ± 1.8 2.61 ± 2.1	2.27 ± 1.7 1.88 ± 1	2.89 ± 1.3 3.14 ± 2.3	6.03 ± 4.8 4.3 ± 5.5	4.59 ± 3.3 3.5 ± 1.6	2.34 ± 0.75 2.1 ± 0.57	87.4 ± 81 78 ± 117	2.59 ± 0.58 2.75 ± 1	9.78 ± 33 2.2 ± 2.2
La	μg·L^{-1}	0.032 ± 0.03 0.017 ± 0.03	0.022 ± 0.03 0.01 ± 0.006	0.011 ± 0.005 0.01 ± 0.006	0.236 ± 0.55 0.092 ± 0.11	0.028 ± 0.012 0.027 ± 0.01	0.026 ± 0.017 0.023 ± 0.02	0.04 ± 0.039 0.03 ± 0.02	0.57 ± 0.04 0.035 ± 0.05	0.055 ± 0.06 0.033 ± 0.03	0.064 ± 0.03 0.05 ± 0.02	0.123 ± 0.11 0.09 ± 0.06	0.155 ± 0.11 0.14 ± 0.09	0.072 ± 0.18 0.03 ± 0.06
Ce	μg·L^{-1}	0.05 ± 0.06 0.023 ± 0.033	0.022 ± 0.012 0.019 ± 0.008	0.022 ± 0.014 0.018 ± 0.01	0.024 ± 0.014 0.022 ± 0.02	0.06 ± 0.03 0.069 ± 0.04	0.058 ± 0.042 0.04 ± 0.05	0.062 ± 0.06 0.04 ± 0.07	0.262 ± 0.046 0.149 ± 0.22	0.129 ± 0.12 0.078 ± 0.09	0.154 ± 0.07 0.14 ± 0.03	0.268 ± 0.26 0.184 ± 0.17	0.357 ± 0.28 0.297 ± 0.24	0.12 ± 0.2 0.05 ± 0.12
Pr	μg·L^{-1}	0.003 ± 0.002 0.002 ± 0.002	0.002 ± 0.002 0.002 ± 0.001	0.003 ± 0.002 0.002 ± 0.002	0.002 ± 0.001 0.002 ± 0.002	0.007 ± 0.004 0.007 ± 0.006	0.007 ± 0.005 0.05 ± 0.007	0.006 ± 0.005 0.004 ± 0.007	0.012 ± 0.011 0.008 ± 0.01	0.016 ± 0.015 0.01 ± 0.01	0.02 ± 0.01 0.017 ± 0.005	0.035 ± 0.03 0.024 ± 0.02	0.048 ± 0.035 0.04 ± 0.03	0.013 ± 0.02 0.006 ± 0.01
Nd	μg·L^{-1}	0.012 ± 0.008 0.011 ± 0.007	0.01 ± 0.007 0.008 ± 0.003	0.011 ± 0.009 0.01 ± 0.008	0.009 ± 0.005 0.009 ± 0.009	0.029 ± 0.016 0.03 ± 0.02	0.028 ± 0.021 0.02 ± 0.028	0.024 ± 0.02 0.018 ± 0.03	0.046 ± 0.04 0.031 ± 0.04	0.065 ± 0.06 0.041 ± 0.05	0.082 ± 0.04 0.07 ± 0.02	0.141 ± 0.136 0.101 ± 0.09	0.203 ± 0.15 0.166 ± 0.14	0.054 ± 0.08 0.02 ± 0.06
Sm	μg·L^{-1}	0.007 ± 0.003 0.006 ± 0.004	0.006 ± 0.002 0.006 ± 0.003	0.009 ± 0.003 0.009 ± 0.003	0.004 ± 0.001 0.004 ± 0.001	0.012 ± 0.004 0.012 ± 0.005	0.012 ± 0.005 0.011 ± 0.01	0.009 ± 0.006 0.006 ± 0.007	0.018 ± 0.01 0.013 ± 0.01	0.036 ± 0.042 0.018 ± 0.01	0.023 ± 0.01 0.021 ± 0.07	0.034 ± 0.03 0.024 ± 0.02	0.054 ± 0.045 0.046 ± 0.03	0.018 ± 0.02 0.01 ± 0.01
Eu	μg·L^{-1}	0.001 ± 0.0009 0.001 ± 0.0002	0.0011 ± 0.001 0.001 ± 0.0001	0.001 ± 0.001 0.001 ± 0.0003	0.001 ± 0.0004 0.001 ± 0.001	0.0022 ± 0.001 0.002 ± 0.002	0.002 ± 0.001 0.001 ± 0.002	0.002 ± 0.001 0.001 ± 0.001	0.003 ± 0.002 0.002 ± 0.003	0.003 ± 0.001 0.002 ± 0.001	0.004 ± 0.002 0.004 ± 0.002	0.015 ± 0.009 0.011 ± 0.01	0.017 ± 0.03 0.009 ± 0.007	0.004 ± 0.01 0.002 ± 0.003

Table A1. *Cont.*

Element	Units	Isolated			Sporadic			Discontinuous			Continuous			Average
		Spring	Summer	Autumn	Spring	Summer	Autumn	Spring	Summer	Autumn	Spring	Summer	Autumn	
Gd	$\mu g{\cdot}L^{-1}$	0.003 ± 0.002	0.004 ± 0.002	0.004 ± 0.003	0.0024 ± 0.002	0.009 ± 0.005	0.007 ± 0.004	0.005 ± 0.0048	0.014 ± 0.01	0.02 ± 0.02	0.02 ± 0.01	0.036 ± 0.03	0.05 ± 0.04	0.014 ± 0.02
		0.003 ± 0.002	0.003 ± 0.002	0.004 ± 0.002	0.0021 ± 0.002	0.009 ± 0.008	0.005 ± 0.007	0.005 ± 0.006	0.01 ± 0.017	0.011 ± 0.01	0.017 ± 0.007	0.027 ± 0.02	0.04 ± 0.03	0.006 ± 0.01
Tb	$\mu g{\cdot}L^{-1}$	0.0005 ± 0.0003	0.001 ± 0.0003	0.001 ± 0.0004	0.0003 ± 0.0002	0.002 ± 0.001	0.0012 ± 0.001	0.001 ± 0.0007	0.002 ± 0.001	0.003 ± 0.0025	0.003 ± 0.001	0.005 ± 0.005	0.007 ± 0.005	0.002 ± 0.003
		0.0004 ± 0.0003	0.001 ± 0.0003	0.0005 ± 0.0005	0.0002 ± 0.0003	0.002 ± 0.001	0.001 ± 0.001	0.001 ± 0.001	0.002 ± 0.001	0.001 ± 0.001	0.002 ± 0.001	0.004 ± 0.003	0.006 ± 0.005	0.001 ± 0.002
Dy	$\mu g{\cdot}L^{-1}$	0.002 ± 0.0013	0.004 ± 0.0018	0.003 ± 0.0026	0.0022 ± 0.001	0.01 ± 0.006	0.007 ± 0.005	0.004 ± 0.004	0.013 ± 0.008	0.016 ± 0.02	0.016 ± 0.009	0.03 ± 0.028	0.041 ± 0.03	0.012 ± 0.02
		0.002 ± 0.001	0.003 ± 0.002	0.003 ± 0.002	0.002 ± 0.0017	0.009 ± 0.008	0.006 ± 0.008	0.003 ± 0.005	0.01 ± 0.01	0.008 ± 0.01	0.014 ± 0.01	0.021 ± 0.02	0.033 ± 0.03	0.007 ± 0.01
Ho	$\mu g{\cdot}L^{-1}$	0.0005 ± 0.0004	0.001 ± 0.0004	0.001 ± 0.0005	0.0005 ± 0.0003	0.002 ± 0.001	0.0013 ± 0.001	0.001 ± 0.0007	0.003 ± 0.002	0.003 ± 0.003	0.003 ± 0.002	0.006 ± 0.005	0.008 ± 0.006	0.002 ± 0.003
		0.0003 ± 0.0002	0.001 ± 0.0003	0.001 ± 0.0004	0.0004 ± 0.0005	0.002 ± 0.002	0.001 ± 0.002	0.001 ± 0.0007	0.002 ± 0.002	0.002 ± 0.001	0.003 ± 0.001	0.004 ± 0.003	0.007 ± 0.005	0.001 ± 0.002
Er	$\mu g{\cdot}L^{-1}$	0.001 ± 0.001	0.002 ± 0.001	0.002 ± 0.0016	0.001 ± 0.0008	0.006 ± 0.003	0.004 ± 0.003	0.003 ± 0.002	0.01 ± 0.004	0.01 ± 0.0099	0.01 ± 0.005	0.016 ± 0.01	0.024 ± 0.01	0.007 ± 0.01
		0.001 ± 0.0009	0.002 ± 0.001	0.001 ± 0.001	0.0005 ± 0.001	0.006 ± 0.005	0.003 ± 0.005	0.002 ± 0.003	0.006 ± 0.005	0.005 ± 0.004	0.009 ± 0.003	0.012 ± 0.01	0.02 ± 0.017	0.004 ± 0.007
Tm	$\mu g{\cdot}L^{-1}$	0.0001 ± 0.0002	0.0003 ± 0.0002	0.0003 ± 0.0003	0.0001 ± 0.0001	0.0009 ± 0.0005	0.001 ± 0.0005	0.0004 ± 0.0004	0.001 ± 0.0006	0.0015 ± 0.001	0.0014 ± 0.001	0.002 ± 0.002	0.003 ± 0.002	0.001 ± 0.001
		0.0001 ± 0.0001	0.0002 ± 0.0002	0.0003 ± 0.0002	0.0001 ± 0.0002	0.001 ± 0.0007	0.0004 ± 0.001	0.0002 ± 0.0006	0.001 ± 0.0007	0.001 ± 0.001	0.001 ± 0.0001	0.002 ± 0.001	0.003 ± 0.002	0.0005 ± 0.001
Yb	$\mu g{\cdot}L^{-1}$	0.002 ± 0.0012	0.002 ± 0.0014	0.002 ± 0.0016	0.001 ± 0.0009	0.005 ± 0.003	0.004 ± 0.003	0.003 ± 0.002	0.01 ± 0.004	0.01 ± 0.001	0.01 ± 0.004	0.015 ± 0.01	0.024 ± 0.02	0.007 ± 0.01
		0.001 ± 0.001	0.0015 ± 0.001	0.0013 ± 0.001	0.001 ± 0.001	0.005 ± 0.004	0.003 ± 0.004	0.002 ± 0.003	0.005 ± 0.005	0.004 ± 0.005	0.01 ± 0.004	0.01 ± 0.008	0.02 ± 0.01	0.003 ± 0.007
Lu	$\mu g{\cdot}L^{-1}$	0.0003 ± 0.0002	0.0002 ± 0.0002	0.0003 ± 0.0003	0.0002 ± 0.0002	0.001 ± 0.0004	0.0005 ± 0.0005	0.0004 ± 0.0004	0.0008 ± 0.0006	0.0015 ± 0.001	0.0014 ± 0.001	0.002 ± 0.0017	0.004 ± 0.003	0.001 ± 0.001
		0.0002 ± 0.0002	0.0002 ± 0.0001	0.0002 ± 0.0001	0.0001 ± 0.0003	0.001 ± 0.0006	0.0003 ± 0.0007	0.0003 ± 0.0005	0.001 ± 0.0007	0.001 ± 0.001	0.001 ± 0.0006	0.002 ± 0.001	0.003 ± 0.002	0.0004 ± 0.001
Hf	$\mu g{\cdot}L^{-1}$	0.003 ± 0.002	0.003 ± 0.002	0.003 ± 0.002	0.005 ± 0.003	0.016 ± 0.01	0.014 ± 0.01	0.005 ± 0.004	0.013 ± 0.007	0.029 ± 0.04	0.012 ± 0.01	0.021 ± 0.02	0.035 ± 0.03	0.013 ± 0.018
		0.002 ± 0.0015	0.002 ± 0.001	0.003 ± 0.002	0.005 ± 0.003	0.015 ± 0.01	0.011 ± 0.01	0.003 ± 0.007	0.013 ± 0.01	0.015 ± 0.02	0.01 ± 0.01	0.02 ± 0.01	0.03 ± 0.02	0.008 ± 0.01
Pb	$\mu g{\cdot}L^{-1}$	0.274 ± 0.22	0.19 ± 0.1	0.218 ± 0.11	0.139 ± 0.08	0.257 ± 0.16	0.3 ± 0.1	0.085 ± 0.04	0.2 ± 0.11	0.281 ± 0.14	0.065 ± 0.03	0.067 ± 0.04	0.088 ± 0.09	0.184 ± 0.14
		0.241 ± 0.15	0.162 ± 0.09	0.18 ± 0.12	0.111 ± 0.12	0.222 ± 0.13	0.316 ± 0.25	0.085 ± 0.04	0.184 ± 0.1	0.245 ± 0.21	0.06 ± 0.03	0.05 ± 0.06	0.06 ± 0.05	0.15 ± 0.16
Th	$\mu g{\cdot}L^{-1}$	0.002 ± 0.001	0.003 ± 0.001	0.003 ± 0.002	0.003 ± 0.001	0.011 ± 0.01	0.011 ± 0.005	0.006 ± 0.004	0.011 ± 0.01	0.012 ± 0.01	0.01 ± 0.008	0.021 ± 0.02	0.028 ± 0.01	0.01 ± 0.01
		0.002 ± 0.001	0.002 ± 0.001	0.003 ± 0.002	0.003 ± 0.002	0.011 ± 0.01	0.012 ± 0.01	0.003 ± 0.008	0.008 ± 0.009	0.01 ± 0.01	0.006 ± 0.006	0.01 ± 0.01	0.03 ± 0.02	0.01 ± 0.01
U	$\mu g{\cdot}L^{-1}$	0.001 ± 0.0003	0.001 ± 0.0005	0.001 ± 0.0005	0.002 ± 0.001	0.004 ± 0.002	0.003 ± 0.002	0.0012 ± 0.001	0.002 ± 0.0015	0.004 ± 0.006	0.007 ± 0.005	0.017 ± 0.01	0.017 ± 0.01	0.005 ± 0.008
		0.001 ± 0.0004	0.001 ± 0.0006	0.001 ± 0.0009	0.002 ± 0.001	0.004 ± 0.002	0.003 ± 0.002	0.001 ± 0.001	0.001 ± 0.002	0.002 ± 0.001	0.006 ± 0.002	0.01 ± 0.01	0.01 ± 0.01	0.002 ± 0.004

References

1. Colombo, N.; Salerno, F.; Gruber, S.; Freppaz, M.; Williams, M.; Fratianni, S.; Giardino, M. Review: Impacts of permafrost degradation on inorganic chemistry of surface fresh water. *Glob. Planet. Chang.* **2018**, *162*, 69–83. [CrossRef]
2. Kirpotin, S.; Berezin, A.; Bazanov, V.; Polishchuk, Y.; Vorobiov, S.; Mironycheva-Tokoreva, N.; Kosykh, N.; Volkova, I.; Dupré, B.; Pokrovsky, O. Western Siberia wetlands as indicator and regulator of climate change on the global scale. *Int. J. Environ. Stud.* **2009**, *66*, 409–421. [CrossRef]
3. Kirpotin, S.; Polishchuk, Y.; Bryksina, N.; Sugaipova, A.; Kouraev, A.; Zakharova, E.; Pokrovsky, O.S.; Shirokova, L.; Kolmakova, M.; Manassypov, R.; et al. West Siberian palsa peatlands: Distribution, typology, cyclic development, present day climate-driven changes, seasonal hydrology and impact on CO_2 cycle. *Int. J. Environ. Stud.* **2011**, *68*, 603–623. [CrossRef]
4. Hjort, J.; Karjalainen, O.; Aalto, J.; Westermann, S.; Romanovsky, V.E.; Nelson, F.E.; Etzelmüller, B.; Luoto, M. Degrading permafrost puts Arctic infrastructure at risk by mid-century. *Nat. Commun.* **2018**, *9*, 1–9. [CrossRef] [PubMed]
5. Polishchuk, Y.M.; Bogdanov, A.N.; Muratov, I.N.; Polishchuk, V.Y.; Lim, A.; Manasypov, R.M.; Shirokova, L.S.; Pokrovsky, O.S. Minor contribution of small thaw ponds to the pools of carbon and methane in the inland waters of the permafrost-affected part of the Western Siberian Lowland. *Environ. Res. Lett.* **2018**, *13*, 045002. [CrossRef]
6. Polishchuk, Y.M.; Bogdanov, A.N.; Polishchuk, V.Y.; Manasypov, R.M.; Shirokova, L.S.; Kirpotin, S.N.; Pokrovsky, O.S. Size distribution, surface coverage, water, carbon, and metal storage of thermokarst lakes in the permafrost zone of the Western Siberia Lowland. *Water* **2017**, *9*, 228. [CrossRef]
7. Serikova, S.; Pokrovsky, O.S.; Laudon, H.; Krickov, I.V.; Lim, A.G.; Manasypov, R.M.; Karlsson, J. High carbon emissions from thermokarst lakes of Western Siberia. *Nat. Commun.* **2019**, *10*, 1–7. [CrossRef]
8. Marsh, P.; Russell, M.; Pohl, S.; Haywood, H.; Onclin, C. Changes in thaw lake drainage in the Western Canadian Arctic from 1950 to 2000. *Hydrol. Process.* **2009**, *23*, 145–158. [CrossRef]
9. Rautio, M.; Dufresne, F.; Laurion, I.; Bonilla, S.; Vincent, W.F.; Christoffersen, K.S. Shallow freshwater ecosystems of the circumpolar Arctic. *Écoscience* **2011**, *18*, 204–222. [CrossRef]
10. Grosse, G.; Goetz, S.; McGuire, A.D.; Romanovsky, V.E.; Schuur, E.A.G. Changing permafrost in a warming world and feedbacks to the Earth system. *Environ. Res. Lett.* **2016**, *11*, 040201. [CrossRef]
11. Vonk, J.E.; Tank, S.E.; Bowden, W.B.; Laurion, I.; Vincent, W.F.; Alekseychik, P.; Amyot, M.; Billet, M.F.; Canário, J.; Cory, R.M.; et al. Reviews and syntheses: Effects of permafrost thaw on Arctic aquatic ecosystems. *Biogeosciences* **2015**, *12*, 7129–7167. [CrossRef]
12. Bouchard, F.; Turner, K.W.; MacDonald, L.A.; Deakin, C.; White, H.; Farquharson, N.; Medeiros, A.S.; Wolfe, B.B.; Hall, R.I.; Pienitz, R.; et al. Vulnerability of shallow subarctic lakes to evaporate and desiccate when snowmelt runoff is low. *Geophys. Res. Lett.* **2013**, *40*, 6112–6117. [CrossRef]
13. Manasypov, R.M.; Pokrovsky, O.S.; Kirpotin, S.N.; Shirokova, L.S. Thermokarst lake waters across the permafrost zones of western Siberia. *Cryosphere* **2014**, *8*, 1177–1193. [CrossRef]
14. Arsenault, J.; Talbot, J.; Moore, T.R. Environmental controls of C, N and P biogeochemistry in peatland pools. *Sci. Total Environ.* **2018**, *631–632*, 714–722. [CrossRef]
15. Arsenault, J.; Talbot, J.; Moore, T.R.; Beauvais, M.-P.; Franssen, J.; Roulet, N.T. The spatial heterogeneity of vegetation, hydrology and water chemistry in a peatland with open-water pools. *Ecosystems* **2019**, *22*, 1352–1367. [CrossRef]
16. Pokrovsky, O.S.; Manasypov, R.M.; Loiko, S.V.; Krickov, I.A.; Kopysov, S.G.; Kolesnichenko, L.G.; Vorobyev, S.N.; Kirpotin, S.N. Trace element transport in western Siberian rivers across a permafrost gradient. *Biogeosciences* **2016**, *13*, 1877–1900. [CrossRef]
17. Kokelj, S.V.; Zajdlik, B.; Thompson, M.S. The impacts of thawing permafrost on the chemistry of lakes across the subarctic boreal-tundra transition, Mackenzie Delta region, Canada. *Permafr. Periglac. Process.* **2009**, *20*, 185–199. [CrossRef]
18. Pokrovsky, O.S.; Shirokova, L.S.; Kirpotin, S.N.; Audry, S.; Viers, J.; Dupré, B. Effect of permafrost thawing on organic carbon and trace element colloidal speciation in the thermokarst lakes of western Siberia. *Biogeosciences* **2011**, *8*, 565–583. [CrossRef]

19. Shirokova, L.S.; Pokrovsky, O.S.; Kirpotin, S.N.; Desmukh, C.; Pokrovsky, B.G.; Audry, S.; Viers, J. Biogeochemistry of organic carbon, CO_2, CH_4, and trace elements in thermokarst water bodies in discontinuous permafrost zones of Western Siberia. *Biogeochemistry* **2013**, *113*, 573–593. [CrossRef]
20. Shevchenko, V.P.; Pokrovsky, O.S.; Vorobyev, S.N.; Krickov, I.V.; Manasypov, R.M.; Politova, N.V.; Kopysov, S.G.; Dara, O.M.; Auda, Y.; Shirokova, L.S.; et al. Impact of snow deposition on major and trace element concentrations and elementary fluxes in surface waters of the Western Siberian Lowland across a 1700 km latitudinal gradient. *Hydrol. Earth Syst. Sci.* **2017**, *21*, 5725–5746. [CrossRef]
21. Manasypov, R.M.; Shirokova, L.S.; Pokrovsky, O.S. Experimental modeling of thaw lake water evolution in discontinuous permafrost zone: Role of peat, lichen leaching and ground fire. *Sci. Total Environ.* **2017**, *580*, 245–257. [CrossRef]
22. Osterkamp, T.E.; Viereck, L.; Shur, Y.; Jorgenson, M.T.; Racine, C.; Doyle, A.; Boone, R.D. Observations of thermokarst and its impact on boreal forests in Alaska, U.S.A. *Arctic Antarct. Alp. Res.* **2000**, *32*, 303–315. [CrossRef]
23. Luoto, M.; Seppälä, M. Thermokarst ponds as indicators of the former distribution of palsas in Finnish Lapland. *Permafr. Periglac. Process.* **2003**, *14*, 19–27. [CrossRef]
24. Bouchard, F.; Francus, P.; Pienitz, R.; Laurion, I. Sedimentology and geochemistry of thermokarst ponds in discontinuous permafrost, subarctic Quebec, Canada. *J. Geophys. Res. Biogeosci.* **2011**, *116*, G00M04. [CrossRef]
25. Bouchard, F.; Francus, P.; Pienitz, R.; Laurion, I.; Feyte, S. Subarctic thermokarst ponds: Investigating recent landscape evolution and sediment dynamics in thawed permafrost of Northern Québec (Canada). *Arct. Antarct. Alpine Res.* **2014**, *46*, 251–271. [CrossRef]
26. Bouchard, F.; MacDonald, L.A.; Turner, K.W.; Thienpont, J.R.; Medeiros, A.S.; Biskaborn, B.K.; Korosi, J.; Hall, R.I.; Pienitz, R.; Wolfe, B.B. Paleolimnology of thermokarst lakes: A window into permafrost landscape evolution. *Arctic Sci.* **2017**, *3*, 91–117. [CrossRef]
27. Bouchard, F.; Proult, V.; Pienitz, R.; Antoniades, D.; Tremblay, R.; Vincent, W.F. Periphytic diatom community structure in thermokarst ecosystems of Nunavik (Québec, Canada). *Arctic Sci.* **2018**, *4*, 110–129. [CrossRef]
28. Coulombe, O.; Bouchard, F.; Pienitz, R. Coupling of sedimentological and limnological dynamics in subarctic thermokarst ponds in Northern Québec (Canada) on an interannual basis. *Sediment. Geol.* **2016**, *340*, 15–24. [CrossRef]
29. Yi, S.; Wischnewski, K.; Langer, M.; Muster, S.; Boike, J. Freeze/thaw processes in complex permafrost landscapes of northern Siberia simulated using the TEM ecosystem model: Impact of thermokarst ponds and lakes. *Geosci. Model. Dev.* **2014**, *7*, 1671–1689. [CrossRef]
30. Czudek, T.; Demek, J. Thermokarst in Siberia and its influence on the development of lowland relief. *Quat. Res.* **1970**, *1*, 103–120. [CrossRef]
31. Selroos, J.-O.; Cheng, H.; Vidstrand, P.; Destouni, G. Permafrost thaw with thermokarst wetland-lake and societal-health risks: Dependence on local soil conditions under large-scale warming. *Water* **2019**, *11*, 574. [CrossRef]
32. Audry, S.; Pokrovsky, O.S.; Shirokova, L.S.; Kirpotin, S.N.; Dupré, B. Organic matter mineralization and trace element post-depositional redistribution in Western Siberia thermokarst lake sediments. *Biogeosciences* **2011**, *8*, 3341–3358. [CrossRef]
33. Jorgenson, M.T.; Shur, Y. Evolution of lakes and basins in northern Alaska and discussion of the thaw lake cycle. *J. Geophys. Res. Earth Surface* **2007**, *112*, F02S17. [CrossRef]
34. Hinkel, K.M.; Eisner, W.R.; Bockheim, J.G.; Nelson, F.E.; Peterson, K.M.; Dai, X. Spatial extent, age, and carbon stocks in drained thaw lake basins on the Barrow Peninsula, Alaska. *Arct. Antarct. Alpine Res.* **2003**, *35*, 291–300. [CrossRef]
35. Matthews, J.A.; Dahl, S.O.; Berrisford, M.S.; Nesje, A. Cyclic development and thermokarstic degradation in the mid-alpine zone at Leirpullan, Dovrefjel, southern Norway. *Permafr. Periglac. Process.* **1997**, *8*, 107–122. [CrossRef]
36. Ellis, C.J.; Rochefort, L.; Gauthier, G.; Pienitz, R. Paleoeco-logical evidence for transitions between contrasting landformsin a polygon-patterned High Arctic wetland. *Arct. Antarct. Alpine Res.* **2008**, *40*, 624–637. [CrossRef]
37. Calmels, F.; Allard, M.; Delisle, G. Development and decay of a lithalsa in Northern Québec: A geomorphological history. *Geomorphology* **2008**, *97*, 287–299. [CrossRef]

38. Grosse, G.; Jones, B.; Arp, C. Thermokarst Lakes, Drainage, and Drained Basins. In *Treatise on Geomorphology*, 3rd ed.; Shroder, J., Giardino, R., Harbor, J., Eds.; Academic Press: San Diego, CA, USA, 2013; Volume 8, pp. 325–353.
39. Lenz, J.; Wetterich, S.; Jones, B.M.; Meyer, H.; Bobrov, A.; Grosse, G. Evidence of multiple thermokarst lake generations from an 11 800-year-old permafrost core on the northern Seward Peninsula, Alaska. *Boreas* **2016**, *45*, 584–603. [CrossRef]
40. Manasypov, R.M.; Vorobyev, S.N.; Loiko, S.V.; Kritzkov, I.V.; Shirokova, L.S.; Shevchenko, V.P.; Kirpotin, S.N.; Kulizhsky, S.P.; Kolesnichenko, L.G.; Zemtzov, V.A.; et al. Seasonal dynamics of organic carbon and metals in thermokarst lakes from the discontinuous permafrost zone of western Siberia. *Biogeosciences* **2015**, *12*, 3009–3028. [CrossRef]
41. Walter Anthony, K.M.; Anthony, P.; Grosse, G.; Chanton, J. Geologic methane seeps along boundaries of Arctic permafrost thaw and melting glaciers. *Nat. Geosci.* **2012**, *5*, 419–426. [CrossRef]
42. Pokrovsky, O.S.; Shirokova, L.S.; Kirpotin, S.N.; Kulizhsky, S.P.; Vorobiev, S.N. Impact of western Siberia heat wave 2012 on greenhouse gases and trace metal concentration in thaw lakes of discontinuous permafrost zone. *Biogeosciences* **2013**, *10*, 5349–5365. [CrossRef]
43. Pokrovsky, O.S.; Shirokova, L.S.; Manasypov, R.M.; Kirpotin, S.N.; Kulizhsky, S.P.; Kolesnichenko, L.G.; Loiko, S.V.; Vorobyev, S.N. Thermokarst lakes of Western Siberia: A complex biogeochemical multidisciplinary approach. *Int. J. Environ. Stud.* **2014**, *71*, 733–748. [CrossRef]
44. Loiko, S.V.; Pokrovsky, O.S.; Raudina, T.V.; Lim, A.; Kolesnichenko, L.G.; Shirokova, L.S.; Vorobyev, S.N.; Kirpotin, S.N. Abrupt permafrost collapse enhances organic carbon, CO_2, nutrient and metal release into surface waters. *Chem. Geol.* **2017**, *471*, 153–165. [CrossRef]
45. Laudon, H.; Köhler, S.; Buffam, I. Seasonal TOC export from seven boreal catchments in northern Sweden. *Aquat. Sci.* **2004**, *66*, 223–230. [CrossRef]
46. Pokrovsky, O.S.; Karlsson, J.; Giesler, R. Freeze-thaw cycles of Arctic thaw ponds remove colloidal metals and generate low-molecular-weight organic matter. *Biogeochemistry* **2018**, *137*, 321–336. [CrossRef]
47. Ala-aho, P.; Soulsby, C.; Pokrovsky, O.S.; Kirpotin, S.N.; Karlsson, J.; Serikova, S.; Vorobyev, S.N.; Manasypov, R.M.; Loiko, S.; Tetzlaff, D. Using stable isotopes to assess surface water source dynamics and hydrological connectivity in a high-latitude wetland and permafrost influenced landscape. *J. Hydrol.* **2018**, *556*, 279–293. [CrossRef]
48. Erwin, K.L. Wetlands and global climate change: The role of wetland restoration in a changing world. *Wetl. Ecol. Manag.* **2009**, *17*, 71–84. [CrossRef]
49. Raudina, T.V.; Loiko, S.V.; Lim, A.G.; Krickov, I.V.; Shirokova, L.S.; Istigechev, G.I.; Kuzmina, D.M.; Kulizhsky, S.P.; Vorobyev, S.N.; Pokrovsky, O.S. Dissolved organic carbon and major and trace elements in peat porewater of sporadic, discontinuous, and continuous permafrost zones of western Siberia. *Biogeosciences* **2017**, *14*, 3561–3584. [CrossRef]
50. Raudina, T.V.; Loiko, S.V.; Lim, A.; Manasypov, R.M.; Shirokova, L.S.; Istigechev, G.I.; Kuzmina, D.M.; Kulizhsky, S.P.; Vorobyev, S.N.; Pokrovsky, O.S. Permafrost thaw and climate warming may decrease the CO_2, carbon, and metal concentration in peat soil waters of the Western Siberia Lowland. *Sci. Total Environ.* **2018**, *634*, 1004–1023. [CrossRef]
51. Pokrovsky, O.S.; Manasypov, R.M.; Loiko, S.; Shirokova, L.S.; Krickov, I.A.; Pokrovsky, B.G.; Kolesnichenko, L.G.; Kopysov, S.G.; Zemtzov, V.A.; Kulizhsky, S.P.; et al. Permafrost coverage, watershed area and season control of dissolved carbon and major elements in western Siberian rivers. *Biogeosciences* **2015**, *12*, 6301–6320. [CrossRef]
52. Krickov, I.V.; Lim, A.G.; Manasypov, R.M.; Loiko, S.V.; Shirokova, L.S.; Kirpotin, S.N.; Karlsson, J.; Pokrovsky, O.S. Riverine particulate C and N generated at the permafrost thaw front: Case study of western Siberian rivers across a 1700-km latitudinal transect. *Biogeosciences* **2018**, *15*, 6867–6884. [CrossRef]
53. Frey, K.E.; McClelland, J.W.; Holmes, R.M.; Smith, L.C. Impacts of climate warming and permafrost thaw on the riverine transport of nitrogen and phosphorus to the Kara Sea. *J. Geophys. Res. Biogeosci.* **2007**, *112*, G04S58. [CrossRef]
54. Frey, K.E.; Smith, L.C. Amplified carbon release from vast West Siberian peatlands by 2100. *Geophys. Res. Lett.* **2005**, *32*, L09401. [CrossRef]

55. Vorobyev, S.N.; Pokrovsky, O.S.; Serikova, S.; Manasypov, R.M.; Krickov, I.V.; Shirokova, L.S.; Lim, A.; Kolesnichenko, L.G.; Kirpotin, S.N.; Karlsson, J. Permafrost boundary shift in Western Siberia may not modify dissolved nutrient concentrations in rivers. *Water* **2017**, *9*, 985. [CrossRef]
56. Blois, J.L.; Williams, J.W.; Fitzpatrick, M.C.; Jackson, S.T.; Ferrier, S. Space can substitute for time in predicting climate-change effects on biodiversity. *Proc. Natl. Acad. Sci. USA* **2013**, *110*, 9374–9379. [CrossRef]
57. Masing, V.; Botch, M.; Läänelaid, A. Mires of the former Soviet Union. *Wetl. Ecol Manag.* **2010**, *18*, 397–433. [CrossRef]
58. Trofimova, I.E.; Balybina, A.S. Classification of climates and climatic regionalization of the West-Siberian plain. *Geogr. Nat. Resour.* **2014**, *35*, 114–122. [CrossRef]
59. Serikova, S.; Pokrovsky, O.S.; Ala-Aho, P.; Kazantsev, V.; Kirpotin, S.N.; Kopysov, S.G.; Krickov, I.V.; Laudon, H.; Manasypov, R.M.; Shirokova, L.S.; et al. High riverine CO_2 emissions at the permafrost boundary of Western Siberia. *Nat. Geosci.* **2018**, *11*, 825–829. [CrossRef]
60. Pokrovsky, O.S.; Manasypov, R.M.; Loiko, S.V.; Shirokova, L.S. Organic and organo-mineral colloids in discontinuous permafrost zone. *Geochim. Cosmochim. Acta* **2016**, *188*, 1–20. [CrossRef]
61. Brown, J.; Ferrians, O.J., Jr.; Heginbottom, J.A.; Melnikov, E.S. *Circum-Arctic Map of Permafrost and Ground Ice Conditions*; National Snow and Ice Data Center/World Data Center for Glaciology Digital Media: Boulder, CO, USA, 2001; pp. 28–44.
62. Savchenko, N.V. Nature of lakes in subarctic of West Siberia. *Geogr. Prir. Resur.* **1992**, *1*, 85–92. (In Russian)
63. Vasyukova, E.V.; Pokrovsky, O.S.; Viers, J.; Oliva, P.; Dupré, B.; Martin, F.; Candaudap, F. Trace elements in organic- and iron-rich surficial fluids of the boreal zone: Assessing colloidal forms via dialysis and ultrafiltration. *Geochim. Cosmochim. Acta* **2010**, *74*, 449–468. [CrossRef]
64. Yeghicheyan, D.; Bossy, C.; Coz, M.B.L.; Douchet, C.; Granier, G.; Heimburger, A.; Lacan, F.; Lanzanova, A.; Rousseau, T.C.C.; Seidel, J.-L.; et al. A compilation of silicon, rare earth element and twenty-one other trace tlement concentrations in the natural river water reference material SLRS-5 (NRC-CNRC). *Geostand. Geoanalytical Res.* **2013**, *37*, 449–467. [CrossRef]
65. De la Cruz, O.; Holmes, S. The duality diagram in data analysis: Examples of modern applications. *Ann. Appl. Stat.* **2011**, *5*, 2266–2277. [CrossRef] [PubMed]
66. Bastviken, D.; Cole, J.; Pace, M.; Tranvik, L. Methane emissions from lakes: Dependence of lake characteristics, two regional assessments, and a global estimate. *Glob. Biogeochem. Cycles* **2004**, *18*, GB4009. [CrossRef]
67. Juutinen, S.; Rantakari, M.; Kortelainen, P.; Huttunen, J.T.; Larmola, T.; Alm, J.; Silvola, J.; Martikainen, P.J. Methane dynamics in different boreal lake types. *Biogeosciences* **2009**, *6*, 209–223. [CrossRef]
68. Wauthy, M.; Rautio, M.; Christoffersen, K.S.; Forsström, L.; Laurion, I.; Mariash, H.L.; Peura, S.; Vincent, W.F. Increasing dominance of terrigenous organic matter in circumpolar freshwaters due to permafrost thaw. *Limnol. Oceanogr. Lett.* **2018**, *3*, 186–198. [CrossRef]
69. Peura, S.; Wauthy, M.; Simone, D.; Eiler, A.; Einarsdóttir, K.; Rautio, M.; Bertilsson, S. Ontogenic succession of thermokarst thaw ponds is linked to dissolved organic matter quality and microbial degradation potential. *Limnol. Oceanogr.* **2020**, *65*, S248–S263. [CrossRef]
70. Shirokova, L.S.; Chupakov, A.V.; Zabelina, S.A.; Neverova, N.V.; Payandi-Rolland, D.; Causserand, C.; Karlsson, J.; Pokrovsky, O.S. Humic surface waters of frozen peat bogs (permafrost zone) are highly resistant to bio- and photodegradation. *Biogeosciences* **2019**, *16*, 2511–2526. [CrossRef]
71. Babkina, E.A.; Leibman, M.O.; Dvornikov, Y.A.; Fakashchuk, N.Y.; Khairullin, R.R.; Khomutov, A.V. Activation of cryogenic processes in Central Yamal as a result of regional and local change in climate and thermal state of permafrost. *Russ. Meteorol. Hydrol.* **2019**, *44*, 283–290. [CrossRef]
72. Xu, N.; Braida, W.; Christodoulatos, C.; Chen, J. A review of molybdenum adsorption on soils/bed sediments: Speciation, mechanisms and model applications. *Soil Sediment Contam. Int. J.* **2013**, *22*, 912–929. [CrossRef]
73. Tournassat, C.; Tinnacher, R.M.; Grangeon, S.; Davis, J.A. Modeling uranium (VI) adsorption onto montmorillonite under varying carbonate concentrations: A surface complexation model accounting for the spillover effect on surface potential. *Geochim. Cosmochim. Acta* **2018**, *220*, 291–308. [CrossRef]
74. Payandi-Rolland, D.; Shirokova, L.S.; Nakhle, P.; Tesfa, M.; Abdou, A.; Causserand, C.; Lartiges, B.; Rols, J.-L.; Guérin, F.; Bénézeth, P.; et al. Aerobic release and biodegradation of dissolved organic matter from frozen peat: Effects of temperature and heterotrophic bacteria. *Chem. Geol.* **2020**, *536*, 119448. [CrossRef]

75. Chupakov, A.V.; Pokrovsky, O.S.; Moreva, O.Y.; Shirokova, L.S.; Neverova, N.V.; Chupakova, A.A.; Kotova, E.I.; Vorobyeva, T.Y. High resolution multi-annual riverine fluxes of organic carbon, nutrient and trace element from the largest European Arctic river, Severnaya Dvina. *Chem. Geol.* **2020**, *538*, 119491. [CrossRef]
76. Vorobyev, S.N.; Pokrovsky, O.S.; Kolesnichenko, L.G.; Manasypov, R.M.; Shirokova, L.S.; Karlsson, J.; Kirpotin, S.N. Biogeochemistry of dissolved carbon, major, and trace elements during spring flood periods on the Ob River. *Hydrol. Process.* **2019**, *33*, 1579–1594. [CrossRef]
77. Schuur, E.A.G.; McGuire, A.D.; Schädel, C.; Grosse, G.; Harden, J.W.; Hayes, D.J.; Hugelius, G.; Koven, C.D.; Kuhry, P.; Lawrence, D.M.; et al. Climate change and the permafrost carbon feedback. *Nature* **2015**, *520*, 171–179. [CrossRef]
78. Biskaborn, B.K.; Smith, S.L.; Noetzli, J.; Matthes, H.; Vieira, G.; Streletskiy, D.A.; Schoeneich, P.; Romanovsky, V.E.; Lewkowicz, A.G.; Abramov, A.; et al. Permafrost is warming at a global scale. *Nat. Commun.* **2019**, *10*, 1–11. [CrossRef]
79. Turetsky, M.R.; Abbott, B.W.; Jones, M.C.; Anthony, K.W.; Olefeldt, D.; Schuur, E.A.G.; Koven, C.; McGuire, A.D.; Grosse, G.; Kuhry, P.; et al. Permafrost collapse is accelerating carbon release. *Nature* **2019**, *569*, 32–34. [CrossRef]
80. Anisimov, O.A.; Shiklomanov, N.I.; Nelson, F.E. Variability of seasonal thaw depth in permafrost regions: A stochastic modeling approach. *Ecol. Model.* **2002**, *153*, 217–227. [CrossRef]
81. Pavlov, A.V.; Moskalenko, N.G. The thermal regime of soils in the north of Western Siberia. *Permafr. Periglac. Process.* **2002**, *13*, 43–51. [CrossRef]
82. Anisimov, O.; Reneva, S. Permafrost and changing climate: The Russian Perspective. *AMBIO A J. Human Environ.* **2006**, *35*, 169–175. [CrossRef]
83. Moskalenko, N.G. Permafrost and vegetation changes in the Nadym region of West Siberian northern taiga due to the climate change and technogenesis. *Kriosfera Zemli* **2009**, *8*, 18–23. (In Russian)
84. Frey, K.E.; McClelland, J.W. Impacts of permafrost degradation on arctic river biogeochemistry. *Hydrol. Process.* **2009**, *23*, 169–182. [CrossRef]
85. Anisimov, O.; Kokorev, V.; Zhil'tsova, Y. Temporal and spatial patterns of modern climatic warming: Case study of Northern Eurasia. *Clim. Chang.* **2013**, *118*, 871–883. [CrossRef]
86. Walvoord, M.A.; Kurylyk, B.L. Hydrologic impacts of thawing permafrost—A review. *Vadose Zone J.* **2016**, *15*, 20. [CrossRef]
87. Karlsson, J.M.; Lyon, S.W.; Destouni, G. Temporal behavior of lake size-distribution in a thawing permafrost landscape in Northwestern Siberia. *Remote Sens.* **2014**, *6*, 621–636. [CrossRef]
88. Tank, S.E.; Lesack, L.F.W.; Hesslein, R.H. Northern delta lakes as summertime CO_2 absorbers within the arctic landscape. *Ecosystems* **2009**, *12*, 144–157. [CrossRef]

water

MDPI

Article

Impact of Permafrost Thaw and Climate Warming on Riverine Export Fluxes of Carbon, Nutrients and Metals in Western Siberia

Oleg S. Pokrovsky [1,2,3,*], Rinat M. Manasypov [1], Sergey G. Kopysov [4], Ivan V. Krickov [1], Liudmila S. Shirokova [2,3], Sergey V. Loiko [1], Artem G. Lim [1], Larisa G. Kolesnichenko [1], Sergey N. Vorobyev [1] and Sergey N. Kirpotin [1]

1 BIO-GEO-CLIM Laboratory, Tomsk State University, 36 Lenina ave, 634050 Tomsk, Russia; rmmanassypov@gmail.com (R.M.M.); krickov_ivan@mail.ru (I.V.K.); s.loyko@yandex.ru (S.V.L.); lim_artyom@mail.ru (A.G.L.); klg77777@mail.ru (L.G.K.); soil@green.tsu.ru (S.N.V.); kirp@mail.tsu.ru (S.N.K.)

2 Geosciences and Environment Toulouse, UMR 5563 CNRS, 14 Avenue Edouard Belin, 31400 Toulouse, France; liudmila.shirokova@get.omp.eu

3 N. Laverov Federal Center for Integrated Arctic Research, Russian Academy of Sciences, 23 Naberezhnaya Sev. Dviny, 163000 Arkhangelsk, Russia

4 Institute of Monitoring of Climatic and Ecological Systems SB RAS, 10/3 Academichesky ave, 634055 Tomsk, Russia; wosypok@mail.ru

* Correspondence: oleg.pokrovsky@get.omp.eu

Received: 30 April 2020; Accepted: 22 June 2020; Published: 24 June 2020

Abstract: The assessment of riverine fluxes of carbon, nutrients, and metals in surface waters of permafrost-affected regions is crucially important for constraining adequate models of ecosystem functioning under various climate change scenarios. In this regard, the largest permafrost peatland territory on the Earth, the Western Siberian Lowland (WSL) presents a unique opportunity of studying possible future changes in biogeochemical cycles because it lies within a south–north gradient of climate, vegetation, and permafrost that ranges from the permafrost-free boreal to the Arctic tundra with continuous permafrost at otherwise similar relief and bedrocks. By applying a "substituting space for time" scenario, the WSL south-north gradient may serve as a model for future changes due to permafrost boundary shift and climate warming. Here we measured export fluxes (yields) of dissolved organic carbon (DOC), major cations, macro- and micro- nutrients, and trace elements in 32 rivers, draining the WSL across a latitudinal transect from the permafrost-free to the continuous permafrost zone. We aimed at quantifying the impact of climate warming (water temperature rise and permafrost boundary shift) on DOC, nutrient and metal in rivers using a "substituting space for time" approach. We demonstrate that, contrary to common expectations, the climate warming and permafrost thaw in the WSL will likely decrease the riverine export of organic C and many elements. Based on the latitudinal pattern of riverine export, in the case of a northward shift in the permafrost zones, the DOC, P, N, Si, Fe, divalent heavy metals, trivalent and tetravalent hydrolysates are likely to decrease the yields by a factor of 2–5. The DIC, Ca, SO_4, Sr, Ba, Mo, and U are likely to increase their yields by a factor of 2–3. Moreover, B, Li, K, Rb, Cs, N-NO_3, Mg, Zn, As, Sb, Rb, and Cs may be weakly affected by the permafrost boundary migration (change of yield by a factor of 1.5 to 2.0). We conclude that modeling of C and element cycle in the Arctic and subarctic should be region-specific and that neglecting huge areas of permafrost peatlands might produce sizeable bias in our predictions of climate change impact.

Keywords: river flux; weathering; organic matter; permafrost; trace element; river

1. Introduction

Arctic warming is anticipated to result in massive carbon (C) mobilization from permafrost peat to atmosphere, rivers and lakes, thereby potentially worsening global warming via greenhouse gases (GHG) emissions [1,2]. Permafrost peatlands cover roughly 2.8 million km^2 or 14% of permafrost-affected areas, mostly in Northern Eurasia (Bolshezemelskaya Tundra, 0.25×10^6 km^2; western Siberia, 1.05×10^6 km^2; Northern Siberia and Eastern Siberia lowlands, 0.5×10^6 km^2) and contain a huge amount of highly vulnerable carbon in soil and surface waters [3]. Except for several regional studies of peatland lakes and small streams in Canada [4–6] and northern Sweden [7], the control factors, timing and reality of C and related elements release from soil and sediments are largely unknown, in part because element export by rivers across the permafrost peatlands is still poorly quantified.

One of the largest peatlands in the world is the Western Siberia Lowland (WSL) which exhibits a disproportionally high contribution to C storage and exchange fluxes [8] and presents a prominent exception to well-studied aquatic and soil ecosystems in mountainous territories of Northern America and Scandinavia. Specific features of the WSL are: (1) developed on flat, low runoff terrain with long water residence time in both lentic and lotic systems; (2) acts as important C stock in the form of organic-rich histosols (peat soils); (3) emits substantial amount of CO_2 and CH_4 to the atmosphere from the surface of inland waters; and (4) exhibits high dissolved organic carbon (DOC) concentration and low pH in waters in contact with peat soils. These factors determine rather unique and still poorly known aquatic communities in dystrophic to mesotrophic peatland waters. Due to the lack of nutrients, shallow photic layer and high sensitivity to water warming, aquatic ecosystems of frozen peatlands are highly vulnerable to the permafrost thaw and can respond unexpectedly to ongoing climate warming in terms in their C, nutrient, and metal storage and export fluxes as well biodiversity and organisms adaptation strategies.

Western Siberia is a key region for biogeochemical studies in the Arctic [2] as this region includes the continuous-discontinuous permafrost transition and experiences substantial thermokarst. As a result, this high-priority region is most susceptible to thaw-induced change in solute transport by rivers and its export to the ocean and atmosphere, notably via activation of deep underground flow [9] and exposure of large volumes of previously frozen peat and mineral soils [10–12]. Raised by incontestable Arctic amplification of overall climate change, the fate of carbon, nutrient and metal in Arctic rivers is at the forefront of field and modeling studies [13–20]. In particular, the Arctic Great Rivers Observatory (GRO) program assessed concentrations and export fluxes of 6 largest Arctic Rivers including the Ob River, draining sizeable part of the WSL [21,22]. However, widespread loss of hydrological monitoring in the beginning of 21st century was especially pronounced for small and medium-size rivers of the permafrost-affected part of the WSL [23]. At the same time, the Ob River is not suitable for modeling of possible changes in western Siberia because (1) it is strongly influenced by its largest permafrost-free tributary, the Irtysh River; and (2) it cannot be used for assessing the fluxes of northern (permafrost) zones as it integrates vast territory of variable permafrost coverage (20% in average) and landscape parameters. For these reasons, the study of small rivers at the WSL territory is more suitable for assessing both mechanisms of flux formation and its possible future changes.

Currently, a dominant paradigm is that riverine fluxes of C and inorganic nutrients are increasing in virtually all permafrost-affected rivers [14,24,25]. In particular, the increase in suspended versus dissolved transport of elements may be due to abrasion of lake shores and riverbanks. At the same time, enhanced groundwater input will lead to an increase in the transport of truly dissolved forms. Recently, following the pioneering work of Frey et al. [26–28], the concentrations of dissolved, particulate and colloidal carbon, nutrient and metals in WSL rivers, lakes and soil waters have been studied over a sizable latitudinal gradient [29–38]. These results allowed a first-order assessment of the consequences of climate warming and permafrost thaw on river water concentrations of dissolved and particulate forms of elements from western Siberia to the Arctic Ocean and C emission to the atmosphere. For this, a "substituting space for time" approach was employed, which postulates, in a broad context, that spatial phenomena which are observed today can be used to describe past and future events [39].

Thus, a concentration pattern of major and trace elements (TE) in WSL river suspended matter implies that, upon a progressive shift of the permafrost boundary northward, there will be a sizeable decrease in concentrations of alkalis, alkaline-earths, divalent heavy metals, and trivalent and tetravalent hydrolysates in northern rivers in currently discontinuous and continuous permafrost zones [31]. This decrease may be by a factor of 2–5 from the position of the minimum elemental concentration in sporadic to discontinuous permafrost zones relative to the continuous permafrost zone. Concerning the dissolved element concentration and potential transport in the WSL rivers, the following results were achieved implying the substituting space for time scenario. From the permafrost thaw perspective, the increase in depth of the active layer and connectivity of a river with underground water reservoirs may decrease the colloidal fraction (1 kDa–0.45 μm) of OC, Fe, Al and number of divalent metals as well as low-soluble trivalent and tetravalent hydrolysates in the sporadic and isolated permafrost zone as it becomes permafrost-free [30]. The forestation of wetlands and lake drainage may slightly diminish colloidal transport of DOC and metals at the expense of low molecular weight forms. Overall, given the significant role of seasonal and forestation effects on colloidal forms of OC and TE in WSL rivers, major changes in the speciation of riverine C and nutrients in the WSL may occur due to changes in vegetation rather than in temperature and precipitation [30]. A northward permafrost boundary shift with increase in air and water temperature may decrease or maintain, rather than increase of major nutrient (K, P, N, Si) and DOC concentrations in rivers draining through continuous permafrost zone [37].

All these predictions described above were made based solely on evolution of concentrations of soluble and particulate C, metals and nutrients in WSL rivers, without taking into account the hydrological flux. The latter could not be addressed until now due to lack of reliable information on river discharge over different seasons. The present work aims to fill this gap by quantifying the export fluxes of ~30 WSL rivers of various size, combined with new hydrological modeling of region and season-specific river runoff of the WSL territory. The chosen rivers encompass a large gradient of climate, vegetation and permafrost distribution and thus enables a substituting space for time scenario to provide a tentative prediction of possible changes in riverine fluxes for short- and long-term prospectives. The present work is built on seasonally resolved sampling performed in 2015 and 2016, and incorporates thorough hydrological modeling to calculate the seasonal discharge. Building upon our previous studies of the WSL rivers [29–31,33,34,37], here we assess for the first time, element export fluxes (yields) across a large permafrost-affected territory and large number of rivers. The obtained elementary yields are essential for further modeling of biogeochemical cycles in the permafrost regions.

2. Study Site and Methods

2.1. Rivers of Western Siberian Lowland, their Sampling and Analyses

The 32 rivers of the Ob, Pur and Taz watersheds in the WSL were sampled (Figure 1). Detailed climatic, lithological and physio-geographic characteristics for the WSL are presented elsewhere [29,33,34]. The dominant lithology is Quaternary deposits (silt, clay, sand) overlaid by peat. The climate gradient of sampled rivers presents a decrease in mean annual air temperature (MAAT) from −0.5 °C in the south (Tomsk region) to −9.5 °C in the north (Arctic coast). Annual precipitation is fairly constant ranging between 550 mm in the south and 600 mm at the lower reaches of the Taz River. The river runoff ranges from 190 ± 30 mm y^{-1} in the south to 300 ± 20 mm y^{-1} in the north [40]. The distribution of the permafrost reflects the south-north MAAT gradient and changes from isolated and sporadic in the south to discontinuous and continuous in the north. In 2015, we sampled rivers in spring (18 May–25 June) and summer (25 July–21 August). In 2016, sampling was performed in spring (17 May–15 June 2016), summer (1–29 August 2016), and autumn (24 September–13 October 2016). We followed the change of seasons during our sampling campaign and moved from the south to the north in spring and from the north to the south in autumn thus collecting the river water at approximately the same period of the discharge. Note that more frequent sampling would be desirable to accurately evaluate the annual export flux, but rather harsh environment and logistical issues constrained sampling. In contrast, route sampling is a common way to assess chemical weathering in extreme environments [41,42], and it is accepted that single sampling during high flow season provides the best agreement with time-series estimates [43].

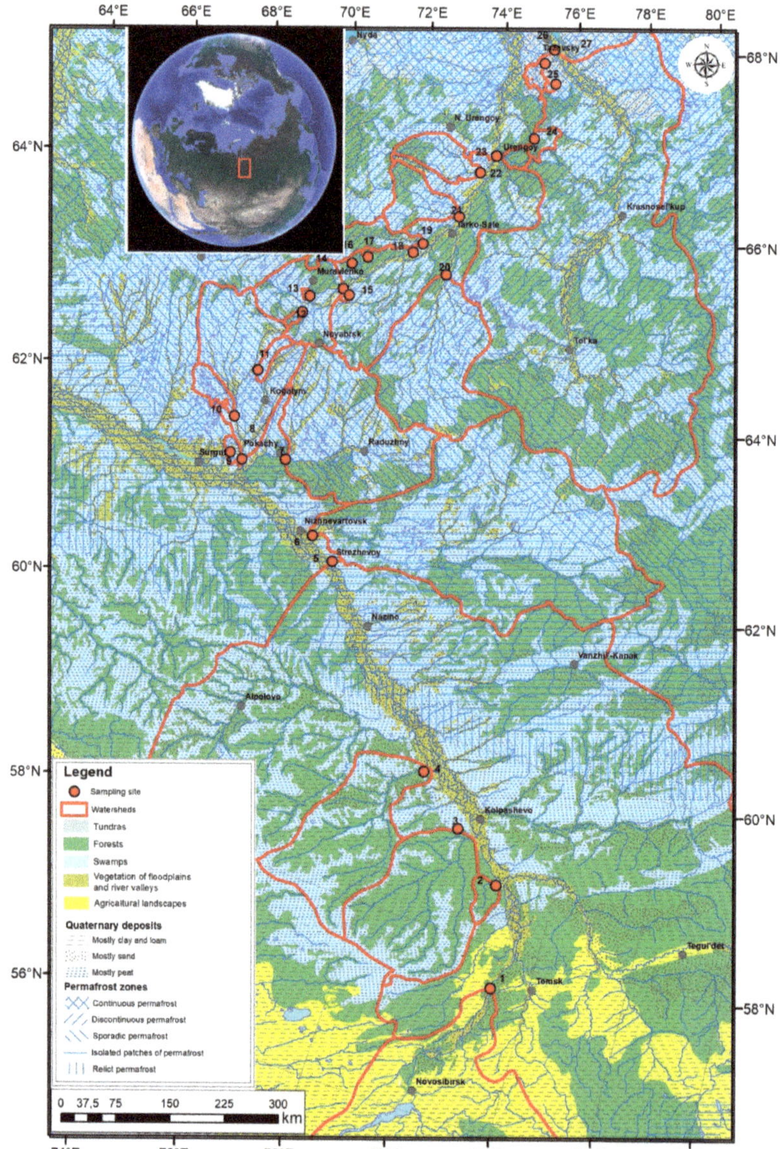

Figure 1. The WSL river sampling points. The numbers on the map correspond to the following rivers: (1) Ob′ (Pobeda), (2) Maliy Tatosh, (3) Chaya, (4) Vyalovka, (5) Ob′ (Aleksandrovskoye), (6) Vakh, (7) Agan, (8) Tromyegan. (9) Vach-Yagun, (10) Vachinguriyagun, (11) Pintyr'yagun, (12) Kamgayakha, (13) Khatytayakha, (14) Pyakupur, (15) Lymbyd'yakha, (16) Apoku-Yakha, (17) Etu-Yakha, (18) Seryareyakha, (19) Purpe, (20) Aivasedapur, (21) Tydylyakha, (22) Yamsovey, (23) Pur, (24) Ngarka Khadyta-Yakha, (25) M. Kheyakha, (26) Nuny-Yakha, (27) Taz. The upper left insert is from Google Maps®.

The two sampled years contrasted in mean monthly and annual temperature and precipitation. The summer 2015 was warm (+0.3 °C above normal for July), while summer 2016 was cool (−3.1° below normal for July). The summer 2015 was wet (180% of normal precipitation for July), and the summer 2016 was less wet (but still 130% of normal precipitation for July). The winter precipitations of 2014–2015 and 2015–2016 were rather similar and sizably higher (ca. 140%) than the normal winter values. The normal values here are defined as between 1970 to 2000 based on the Roshydromet archives [44].

The surface water was collected in a polypropylene 1-L container. Samples were filtered through 0.45 μm cellulose acetate filters using a Nalgene 250 mL filter unit combined with a vacuum pump. All filtrations were run on site, in a protected environment, within 2 h of river water collection. Immediately after filtration, samples for DOC, DIC, major and trace elements were stored in the refrigerator during 1–2 months prior to the analyses, while the samples for nutrients were kept frozen. Dissolved (<0.45 μm) concentrations of nutrients, major and trace elements in WSL river waters were analyzed as described elsewhere [33,34,37,45]. Major anion concentrations (Cl^- and SO_4^{2-}) were measured by ion chromatography (HPLC, Dionex ICS 2000) with an uncertainty of 2%. DOC and DIC were analyzed using a Shimadzu TOC 6000 with an uncertainty of 3–5% [46]. For all major and most trace elements, analyzed by ICP MS, the concentrations in the blanks were below analytical detection limits (≤0.1–1 ng/L for Cd, Ba, Y, Zr, Nb, rare earth elements (REEs), Hf, Pb, Th, U; 1 ng/L for Ga, Ge, Rb, Sr, Sb; ~10 ng/L for Ti, V, total P (P_{tot}), Cr, Mn, Fe, Co, Ni, Cu, Zn, As). The international certified reference material SLRS-5 (Riverine Water Reference Material for Trace Metals) was used to validate the analysis. Further details of analytical uncertainties and detection limits for TE are provided elsewhere [30,34,45].

2.2. Hydrological Parameters, River Discharge Modeling, and Element Export Flux Calculations

The runoff was longitudinally modelled using average altitude of the watershed for gauged rivers [47]. For small and medium rivers draining palsa and polygonal bogs of the permafrost zone, we used empirical equations accounting for hydrological characteristics of these watersheds [48]. In this work, only open water period (May to October) was considered. The contribution of winter time export in large rivers typically does not exceed 10% of mean annual runoff [49]; moreover, small (<1000–10,000 km^2 watershed) WSL rivers freeze solid in winter [33]. We interpolated discharge by using watershed area change along the river mainstem. The river hydrographs were modelled using HBV Light program package (https://www.geo.uzh.ch/en/units/h2k/Services/HBV-Model.html) as described in refs [50–52]. For the gauged rivers, we modelled daily runoff based using mean water temperature and precipitation from adjacent meteostations. The data for reproducing the water discharge at key sites were taken from hydrological yearbooks and automatic information system of State monitoring of water bodies [53]. The daily air temperature and precipitation were taken from Russian Hydrometeocenter (URL: http://aisori.meteo.ru/ClimateR; [44]). The calibration of model parameters demonstrated that the HBV model adequately reproduces the seasonal dynamics of runoff (quality criteria of 0.75 to 0.90). The uncertainties of modelled discharge at ungauged rivers stem from several factors. First, there is uncertainties in characterizing watershed boundaries for the very flat, but poorly resolved, WSL territory. For small rivers, this uncertainty can amount to 30% due to lack of precise topographic information. For large rivers, it can be up to 7% due to difficulties in determination of exact position of the watershed divide on flat bog-lake landscapes. The second cause of uncertainty is a discrepancy between the model parameters of analogous rivers due to intrinsic differences in the conditions of runoff formation (up to 25%). Finally, the amount of precipitation obtained from the nearest meteostations might substantially differ from the actual precipitation at a given watershed, which is reflected, in particular, by low quality of the rain-flood modeling. The maximal overall uncertainty of daily discharges is determined by the most probable value of 30%. However, because we used two-month averaged discharges for calculating elementary yields, the real uncertainty of export fluxes is determined by the uncertainty of watershed delineations and ranges from 7 to 30%.

In order to calculate element export fluxes, we defined six latitudinal belts (56–58° N, 58–60° N, 60–62° N, 62–64° N, 64–66° N and 66–68° N). These latitude belts were selected based on: i) the permafrost map of the WSL, where the permafrost zones roughly follow latitude, and ii) necessity to integrate sufficient number of rivers in each permafrost zone and for statistical comparison. For statistical comparisons, we separated the years 2015 and 2016, because the autumn was sampled only in 2016. The number of rivers used for the latitudinal-average flux calculation ranged from 2 to 10 in 2016 and from 2 to 16 in 2015 for each latitudinal range.

3. Results

3.1. Impact of the Watershed Size and Season on Element Export Fluxes

The yields of DOC, representative major solutes (DIC, Ca, Mg) and nutrients (Si, P_{tot}, K, Fe) are illustrated in Supplementary Figure S1. Flux magnitudes exhibited sizable variation, ranging over one (DOC, Si and Mg) to two (DIC, Ca, K, Fe and P_{tot}) orders of magnitude. Generally, the variations were the highest for rivers having watersheds between 100 and 1000 km^2, for both permafrost-affected and permafrost-free. The river watershed area ($S_{watershed}$) exhibited rather weak control on elementary yields, as also illustrated by statistical treatment of individual seasons and full data set for both years (Supplementary Table S1). Overall, the watershed size exhibited more pronounced correlations with element yield in the permafrost-free zone compared to the permafrost zone. The correlations were quite low during spring flood but become pronounced during summer and autumn baseflow. The highest correlations were observed in summer, when Cl, SO_4, Fe, Cu, Y, Mo, Sb, REEs and Th were positively linked to $S_{watershed}$ ($R_{Spearman} > 0.58$, $p < 0.01$) in the permafrost-free zone.

Partial contribution of spring, summer, and autumn (2 months each) to the overall export of elements during the six-month open-water period (May to October) in 2016 demonstrated the dominant role of spring for DOC, Al and Fe (>60%), for both permafrost-free and permafrost-affected rivers (Figure 2). In contrast, soluble highly mobile elements (DIC, Na, K, Ca, Mg, Si) had less than 40% and 50% contribution of spring in the permafrost-free and permafrost zone, respectively, and a 2–3 times higher contribution during summer baseflow period compared to during the spring flood period. Nutrients (e.g., K, P_{tot}) and metals (e.g., Mn) presented an intermediate case with a half of yield occurring during the spring flood period and with negligible (<10%) role of autumn period in permafrost-free zone but comparable contribution of summer and autumn in the permafrost-affected part of the WSL.

Element export as a function of watershed area (Supplementary Figure S1) and season (Supplementary Figure S2) represent two main groups, namely (1) soluble highly mobile elements (Cl, SO_4, Sr, Rb, Ba and As, Sb, Mo, U (permafrost-free zone only) and (2) DOC and low soluble elements, typically present in the form of organic- and organo-mineral colloids especially in the permafrost zone [30], such as micronutrients (Fe, Co, Ni, Cu, Zn, V), toxic metals (Cr, Cd, Pb), Nb, and trivalent (Al, Ga, Y, REE) and tetravalent (Ti, Hf, Th) hydrolysates. Generally, the share of spring flood contribution in overall open-water export of elements was 20 ± 10% higher in permafrost-affected rivers than in rivers of the permafrost-free zone.

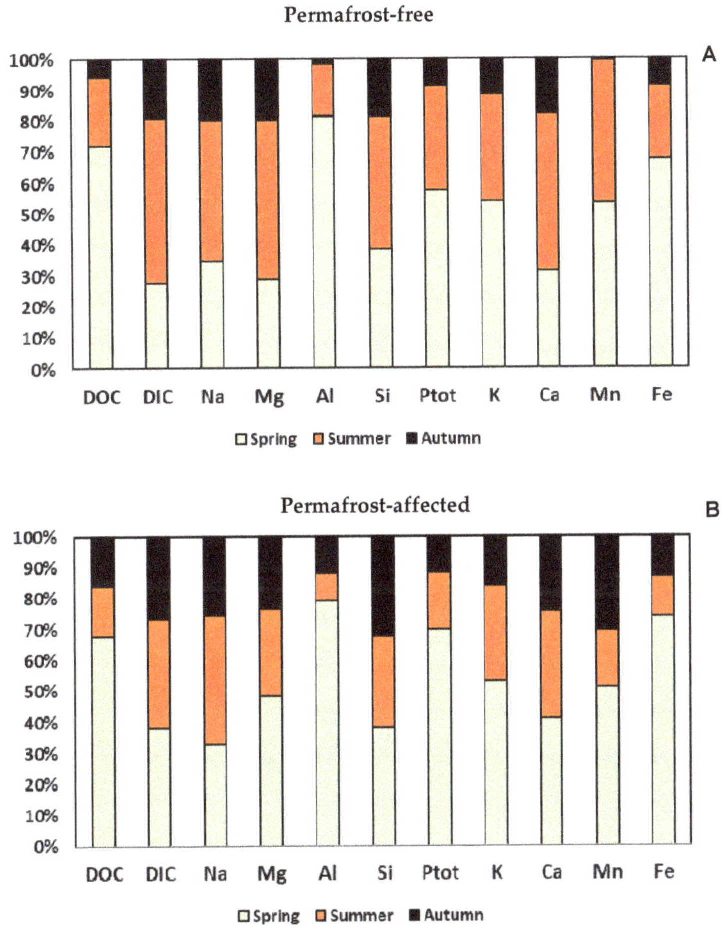

Figure 2. Partial contribution of spring, summer and autumn 2016 to overall open-water period export of elements by WSL rivers, located in the permafrost-free (**A**) and permafrost-affected (**B**) regions.

3.2. Riverine Element Export Fluxes Across the Latitudinal Profile (Permafrost and Climate Gradient)

Because both concentrations [33,34,37] and seasonal fluxes (Supplementary Figure S1 and Table S1) of elements were not strongly impacted by the river size, the watershed-averaged fluxes can be calculated as a function of the watershed latitude. Furthermore, we used ternary molar diagrams (Ca – Mg – (Na + K) and Cl – SO_4 – HCO_3) to reveal the role of season, river size and permafrost coverage of major elementary composition (Supplementary Figure S3), following traditional geochemical classifications [54]. Regardless of the season, water chemical composition from permafrost-free rivers was distinctly different from water from permafrost-affected rivers and strongly enriched in Ca and HCO_3^-. As a result, the latitude (which corresponds with the permafrost zonation) can be considered as the main factor controlling major cation concentrations in the WSL rivers. The average (±2 s.d.) fluxes of dissolved (<0.45 μm) major element export from the WSL watersheds for each latitudinal belt in 2016 and 2015 are shown in Figure 3 and Supplementary Figure S4, respectively. For the sake of scientific rigor, we illustrate the yields separately for 2015 and 2016, but the two-year average values of element export during spring and summer are given in Table 1.

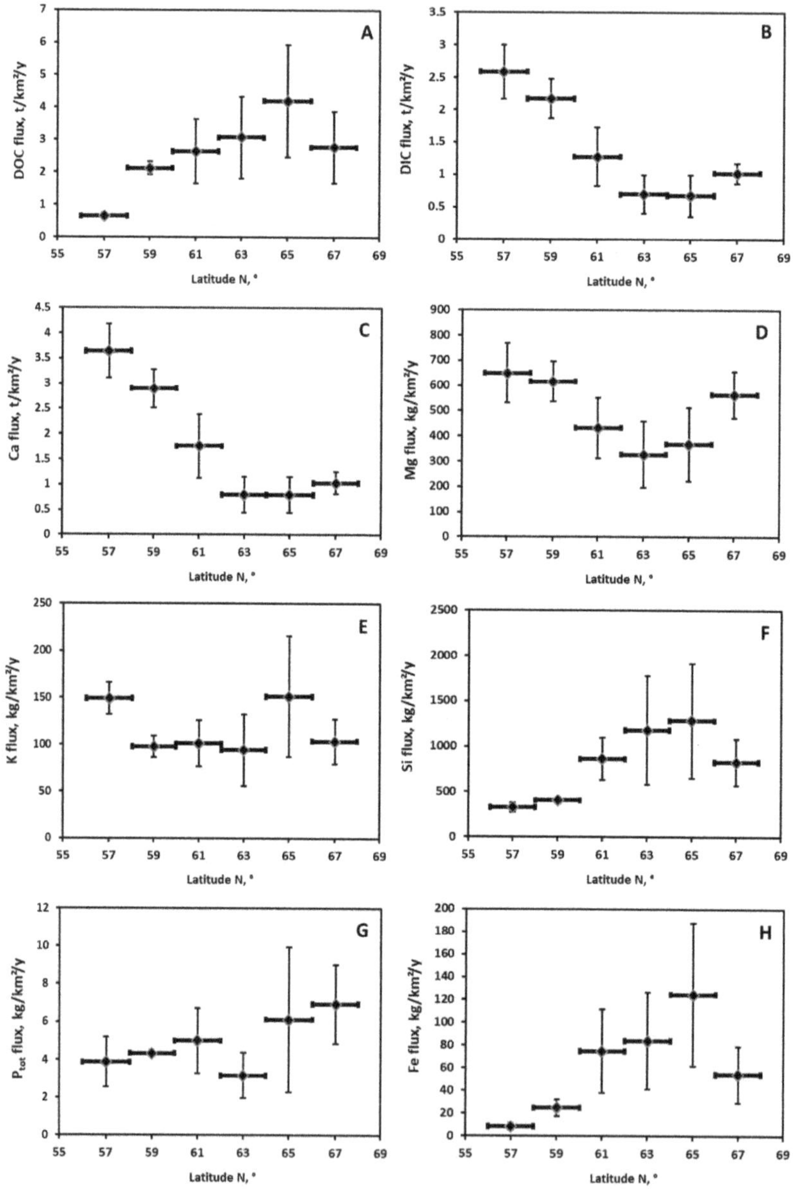

Figure 3. Yields (watershed-area normalized export fluxes) of DOC (**A**), DIC (**B**), Ca (**C**), Mg (**D**), K (**E**), Si (**F**), P_{tot} (**G**) and Fe (**H**) during May-October 2016 in 33 WSL rivers across the latitudinal gradient. The vertical uncertainties represent the 2 s.d. of several rivers belonging to the same latitudinal belt. Thick horizontal bars represent the 2° latitudinal belts. The year is represented by open-water period (May to October), neglecting winter (November to April), when small rivers freeze solid in the north.

Table 1. Mean (±SD) export fluxes of major and trace elements by WSL rivers in the permafrost-free and the permafrost zone.

Element kg/km²/y	Biennial Flux, 2015 + 2016 (May—August)		Annual Flux, 2016 (May—October)	
	Permafrost-Free	Permafrost	Permafrost-Free	Permafrost
DOC	1192 ± 310	6340 ± 1858	1165 ± 202	3399 ± 1279
Major anions				
DIC	2122 ± 389	1150 ± 316	2278 ± 403	839 ± 280
Cl	132 ± 50	2092 ± 1052	160 ± 84	1150 ± 692
SO_4	409 ± 201	308 ± 112	486 ± 319	234 ± 116
Major cations				
Na	452 ± 115	1784 ± 769	528 ± 163	1116 ± 592
Mg	554 ± 103	615 ± 174	602 ± 107	414 ± 125
Ca	3045 ± 545	1585 ± 510	3125 ± 557	1004 ± 368
K	136 ± 21	175 ± 56	130 ± 20	108 ± 36
Macro-nutrients				
P_{tot}	6.41 ± 2.4	22.4 ± 8.2	3.66 ± 1	5.71 ± 2.3
P-PO_4	1.99 ± 0.67	10.7 ± 3.8	-	-
N-NO_3^-	7.4 ± 1.8	8.12 ± 3.4	-	-
N-NH_4^+	3.57 ± 0.8	14.6 ± 6.1	-	-
Si	409 ± 73	1528 ± 415	426 ± 112	1076 ± 430
Micro-nutrients				
B	2.26 ± 0.43	5.63 ± 2.1	2.23 ± 0.39	3.82 ± 1.8
Mn	3.46 ± 1.8	36.7 ± 13	1.4 ± 0.69	16.9 ± 11
Fe	32.9 ± 14.3	292.7 ± 93.8	17.3 ± 7.5	92.2 ± 42
Co	0.012 ± 0.005	0.15 ± 0.06	0.009 ± 0.002	0.066 ± 0.04
Ni	0.126 ± 0.03	0.599 ± 0.22	0.118 ± 0.018	0.322 ± 0.15
Cu	0.159 ± 0.06	0.283 ± 0.12	0.128 ± 0.041	0.177 ± 0.087
Zn	1.31 ± 1	4.55 ± 3	0.203 ± 0.091	1.42 ± 0.89
Rb	0.092 ± 0.016	0.261 ± 0.098	0.078 ± 0.014	0.159 ± 0.059
Mo	0.023 ± 0.01	0.011 ± 0.004	0.029 ± 0.017	0.006 ± 0.002
Ba	3.32 ± 0.52	7.89 ± 3.4	2.87 ± 0.37	2.95 ± 1.4
Toxicants				
As	0.098 ± 0.021	0.337 ± 0.11	0.093 ± 0.026	0.143 ± 0.047
Cd	0.0009 ± 0.0004	0.005 ± 0.002	0.0008 ± 0.0003	0.003 ± 0.001
Sb	0.008 ± 0.003	0.016 ± 0.005	0.008 ± 0.003	0.009 ± 0.004
Pb	0.023 ± 0.009	0.065 ± 0.031	0.005 ± 0.002	0.024 ± 0.012
Geochemical traces				
Li	0.193 ± 0.059	0.57 ± 0.17	0.272 ± 0.094	0.455 ± 0.18
Al	5.37 ± 2.8	41.5 ± 18	5.86 ± 2.6	20.4 ± 11
Ti	0.223 ± 0.075	1.58 ± 0.56	0.512 ± 0.1	1.4 ± 0.52
Cr	0.041 ± 0.01	0.379 ± 0.12	0.024 ± 0.009	0.204 ± 0.08
Ga	0.002 ± 0.001	0.007 ± 0.002	0.0009 ± 0.0004	0.004 ± 0.002
Sr	15.7 ± 3	17.2 ± 8.8	16.6 ± 3.4	9.12 ± 4.8
Y	0.015 ± 0.006	0.11 ± 0.05	0.013 ± 0.003	0.056 ± 0.03
La	0.036 ± 0.029	0.085 ± 0.035	0.012 ± 0.003	0.05 ± 0.026
Ce	0.059 ± 0.048	0.177 ± 0.082	0.02 ± 0.006	0.091 ± 0.051
Pr	0.003 ± 0.001	0.023 ± 0.01	0.003 ± 0.0008	0.012 ± 0.007
Nd	0.013 ± 0.006	0.096 ± 0.044	0.012 ± 0.003	0.049 ± 0.028
Sm	0.003 ± 0.001	0.022 ± 0.01	0.003 ± 0.0008	0.012 ± 0.006
Eu	0.0009 ± 0.0003	0.006 ± 0.002	0.001 ± 0.0002	0.003 ± 0.002
Tb	0.0006 ± 0.0005	0.012 ± 0.006	0.003 ± 0.0007	0.012 ± 0.007
Gd	0.003 ± 0.001	0.014 ± 0.006	0.0007 ± 0.0002	0.002 ± 0.0009
Dy	0.003 ± 0.001	0.02 ± 0.009	0.003 ± 0.0007	0.01 ± 0.006
Ho	0.0005 ± 0.0002	0.004 ± 0.002	0.0005 ± 0.0001	0.002 ± 0.001
Er	0.002 ± 0.0007	0.012 ± 0.005	0.001 ± 0.0004	0.006 ± 0.003
Tm	0.0002 ± 0.0001	0.002 ± 0.0007	0.0002 ± 0.00005	0.0009 ± 0.0005
Yb	0.001 ± 0.0006	0.011 ± 0.005	0.001 ± 0.0003	0.006 ± 0.003
Lu	0.0002 ± 0.0001	0.002 ± 0.0007	0.0002 ± 0.00005	0.0009 ± 0.0005

Table 1. *Cont.*

Element kg/km²/y	Biennial Flux, 2015 + 2016 (May—August)		Annual Flux, 2016 (May—October)	
	Permafrost-Free	Permafrost	Permafrost-Free	Permafrost
Zr	0.901 ± 0.33	1.47 ± 0.31	1.27 ± 0.46	1.57 ± 0.6
Cs	0.0003 ± 0.0001	0.0021 ± 0.0009	0.0003 ± 0.0001	0.001 ± 0.0005
Nb	0.0007 ± 0.0002	0.004 ± 0.002	0.0009 ± 0.0003	0.002 ± 0.001
Hf	0.0007 ± 0.0004	0.008 ± 0.003	0.001 ± 0.0004	0.005 ± 0.003
Th	0.002 ± 0.0009	0.009 ± 0.004	0.002 ± 0.0005	0.005 ± 0.003
U	0.021 ± 0.01	0.005 ± 0.002	0.03 ± 0.019	0.003 ± 0.001

Although the maximal uncertainties on discharge do not exceed 30% (see Section 2.2), the regional element yield assessments are subjected to high variation among rivers (Figure S1) with s.d. often at 50% (see Figure 3, Supplementary Figure S4). Among possible causes of these sizable uncertainties are non-linear and hysteretic relationships between concentrations and discharge which could not be resolved due to low frequency of sampling. Another reason of this variability could be highly dynamic behavior of element concentration during summer and autumn baseflow and spring flood, reflecting source limitation, chemostatis, or transport limitation (i.e., refs. [55–57]).

The DOC yield was minimal in rivers south to 59°N and exhibited a maximum at 61–65° N and 63–65° N in 2015 and 2016, respectively, with overall magnitude of variation by a factor of 4 (Figure S4A and Figure 3A, respectively). The yields of DIC and Ca decreased more than five-fold from south to north and achieved minimal values in the 63–65° N belt, for both years of observation (Figure 3B,C and Supplementary Figure S4B,C). Magnesium showed a weak minimum of yield in the 63–65° N which was however pronounced only in 2016; the overall variations were less than a factor of 2 (Figure 3D and Supplementary Figure S4D). Potassium yield remained fairly constant across the latitudinal profile with overall variations less than a factor of 1.5 to 2.0 between various latitudinal belts (Figure 3E and Supplementary Figure S4E). Silicon showed a three-fold increase in export fluxes from the south to the north, quite similar for two years of monitoring (Figure 3F and Supplementary Figure S4F).

The latitudinal pattern of other major and trace element export fluxes followed these three main types of behavior described above: (1) minimal in the south and a northward increase by a factor of 3 to 5 (Si, POC, P-PO_4; N-NH_4, Ni, REEs, Zr, Nb, Hf, and Th), with a local maximum at 63–65° N (DOC, Al, Ti, Cr, Mn, Fe, Co, Ni, Cu, Zn, Ga, Rb, Cd, Cs, and Pb); (2) northward decrease by a factor of 3 to 5 (DIC, Ca, SO_4, Sr, Mo, W, and U), and (3) overall independence of yield on latitude, or the latitudinal variations were less than a factor of 2 (Cl, N-NO_3, Li, B, Mg, P_{tot}, K, As, Zr, Sb, and Ba). Note that some elements (Mg, Zn, As, Rb) may belong to one or another group depending on the year of observation.

The yields of all 51 dissolved elements are listed in Table 1, which presents the export fluxes of two-year average (May to August) and open-water period (May to October 2016) in permafrost-free and permafrost-affected zones. The ratio of element flux in the non-permafrost zone to that in the permafrost zone ($R_{absent/permafrost}$, Figure 4) demonstrates a distinct order of elements whose export occurs preferentially in southern or northern part of the WSL. There are three main groups of elements - those exhibiting the highest (a factor of 2 to 10) yield in the permafrost-free zone (DIC, SO_4, Ca, Mo and U) and those showing a maximum in the permafrost-affected zone (As, P_{tot}, Li, B, Rb, Na, Si with $0.5 \leq R_{absent/permafrost} \leq 0.25$, and Ni, Th, Cd, DOC, Cs, Nb, Ti, Al, Fe, Co, Cr, Mn, Hf, Co, and REEs with $R_{absent/permafrost} < 0.25$). The elements of an intermediate group showed comparable (±30%) yields in permafrost-free and permafrost-affected parts of the WSL (Sr, Mg, K, Ba, Sb, Zr, Cu).

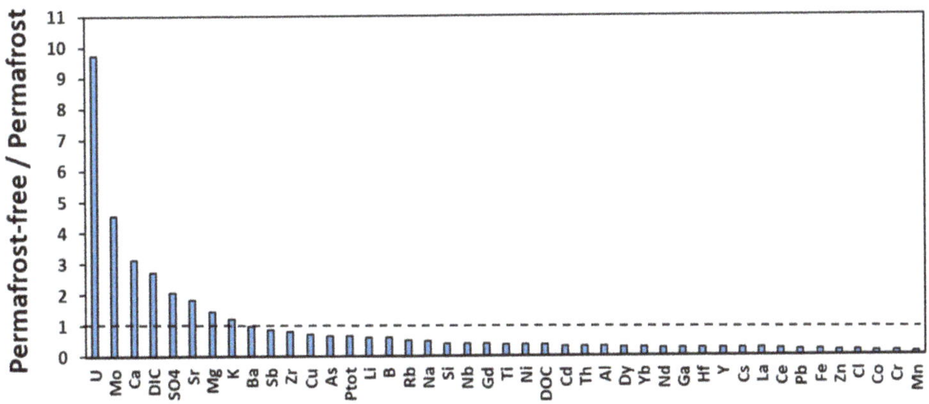

Figure 4. A histogram of the ratio of element average yield in the permafrost-free rivers to that in the permafrost-affected rivers for 6 open-water months (May to October) of 2016 (based on the data listed in Table 1).

4. Discussion

4.1. Effect of Seasons, Watershed Size, Latitude and Permafrost Coverage on Elementary Yields

There are several distinct groups of elements defined according to the latitudinal patterns of their yields. These groups reflect the relative mobility of elements, consistent with general knowledge of river hydrochemistry in the high latitude regions [58]. The DOC, organically-bound metals (V, Cr, Mn, Fe, Co, Ni, Cu, Zn, Cd and Pb) and many low soluble TE–geochemical traces (Al, REEs, Nb, Ti, Zr, Hf, and Th) exhibited similar 1st-type latitudinal pattern (minimum in the south and maximum in the permafrost-affected zone). The overall open-water period transport of these elements was dominated by spring flood, as it was also observed in other boreal and subarctic rivers (i.e., the Severnaya Dvina River, [59]). The nutrients (K, Si, P-PO_4, P_{tot}, and N-NH_4^+) and TE—analogous of macronutrients (Rb)—exhibited 50 ± 10% share of annual export during spring flood and a northward increase. Finally, the yield of geochemically mobile elements (DIC, Cl^-, SO_4, Li, B, Mg, Ca, Sr, Mo, and U) decreased northward, with less than 40–50% of overall contribution provided by spring flood period.

The division of elements into these groups depends on (1) the share of different seasons in overall 6-month open water period export, (2) the shape of the latitudinal pattern (Figure 3 and Figure S4), and (3) the difference in total 2-year averaged riverine yields between permafrost-free and permafrost-affected zones (Table 1, Figure 4). Such a distinction is consistent with two main factors controlling the element export (transport) in the WSL rivers. First, this is source-limitation, when the input of elements from soils to the river is controlled by river connectivity to deep and shallow groundwater reservoirs [27]. Due to the presence of carbonate mineral concretions in clay-silt bedrocks, especially in the south of the WSL [33], the groundwater is enriched in soluble elements such as alkalis and alkaline-earth metals, oxyanions and U(VI). The second is transport-limitation, when the export of an element is controlled by the availability of its carrier such as organic and organo-ferric, organo-aluminum colloids [30,34,45]. This factor reflects the superposition of surface source (topsoil, vegetation) providing organic colloids, and deep soil (mineral horizons), together with Fe(II)-rich groundwaters, providing soluble TE. In addition to these two main groups of elements, the mineral nutrients are not limited by transport and exhibit quite complex pattern which reflects their deep (groundwater, bedrock lithology) and surface (plant litter, atmospheric deposition) sources. These elements are strongly impacted by their biotic uptake in the river channel or seasonal release from decaying aquatic macrophytes, plankton and periphyton [37,60]. Various internal factors, operating in soils and riparian zone of the river, are capable of modifying the export of nutrients. For example, low export fluxes of both total P (phosphate and organic P) and P-PO_4 in southern (56–60° N) latitudinal belt compared to northern, permafrost-affected rivers (see Figure 3G, Figure S4G) may be due P retention via adsorption onto and coprecipitation with Al, Fe hydroxides in the deeper part of soil profile [61,62] as well as P removal in the form of Ca phosphate minerals [63] occurring in the unfrozen mineral soils, exposed to surface fluids. Furthermore, P uptake by abundant terrestrial and aquatic vegetation is most pronounced in the south. In the north, the mineral soils are essentially frozen, and the requirements of aquatic and terrestrial biota for this nutrient are much lower [64].

The effect of watershed size on DOC, DIC, cations, and Si export fluxes was rather minor. This is at odds with strong river size control on element yields as it is known in temperate and mountainous regions [65,66]. In particular, small catchments of wetlands exhibit generally lower runoff than the medium and large rivers, and the runoff is one of the major controlling factors of chemical weathering and element export [46,67–70]. Season also played a secondary role in determining element yield pattern. However, sizable correlations of element fluxes with $S_{watershed}$ in summer and autumn, observed solely in the permafrost-free zone (Supplementary Table S1) can be explained by more pronounced impact of deep underground waters in large rivers compared to small ones. These waters typically contain a high concentration of soluble, labile elements (e.g., Cl, SO_4, Na, Mo, As) [33,34] but also Fe(II), those oxidation in the riparian and hyporheic zones create large amount of organo-ferric colloids [45] capable to provide enhanced concentrations of typically insoluble low mobile elements such as trivalent and tetravalent hydrolysates.

The latitude was revealed to be the primary governing parameter of elementary yields and clearly marked the difference between permafrost-free, discontinuous and continuous permafrost regions. The most northern regions of the WSL exhibited rather high yields of DOC, DIC, Si and cations. It is possible that in continuous permafrost zone of frozen peat bogs, the underlining mineral layer is protected by the permafrost. As a result, the active (seasonally unfrozen) layer is located within the organic layer which comprises live vegetation, plant litter, and upper peat layer. This organic matrix is extremely reactive, and capable of releasing sizable amount of DOC, P, Si, Ca and nutrients over very short periods of time during contact with surface waters [71,72]. Water temperature (between 4 and 25 °C) has only minor effect on C and element release from both thawed and frozen peat [72]. Therefore, even short-term contact of water with surface peat and vegetation is capable mobilizing sizable amount of DOC and nutrients, despite low temperature in the northern regions. In this regard, the elementary export by the WSL rivers is strongly controlled by dynamics of peat formation/decay across the territory. The particularity of the WSL is that, currently, this region is recovering from the last glaciation. As a result, the ecosystems are highly non-stationary: the peat actively accumulates in the south [73], while in the north, the frozen peat is subjected to thawing and degradation [74–76]. The uptake of elements from groundwater, river and forest tree litter by growing peat in southern mires counteracts with DOC and element release from thawing/degrading peat in the northern palsas. The elements affected by these processes are those that exhibit the highest concentration in peat relative to undelaying mineral (silt, sand) horizons. According to a previous assessment of elementary peat composition in the WSL [77], the peat is sizably enriched in C, (V, Cr), trivalent (TE^{3+}) and tetravalent (TE^{4+}) trace metals (Al, Y, REEs, Ti, Zr, Hf, Th) and U, Zn, Pb and depleted in highly mobile alkalis and alkaline-earths metals, As, and Mo. Therefore, enhanced riverine yield of low-soluble TE^{3+}, TE^{4+}, U, and some divalent metals in the north relative to the south is possibly due to these elements being tightly linked to peatland evolutionary pattern across the WSL. Similarly, depletion of peat relative to underlying mineral horizons in soluble, highly mobile elements is consistent with enhanced export of these elements by southern rivers, where the peat accumulation occurs.

4.2. Comparison of Major Cation, DIC, Si and DOC Export Fluxes in the WSL with Other Boreal Regions

The average total dissolved cation flux (TDS_c = Na + K + Ca + Mg) over May-October 2016 ranged from 4.40 ± 0.55 t km^{-2} y^{-1} in the permafrost-free zone to 2.64 ± 0.59 t km^{-2} y^{-1} in the permafrost-affected zone, which is lower than the fluxes of Central Siberian rivers of the same latitude, draining basaltic rocks (5 to 8 t km^{-2} y^{-1}, [78]), large Siberian rivers such as Yenisey and Lena (6.2 and 6.8 t km^{-2} y^{-1}, respectively, [79]), the Ob River in its middle course (6.0 t km^{-2} y^{-1}, [80]), and the permafrost-free Eurasian Arctic rivers draining sedimentary rocks (Sev. Dvina, 9.5 t km^{-2} y^{-1}, [59]; Pechora, 6.6 t km^{-2} y^{-1}, [79]). The TDS_c yield of permafrost-affected WSL rivers is, however, comparable with previous estimations of that in the middle-size Siberian rivers (2.8, 2.5, and 2.3 t km^{-2} y^{-1} for Kolyma, Indigirka and Anabar, respectively [79]). The main reason for relatively low cationic fluxes of small WSL rivers compared to other large rivers of Northern Eurasia of similar runoff are i) low connectivity of WSL rivers with underlying bedrocks and groundwater, due to thick peat layer and permafrost, and ii) essentially weathered character of silicate rocks (sands, clays) in western Siberia. Note also that the transport of major cations in permafrost-free rivers is strongly pronounced during winter (e.g., 35–40% of total annual yield in the Severnaya Dvina River [59]), so it is possible that cationic fluxes of WSL rivers during May-October are somewhat underestimated relative to annual export. At the same time, the majority of small (<4000 km^2 watershed area) rivers in the northern part of the WSL freeze solid during the winter [33], so the winter flux may be non-negligible only for southern, permafrost-free rivers.

In contrast to major cations, no difference in the export fluxes between small WSL rivers and large and medium size Eurasian rivers was detected for the DIC. The DIC export of small WSL rivers in the permafrost-free zone (2.2 ± 0.4 t km^{-2} y^{-1}) is in agreement with recent estimations of DIC export in the middle course of the Ob River (2.9 t km^{-2} y^{-1}, [80]) and with the mean riverine DIC yield of the entire Eurasian Arctic basin (2.2 t km^{-2} y^{-1}, [79]). However, DIC export by permafrost-affected WSL rivers is somewhat lower (1–2 t km^{-2} y^{-1}) and comparable to medium size rivers of Central and Eastern Siberia (0.6–2.2 t km^{-2} y^{-1}, [79]). A tentative explanation of elevated DIC (but not TDS_c) flux in the WSL rivers is CO_2 and HCO_3^- generation by microbial (and photolytic) processing of peat soil organic carbon (both DOC and POC), which is the main cause of very high CO_2 emission from WSL inland waters [81,82]. The light isotopic composition of DIC in the WSL rivers ($-25 \leq \delta^{13}C_{DIC} < -10$ ‰, [33]) is consistent with this hypothesis.

The riverine Si fluxes in the southern part (<61° N) of the WSL (0.5–1.0 t km^{-2} y^{-1}) are comparable to those of the Ob River (0.62 t km^{-2} y^{-1}, [80] and small rivers of the northern Sweden (0.9 t km^{-2} y^{-1}, [83]). In contrast, the fluxes of the permafrost-affected rivers (1.0–1.5 t km^{-2} y^{-1}, Figure 3F and Figure S4F and Table 1) contradict the expected trend of decreasing flux with the increasing latitude and decreasing temperature, given that the chemical weathering of silicate rocks is much slower in colder climates [67–70]. In fact, we noted that the northern fluxes of WSL are comparable with those of the temperate rivers such as Mississippi and Yangtze [84]. Moreover, if the silicate rock weathering significantly controls element delivery from the soil to the river, such a northward increase would occur for cations (Ca, Mg, Na) as it is known for typical silicate terrains of the boreal zone [85], but this is not observed in the WSL territory. As such, we hypothesize that two- to three-fold increase in Si yield of northern rivers relatively to southern rivers is linked to combination of (1) enhanced mobilization from plant litter via suprapermafrost flow in the permafrost zone, (2) limited silicate secondary mineral formation in shallow, essentially frozen northern soils compared to southern soils, and (3) strong uptake of Si by both terrestrial plants and aquatic macrophytes and periphyton in the southern rivers.

The DOC fluxes in the permafrost-affected regions of the WSL territory were quite high (2 to 6 t km^{-2} y^{-1}) compared to the middle and lower reaches of the Ob River (0.64 and 1.2 t km^{-2} y^{-1}, respectively, [79,80]). At the same time, these fluxes are comparable with those in large boreal river draining permafrost-free wetlands (the Severnaya Dvina River, 4.2 ± 0.8 t C_{org} km^{-2} y^{-1}, [59]) and are the highest among all known rivers flowing to the Arctic Ocean. Indeed, the Ob, Yukon, Lena, Yenisey, and Mackenzie rivers exhibit a DOC yield from 0.5 to 2.5 t km^{-2} y^{-1} ([79,86]). We suggest that the main factor responsible for such high DOC yield in small WSL rivers is high proportion of peatlands on their watershed (i.e., typically from 40 to 60%, according to the GIS data [34]). The peatland-draining Taz (watershed = 150,000 km^2), Pur (112,000 km^2), and Nadym (64,000 km^2) rivers, located entirely in the discontinuous permafrost zone, also have a DOC yield of 1.9, 2.1, and 4.4 t C_{org} km^{-2} y^{-1}, respectively ([79] and calculations based on data from the RHS).

The northward increase in DOC flux possibly reflects strong leaching of OM from the plant litter and organic-rich topsoil (Histosol). In the north, the adsorption of DOM on underlying mineral horizons is minimal because these horizons are frozen. As a result, the riverine DOC export in the permafrost zone of the WSL is controlled by water travel time through the peat layer and underlying mineral horizons and the water residence time necessary for DOC leaching from upper vegetation layer (moss, lichen, litter). However, quantitative modeling of DOM and element reactive transport in the WSL peatlands, on the scale of a small watershed or large river, is beyond the scope of this study.

4.3. Possible Evolution of Western Siberia Rivers Elementary Yields Under Climate Change Scenario

The space for time approach employed in the present work provides some future projections for riverine element behavior. However, it exhibits a number of shortcomings whose analysis goes beyond the scope of this work. In particular, this approach does not address the time scale, necessary for the northern ecosystem to reach the new "more southern" state; it ignores possible shift in the structure of vegetation and soil microbial community that can indirectly impact the carbon and nutrient, in terms of landscape storage and removal via rivers, and it does not include information on the altered seasonality such as extended hydrologic seasons, earlier snowmelt, higher precipitation that will likely occur as a result of climate change. Nevertheless, as a first order empirical assessment, the following predictions can be made. The first consequence of climate warming in western Siberia is thawing of frozen peat and underlying mineral horizons. The thickness of the active layer (ALT) is projected to increase more than 30% during this century across the tundra area in the Northern Hemisphere [87–89]. In the WSL, this increase will be most dramatic in the north, where the peat deposits are thinner than those in the discontinuous permafrost zone [48,90–92]. The main consequences of the ALT increase may be the involvement of mineral (clay) horizons into water infiltration within the soil profile [90]. As a result, the DOC originated from the leaching of the upper peat layers and plant litter degradation will be retained on mineral surfaces via adsorption onto and incorporation into clay interlayers [93–97]. For inorganic solutes, the effect of ALT increase will be lower than that of DOC, given much lower affinity of HCO_3, cations and Si to clay surfaces and the lack of unweathered (primary) silicate rocks underneath the peat soil column. However, if the thawing will open new water paths between deep groundwater reservoirs and the river, this may increase the riverine export of major cations and DIC [9,33].

The second consequence of climate change in western Siberia is a shift of the permafrost zone boundaries further north [98–100]. Within the substituting space for time scenario, such a permafrost boundary change can be considered equivalent to the northward shift of the river latitudes as shown in Figure 3. Based on latitudinal pattern of major and TE (as illustrated in Figure 3 and Figure S4), the following groups of the river water components likely to change their riverine yield over open-water period in case of anticipated shift in the permafrost zones: (1) Elements those yields likely to increase by a factor of 2 to 3: DIC, Ca, SO_4, Sr, Ba, Mo, U; (2) Elements likely to decrease the yields by a factor of 2 to 5: DOC, Fe, Si, Ptot, P-PO_4, N-NH_4, divalent heavy metals, trivalent and tetravalent hydrolysates, and (3) Elements weakly affected by the permafrost boundary change and the ALT increase (change of yield by a factor of 1.5 to 2.0): B, Li, K, Rb, Cs, N-NO_3, Mg, Zn, As, Sb, Rb, Cs).

The impact of climate warming on riverine fluxes in the WSL is not restricted to the shift of permafrost zones and the change of water flow path (deep versus surface). It has to be placed in the context of changing precipitation, plant biomass productivity and modification of the seasonality [101]. Complex evaluation of these factors goes beyond the scope of this study and it requires ecosystem-level regional modeling. Overall, the WSL is likely to have increased lateral export of DIC but decreased export of DOC. However, the changes of both fluxes (+2 and −3 to −4 t C km^{-2} y^{-1} for DIC and DOC, respectively) are dwarfed compared to possible magnitude of C emission from WSL inland waters (rivers and lakes) to the atmosphere: 10–20 t km^{-2} y^{-1} in permafrost-free and isolated zone and 20–40 t C km^{-2} y^{-1} in discontinuous and continuous permafrost zone [102].

5. Conclusions

Based on a two-year seasonal sampling of 32 western Siberian rivers of various size (from 10 to 10^5 km^2 in watershed area) draining through a sizable permafrost gradient, we measured riverine export fluxes of dissolved (<0.45 μm) C, nutrients, major and TE over a six-month open-water period (May to October). The export fluxes of DOC, DIC, major cations, macro- and micro-nutrients, toxicants, and geochemical tracers were weakly dependent on the size of the river. The primary parameter of export fluxes control was latitude, which marked the position of the permafrost zones.

There are several distinct groups of elements defined according to the latitudinal patterns of their yields. These groups reflect the relative mobility of elements, consistent with general knowledge of river hydrochemistry in high latitude regions. The DOC, organically-bound metals (V, Cr, Mn, Fe, Co, Ni, Cu, Zn, Cd and Pb) and many low soluble TE (Al, Ti, Zr, Hf, Th and REEs) exhibited similar latitudinal pattern with a minimum in the south and a maximum in the permafrost zone). A northward increase in Si export flux may be due to a decrease in Si uptake by plants in the north and strong Si retaining by mire and forest vegetation in the south.

An increase in DOC export fluxes from the south to the north (by a factor of 3 to 4 depending on the years of observation) could be due to leaching of OM from the plant litter and organic-rich topsoil (Histosol). The removal of DOC by adsorption on mineral horizons was hypothesized to be very low in the north. As a result, the riverine DOC export in the permafrost zone of the WSL may be strongly controlled by the water residence time necessary for DOC leaching from upper vegetation layer (moss, lichen, litter). This calls a need for quantitative modeling of DOM and element reactive transport in WSL peatlands, both at the scale of small watershed and large rivers. Furthermore, the peculiarity of western Siberia is that the elementary export by WSL rivers is strongly controlled by dynamics of peat formation and decay across the territory. Because the WSL is currently recovering from the last glaciation, this territory is dominated by non-stationary ecosystems with strong latitudinal contrast: the fresh peat is accumulating in the south whereas the old frozen peat is thawing and degrading in the north. As a result, uptake of elements from groundwater, river and forest tree litter by growing peat in southern bog competes with DOC and element release from thawing/degrading peat in the northern palsas. Note that, unlike many other permafrost-affected regions in the world whose CO_2 uptake rate during weathering is likely to increase under climate warming, the WSL may increase its riverine export of DIC but decrease the export of DOC.

Supplementary Materials: The following are available online at http://www.mdpi.com/2073-4441/12/6/1817/s1, Figure S1: Correlation between elementary fluxes and watershed area for permafrost-free and permafrost-affected regions; Figure S2: Partial contribution of spring, summer and autumn 2016 to overall open-water period export of anions and trace elements by WSL rivers; Figure S3: Ternary molar diagrams of major cations and anions in the WSL rivers. Figure S4: Yields of DOC (A), DIC (B), Ca (C), Mg (D), K (E), Si (F), P_{tot} (G) and Fe (H) during May-August 2015 in 33 WSL rivers across the latitudinal gradient. Table S1: Spearman correlation coefficients ($p < 0.05$) between element export flux (yield) and watershed area.

Author Contributions: O.S.P. designed the study and wrote the paper; R.M.M., I.V.K., and S.V.L. performed sampling, analysis of cations and anions, and their interpretation; S.N.V. and S.N.K. were responsible for the choice of sampling objects and statistical treatment; S.V.L. provided the background information on soil, peat, and permafrost active layer; L.S.S. was in charge of DOC, DIC, and anion measurements and their interpretation; L.G.K. provided GIS-based interpretation, mapping, and identification of river watersheds; S.G.K. performed all primary hydrological data collection, and their analysis and interpretation. All 10 authors participated in field expeditions. All authors have read and agreed to the published version of the manuscript.

Funding: Russian Science Foundation: No 18-17-00237 and 18-77-10045. Russian Fund for Fundaental Research: 19-55-15002, 20-05-00729_a.

Acknowledgments: We acknowledge main support from RSF grant No 18-17-00237 and RFBR grants No 19-55-15002, 20-05-00729_a, and RSF grant No 18-77-10045 for field work.

Conflicts of Interest: The authors declare no conflict of interest.

References

1. Turetsky, M.R.; Abbott, B.W.; Jones, M.C.; Anthony, K.W.; Olefeldt, D.; Schuur, E.A.G.; Grosse, G.; Kuhry, P.; Hugelius, G.; Koven, C.; et al. Carbon release through abrupt permafrost thaw. *Nat. Geosci.* **2020**, *13*, 138–143. [CrossRef]
2. Vonk, J.E.; Tank, S.E.; Walvoord, M.A. Integrating hydrology and biogeochemistry across frozen landscapes. *Nat. Commun.* **2019**, *10*, 1–4. [CrossRef] [PubMed]
3. Smith, L.C.; Sheng, Y.; MacDonald, G.M. A first Pan-Arctic assessment of the influence of glaciation, permafrost, topography and peatlands on Northern hemisphere lake distribution. *Permafr. Periglac. Proc.* **2007**, *18*, 201–208. [CrossRef]
4. McEnroe, N.A.; Roulet, N.T.; Moore, T.R.; Garneau, M. Do pool surface area and depth control CO_2 and CH_4 fluxes from an ombrotrophic raised bog, James Bay, Canada? *J. Geophys. Res. Biogeosci.* **2009**, 114. [CrossRef]
5. Olefeldt, D.; Roulet, N.T. Effects of permafrost and permafrost thaw on dissolved organic carbon in a subarctic peatland complex. *J. Geophys. Res. Biogeosci.* **2012**, 117. [CrossRef]
6. Olefeldt, D.; Roulet, N.; Giesler, R.; Persson, A. Total waterborne carbon export and DOC composition from ten nested subarctic peatland catchments—Importance of peatland cover, groundwater influence, and inter-annual variability of precipitation patterns. *Hydrol. Process.* **2013**, *27*, 2280–2294. [CrossRef]
7. Roehm, C.L.; Giesler, R.; Karlsson, J. Bioavailability of terrestrial organic carbon to lake bacteria: The case of a degrading subarctic permafrost mire complex. *J. Geophys. Res. Biogeosci.* **2009**, 114. [CrossRef]
8. Amesbury, M.J.; Gallego-Sala, A.; Loisel, J. Peatlands as prolific carbon sinks. *Nat. Geosci.* **2019**, *12*, 880–881. [CrossRef]
9. Frey, K.E.; McClelland, J.W. Impacts of permafrost degradation on arctic river biogeochemistry. *Hydrol. Process.* **2009**, *23*, 169–182. [CrossRef]
10. Bartsch, A.; Leibman, M.; Strozzi, T.; Khomutov, A.; Widhalm, B.; Babkina, E.; Mullanurov, D.; Ermokhina, K.; Kroisleitner, C.; Bergstedt, H. Seasonal Progression of Ground Displacement Identified with Satellite Radar Interferometry and the Impact of Unusually Warm Conditions on Permafrost at the Yamal Peninsula in 2016. *Remote Sens.* **2019**, *11*, 1865. [CrossRef]
11. Beilman, D.W.; MacDonald, G.M.; Smith, L.C.; Reimer, P.J. Carbon accumulation in peatlands of West Siberia over the last 2000 years. *Glob. Biogeochem. Cycles* **2009**, 23. [CrossRef]
12. Liljedahl, A.K.; Boike, J.; Daanen, R.P.; Fedorov, A.N.; Frost, G.V.; Grosse, G.; Hinzman, L.D.; Iijma, Y.; Jorgenson, J.C.; Matveyeva, N.; et al. Pan-Arctic ice-wedge degradation in warming permafrost and its influence on tundra hydrology. *Nat. Geosci.* **2016**, *9*, 312–318. [CrossRef]
13. Colombo, M.; Brown, K.A.; De Vera, J.; Bergquist, B.A.; Orians, K.J. Trace metal geochemistry of remote rivers in the Canadian Arctic Archipelago. *Chem. Geol.* **2019**, *525*, 479–491. [CrossRef]
14. Drake, T.W.; Tank, S.E.; Zhulidov, A.V.; Holmes, R.M.; Gurtovaya, T.; Spencer, R.G.M. Increasing alkalinity export from large russian Arctic rivers. *Environ. Sci. Technol.* **2018**, *52*, 8302–8308. [CrossRef]
15. Fabre, C.; Sauvage, S.; Tananaev, N.; Noël, G.E.; Teisserenc, R.; Probst, J.L.; Pérez, J.M.S. Assessment of sediment and organic carbon exports into the Arctic ocean: The case of the Yenisei River basin. *Water Res.* **2019**, *158*, 118–135. [CrossRef]
16. Griffin, C.G.; McClelland, J.W.; Frey, K.E.; Fiske, G.; Holmes, R.M. Quantifying CDOM and DOC in major Arctic rivers during ice-free conditions using Landsat TM and ETM+ data. *Remote Sens. Environ.* **2018**, *209*, 395–409. [CrossRef]
17. Guo, L.; Ping, C.-L.; Macdonald, R.W. Mobilization pathways of organic carbon from permafrost to arctic rivers in a changing climate. *Geophys. Res. Lett.* **2007**, 34. [CrossRef]
18. Hindshaw, R.S.; Teisserenc, R.; Le Dantec, T.; Tananaev, N. Seasonal change of geochemical sources and processes in the Yenisei River: A Sr, Mg and Li isotope study. *Geochim. Cosmochim. Acta* **2019**, *255*, 222–236. [CrossRef]
19. Li, M.; Peng, C.; Zhou, X.; Yang, Y.; Guo, Y.; Shi, G.; Zhu, Q. Modeling global riverine DOC flux dynamics from 1951 to 2015. *J. Adv. Modeling Earth Syst.* **2019**, *11*, 514–530. [CrossRef]
20. Murphy, M.J.; Porcelli, D.; Pogge von Strandmann, P.A.E.; Hirst, C.A.; Kutscher, L.; Katchinoff, J.A.; Mörth, C.-M.; Maximov, T.; Andersson, P.S. Tracing silicate weathering processes in the permafrost-dominated Lena River watershed using lithium isotopes. *Geochim. Cosmochim. Acta* **2019**, *245*, 154–171. [CrossRef]

21. Holmes, R.M.; Coe, M.T.; Fiske, G.J.; Gurtovaya, T.; McClelland, J.W.; Shiklomanov, A.I.; Spencer, R.G.M.; Tank, S.E.; Zhulidov, A.V. Climate change impacts on the hydrology and biogeochemistry of Arctic Rivers. In *Climatic Changes and Global warming of Inland Waters; Impacts and Mitigation for Ecosystems and Societies*; Goldman, C.R., Kumagi, M., Robarts, R.D., Eds.; John Wiley and Sons: Hoboken, NJ, USA, 2013; pp. 1–26.
22. McClelland, J.W.; Déry, S.J.; Peterson, B.J.; Holmes, R.M.; Wood, E.F. A pan-arctic evaluation of changes in river discharge during the latter half of the 20th century. *Geophys. Res. Lett.* **2006**, 33. [CrossRef]
23. Shiklomanov, A.I.; Lammers, R.B.; Vörösmarty, C.J. Widespread decline in hydrological monitoring threatens Pan-Arctic research. *Eos Trans. Am. Geophys. Union* **2002**, *83*, 13–20. [CrossRef]
24. Shogren, A.J.; Zarnetske, J.P.; Abbott, B.W.; Iannucci, F.; Frei, R.J.; Griffin, N.A.; Bowden, W.B. Revealing biogeochemical signatures of Arctic landscapes with river chemistry. *Sci. Rep.* **2019**, *9*, 1–11. [CrossRef] [PubMed]
25. Toohey, R.C.; Herman-Mercer, N.M.; Schuster, P.F.; Mutter, E.A.; Koch, J.C. Multidecadal increases in the Yukon River Basin of chemical fluxes as indicators of changing flowpaths, groundwater, and permafrost. *Geophys. Res. Lett.* **2016**, *43*, 12–120. [CrossRef]
26. Frey, K.E.; McClelland, J.W.; Holmes, R.M.; Smith, L.C. Impacts of climate warming and permafrost thaw on the riverine transport of nitrogen and phosphorus to the Kara Sea. *J. Geophys. Res.* **2007**, 112. [CrossRef]
27. Frey, K.E.; Siegel, D.I.; Smith, L.C. Geochemistry of west Siberian streams and their potential response to permafrost degradation. *Water Resour. Res.* **2007**, 43. [CrossRef]
28. Frey, K.E.; Smith, L.C. Amplified carbon release from vast West Siberian peatlands by 2100. *Geophys. Res. Lett.* **2005**, 32. [CrossRef]
29. Krickov, I.; Lim, A.; Manasypov, R.M.; Loiko, S.V.; Shirokova, L.S.; Kirpotin, S.N.; Karlsson, J.; Pokrovsky, O.S. Riverine particulate C and N generated at the permafrost thaw front: Case study of western Siberian rivers across a 1700-km latitudinal transect. *Biogeosciences* **2018**, *15*, 6867–6884. [CrossRef]
30. Krickov, I.V.; Pokrovsky, O.S.; Manasypov, R.M.; Lim, A.G.; Shirokova, L.S.; Viers, J. Colloidal transport of carbon and metals by western Siberian rivers during different seasons across a permafrost gradient. *Geochim. Cosmochim. Acta* **2019**, *265*, 221–241. [CrossRef]
31. Krickov, I.V.; Lim, A.G.; Manasypov, R.M.; Loiko, S.V.; Vorobyev, S.N.; Shevchenko, V.P.; Dara, O.M.; Gordeev, V.V.; Pokrovsky, O.S. Major and trace elements in suspended matter of western Siberian rivers: First assessment across permafrost zones and landscape parameters of watersheds. *Geochim. Cosmochim. Acta* **2020**, *269*, 429–450. [CrossRef]
32. Manasypov, R.M.; Pokrovsky, O.S.; Kirpotin, S.N.; Shirokova, L.S. Thermokarst lake waters across the permafrost zones of western Siberia. *Cryosphere* **2014**, *8*, 1177–1193. [CrossRef]
33. Pokrovsky, O.S.; Manasypov, R.M.; Shirokova, L.S.; Loiko, S.V.; Krickov, I.V.; Kopysov, S.; Zemtzov, V.A.; Kulizhsky, S.P.; Vorobyev, S.N.; Kirpotin, S.N. Permafrost coverage, watershed area and season control of dissolved carbon and major elements in western Siberia rivers. *Biogeosciences* **2015**, *12*, 6301–6320. [CrossRef]
34. Pokrovsky, O.S.; Manasypov, R.M.; Loiko, S.; Krickov, I.A.; Kopysov, S.G.; Kolesnichenko, L.G.; Vorobyev, S.N.; Kirpotin, S.N. Trace element transport in western Siberia rivers across a permafrost gradient. *Biogeosci. Discuss.* **2016**, *13*, 1877–1900. [CrossRef]
35. Polishchuk, Y.M.; Bogdanov, A.N.; Muratov, I.N.; Polishchuk, V.Y.; Lim, A.; Manasypov, R.M.; Shirokova, L.S.; Pokrovsky, O.S. Minor contribution of small thaw ponds to the pools of carbon and methane in the inland waters of the permafrost-affected part of the Western Siberian Lowland. *Environ. Res. Lett.* **2018**, *13*, 045002. [CrossRef]
36. Raudina, T.V.; Loiko, S.V.; Lim, A.G.; Krickov, I.V.; Shirokova, L.S.; Istigichev, G.I.; Kuzmina, D.M.; Kulizhsky, S.P.; Vorobyev, S.N.; Pokrovsky, O.S. Dissolved organic carbon and major and trace elements in peat porewater of sporadic, discontinuous, and continuous permafrost zones of western Siberia. *Biogeosciences* **2017**, *14*, 3561–3584. [CrossRef]
37. Vorobyev, S.N.; Pokrovsky, O.S.; Serikova, S.; Manasypov, R.M.; Krickov, I.V.; Shirokova, L.S.; Lim, A.; Kolesnichenko, L.G.; Kirpotin, S.N.; Karlsson, J. Permafrost boundary shift in Western Siberia may not modify dissolved nutrient concentrations in rivers. *Water* **2017**, *9*, 985. [CrossRef]
38. Raudina, T.V.; Loiko, S.V.; Lim, A.; Manasypov, R.M.; Shirokova, L.S.; Istigechev, G.I.; Kuzmina, D.M.; Kulizhsky, S.P.; Vorobyev, S.N.; Pokrovsky, O.S. Permafrost thaw and climate warming may decrease the CO_2, carbon, and metal concentration in peat soil waters of the Western Siberia Lowland. *Sci. Total Environ.* **2018**, *634*, 1004–1023. [CrossRef]

39. Blois, J.L.; Williams, J.W.; Fitzpatrick, M.C.; Jackson, S.T.; Ferrier, S. Space can substitute for time in predicting climate-change effects on biodiversity. *Proc. Natl. Acad. Sci. USA* **2013**, *110*, 9374–9379. [CrossRef]
40. Nikitin, S.P.; Zemtsov, V.A. *The Variability of Hydrological Parameters of Western Siberia*; Nauka: Novosibirsk, Russia, 1986; 204p. (In Russian)
41. Huh, Y.; Edmond, J.M. The fluvial geochemistry of the rivers of Eastern Siberia: III. Tributaries of the Lena and Anabar draining the basement terrain of the Siberian Craton and the Trans-Baikal Highlands. *Geochim. Cosmochim. Acta* **1999**, *63*, 967–981. [CrossRef]
42. Spence, J.; Telmer, K. The role of sulfur in chemical weathering and atmospheric CO_2 fluxes: Evidence from major ions, $\delta^{13}C_{DIC}$, and $\delta^{34}S_{SO4}$ in rivers of the Canadian Cordillera. *Geochim. Cosmochim. Acta* **2005**, *69*, 5441–5458. [CrossRef]
43. Qin, J.; Huh, Y.; Edmond, J.M.; Du, G.; Ran, J. Chemical and physical weathering in the Min Jiang, a headwater tributary of the Yangtze River. *Chem. Geol.* **2006**, *227*, 53–69. [CrossRef]
44. Bulygina, O.N.; Razuvaev, V.N.; Alexsandrova, T.M. Description of Daily Air Temperature and Precipitation Data set from Russian Meteorological Stations and from some Meteorological Stations over the Former USSR Territory (TTTR) [Electronic Resource]. Available online: http://meteo.ru/english/climate/descrip11.htm (accessed on 17 March 2020).
45. Pokrovsky, O.S.; Manasypov, R.M.; Loiko, S.V.; Shirokova, L.S. Organic and organo-mineral colloids in discontinuous permafrost zone. *Geochim. Cosmochim. Acta* **2016**, *188*, 1–20. [CrossRef]
46. Prokushkin, A.S.; Pokrovsky, O.S.; Shirokova, L.S.; Korets, M.A.; Viers, J.; Prokushkin, S.G.; Amon, R.M.W.; Guggenberger, G.; McDowell, W.H. Sources and the flux pattern of dissolved carbon in rivers of the Yenisey basin draining the Central Siberian Plateau. *Environ. Res. Lett.* **2011**, *6*, 045212. [CrossRef]
47. Rozhdestvensky, A.V.; Buzin, V.A.; Dobroumov, B.M.; Lobanova, A.G.; Lobanov, V.A.; Plitkin, G.A.; Tumanovskaya, S.M. SP 33-101-2003. Svod Pravil. Opredelenie Osnovny Khraschetnykh Gidrologicheskikh Kharakteristik [Set of Rules. Determination of Basic Design Hydrological Characteristics]. Odobren dlya Primeneniya v Kachestve Normativnogo Dokumenta Postanovleniem Gosstroya Rossii No 218 ot 26.12.2003. Available online: https://files.stroyinf.ru/Data2/1/4294815/4294815038.htm (accessed on 18 March 2020).
48. Novikov, S.M.; Moskvin, Y.P.; Trofimov, S.A.; Usova, L.I.; Batuev, V.I.; Tumanovskaya, S.M.; Smirnova, V.P.; Markov, M.L.; Korotkevicth, A.E.; Potapova, T.M. *Hydrology of Bog Territories of the Permafrost Zone of Western Siberia*; BBM Publishing House: St. Petersbourg, Russia, 2009; 535p. (In Russian)
49. McClelland, J.W.; Holmes, R.M.; Peterson, B.J.; Raymond, P.A.; Striegl, R.G.; Zhulidov, A.V.; Zimov, S.A.; Zimov, N.; Tank, S.E.; Spencer, R.G.M.; et al. Particulate organic carbon and nitrogen export from major Arctic rivers. *Glob. Biogeochem. Cycles* **2016**, *30*, 629–643. [CrossRef]
50. Bergström, S. The HBV Model. In *Structure and Applications*; Swedish Meteorological and Hydrological Institute (SMHI): Norrköping, Sweden, 1992; p. 35.
51. Bergström, S. The HBV model. In *Computer Models of Watershed Hydrology*; Singh, V.P., Ed.; Water Resources Publications: Highlands Ranch, CO, USA, 1995; Chapter 13; pp. 443–476.
52. Seibert, J.; Vis, M.J.P. Teaching hydrological modeling with a user-friendly catchment-runoff-model software package. *Hydrol. Earth Syst. Sci.* **2012**, *16*, 3315–3325. [CrossRef]
53. AIS GMVO Avtomatizirovannaya Informatsionnaya Sistema Gosudarstvennogo Monitoring Vodnykh ob"Ektov [Automated Information System of State Monitoring of Water Bodies]. [Electronic Resource]. 2020. Available online: https://gmvo.skniivh.ru/ (accessed on 17 March 2020).
54. Critelli, T.; Vespasiano, G.; Apollaro, C.; Muto, F.; Marini, L.; De Rosa, R. Hydrogeochemical study of an ophiolitic aquifer: A case study of Lago (Southern Italy, Calabria). *Environ. Earth Sci.* **2015**, *74*, 533–543. [CrossRef]
55. Raymond, P.A. Temperature versus hydrologic controls of chemical weathering fluxes from United States forests. *Chem. Geol.* **2015**, *458*, 1–13. [CrossRef]
56. Raymond, P.A.; Hamilton, S.K. Anthropogenic influences on riverine fluxes of dissolved inorganic carbon to the oceans. *Limnol. Oceanogr. Lett.* **2018**, *3*, 143–155. [CrossRef]
57. Stackpoole, S.M.; Stets, E.G.; Striegl, R.G. The impact of climate and reservoirs on longitudinal riverine carbon fluxes from two major water-sheds in the Central and Intermontane West. *J. Geophys. Res. Biogeosci.* **2014**, *119*, 848–863. [CrossRef]

58. Pokrovsky, O.S.; Viers, J.; Dupré, B.; Chabaux, F.; Gaillardet, J.; Audry, S.; Prokushkin, A.S.; Shirokova, L.S.; Kirpotin, S.N.; Lapitsky, S.A.; et al. Biogeochemistry of carbon, major and trace elements in watersheds of Northern Eurasia drained to the Arctic Ocean: The change of fluxes, sources and mechanisms under the climate warming prospective. *Comptes Rendus Geosci.* **2012**, *344*, 663–677. [CrossRef]
59. Chupakov, A.V.; Pokrovsky, O.S.; Moreva, O.Y.; Shirokova, L.S.; Neverova, N.V.; Chupakova, A.A.; Kotova, E.I.; Vorobyeva, T.Y. High resolution multi-annual riverine fluxes of organic carbon, nutrient and trace element from the largest European Arctic river, Severnaya Dvina. *Chem. Geol.* **2020**, 538. [CrossRef]
60. Newcomer Johnson, T.A.; Kaushal, S.S.; Mayer, P.M.; Smith, R.M.; Sivirichi, G.M. Nutrient retention in restored streams and rivers: A global review and synthesis. *Water* **2016**, *8*, 116. [CrossRef]
61. Bloom, P.R. Phosphorus adsorption by an aluminum-peat complex. *Soil Sci. Soc. Am. J.* **1981**, *45*, 267–272. [CrossRef]
62. Richardson, C. Mechanisms controlling phosphorus retention capacity in fresh-water wetlands. *Science* **1985**, *228*, 1424–1427. [CrossRef] [PubMed]
63. Penn, C.J.; Camberato, J.J. A critical review on soil chemical processes that control how soil pH affects phosphorus availability to plants. *Agriculture* **2019**, *9*, 120. [CrossRef]
64. Prokushkin, A.S.; Hagedorn, F.; Pokrovsky, O.S.; Viers, J.; Kirdyanov, A.V.; Masyagina, O.V.; Prokushkina, M.P.; McDowell, W.H. Permafrost regime affects the nutritional status and productivity of larches in Central Siberia. *Forests* **2018**, *9*, 314. [CrossRef]
65. Stets, E.G.; Striegl, R.G. Carbon export by rivers draining the conterminous United States. *Inland Waters* **2012**, *2*, 177–184. [CrossRef]
66. Giesler, R.; Lyon, S.W.; Morth, C.-M.; Karlsson, J.; Karlsson, E.M.; Jantze, E.J.; Destouni, G.; Humborg, C. Catchment-scale dissolved carbon concentrations and export estimates across six subarctic streams in northern Sweden. *Biogeosciences* **2014**, *11*, 525–537. [CrossRef]
67. Zakharova, E.A.; Pokrovsky, O.S.; Dupré, B.; Zaslavskaya, M.B. Chemical weathering of silicate rocks in Aldan Shield and Baikal Uplift: Insights from long-term seasonal measurements of solute fluxes in rivers. *Chem. Geol.* **2005**, *214*, 223–248. [CrossRef]
68. White, A.F.; Blum, A.E. Effects of climate on chemical weathering in watersheds. *Geochim. Cosmochim. Acta* **1995**, *59*, 1729–1747. [CrossRef]
69. Dessert, C.; Dupré, B.; Gaillardet, J.; Francois, L.M.; Allégre, C.J. Basalt weathering laws and the impact of basalt weathering on the global carbon cycle. *Chem. Geol.* **2003**, *202*, 257–273. [CrossRef]
70. Oliva, P.; Viers, J.; Dupré, B. Chemical weathering in granitic environments. *Chem. Geol.* **2003**, *202*, 225–256. [CrossRef]
71. Payandi-Rolland, D.; Shirokova, L.S.; Nakhle, P.; Tesfa, M.; Abdou, A.; Causserand, C.; Lartiges, B.; Rols, J.-L.; Guérin, F.; Bénézeth, P.; et al. Aerobic release and biodegradation of dissolved organic matter from frozen peat: Effects of temperature and heterotrophic bacteria. *Chem. Geol.* **2020**, *536*, 119448. [CrossRef]
72. Fraysse, F.; Pokrovsky, O.S.; Meunier, J.-D. Experimental study of terrestrial plant litter interaction with aqueous solutions. *Geochim. Cosmochim. Acta* **2010**, *74*, 70–84. [CrossRef]
73. Antipina, T.G.; Preis, Y.I.; Zenin, V.N. Dynamics of forest vegetation and climate in the southern taiga of Western Siberia in the Late Holocene according to spore–pollen analysis and AMS dating of the peat bog. *Russ. J. Ecol.* **2019**, *50*, 445–452. [CrossRef]
74. Fotiev, S.M. Arctic peatlands of the Yamal-Gydan province of Western Siberia. *Earth's Cryosphere* **2017**, *XXI*, 3–15. [CrossRef]
75. Loiko, S.; Raudina, T.; Lim, A.; Kuzmina, D.; Kulizhskiy, S.; Pokrovsky, O. Microtopography Controls of Carbon and Related Elements Distribution in the West Siberian Frozen Bogs. *Geosciences* **2019**, *9*, 291. [CrossRef]
76. Panova, N.K.; Trofimova, S.S.; Antipina, T.G.; Zinoviev, E.V.; Gilev, A.V.; Erokhin, N.G. Holocene dynamics of vegetation and ecological conditions in the southern Yamal Peninsula according to the results of comprehensive analysis of a relict peat bog deposit. *Russ. J. Ecol.* **2010**, *41*, 20–27. [CrossRef]
77. Stepanova, V.M.; Pokrovsky, O.S.; Viers, J.; Mironycheva-Tokareva, N.P.; Kosykh, N.P.; Vishnyakova, E.K. Major and trace elements in peat profiles in Western Siberia: Impact of the landscape context, latitude and permafrost coverage. *Appl. Geochem.* **2015**, *53*, 53–70. [CrossRef]
78. Pokrovsky, O.S.; Schott, J.; Kudryavtzev, D.I.; Dupré, B. Basalt weathering in Central Siberia under permafrost conditions. *Geochim. Cosmochim. Acta* **2005**, *69*, 5659–5680. [CrossRef]

79. Gordeev, V.V.; Martin, J.-M.; Sidorov, I.S.; Sidorova, M.V. A reassessment of the Eurasian river input of water, sediment, major elements, and nutrients to the Arctic Ocean. *Am. J. Sci.* **1996**, *296*, 664–691. [CrossRef]
80. Vorobyev, S.N.; Pokrovsky, O.S.; Kolesnichenko, L.G.; Manasypov, R.M.; Shirokova, L.S.; Karlsson, J.; Kirpotin, S.N. Biogeochemistry of dissolved carbon, major, and trace elements during spring flood periods on the Ob River. *Hydrol. Process.* **2019**, *33*, 1579–1594. [CrossRef]
81. Serikova, S.; Pokrovsky, O.S.; Ala-Aho, P.; Kazantsev, V.; Kirpotin, S.N.; Kopysov, S.G.; Krickov, I.V.; Laudon, H.; Manasypov, R.M.; Shirokova, L.S.; et al. High riverine CO_2 emissions at the permafrost boundary of Western Siberia. *Nat. Geosci.* **2018**, *11*, 825–829. [CrossRef]
82. Serikova, S.; Pokrovsky, O.S.; Laudon, H.; Krickov, I.V.; Lim, A.G.; Manasypov, R.M.; Karlsson, J. High carbon emissions from thermokarst lakes of Western Siberia. *Nat. Commun.* **2019**, *10*, 1–7. [CrossRef] [PubMed]
83. Smedberg, E.; Mörth, C.-M.; Swaney, D.P.; Humborg, C. Modeling hydrology and silicon-carbon interactions in taiga and tundra biomes from a landscape perspective: Implications for global warming feedbacks. *Glob. Biogeochem. Cycles* **2006**, 20. [CrossRef]
84. Guo, L.; Zhang, J.-Z.; Guéguen, C. Speciation and fluxes of nutrients (N, P, Si) from the upper Yukon River. *Glob. Biogeochem. Cycles* **2004**, 18. [CrossRef]
85. Zakharova, E.A.; Pokrovsky, O.S.; Dupré, B.; Gaillardet, J.; Efimova, L.E. Chemical weathering of silicate rocks in Karelia region and Kola peninsula, NW Russia: Assessing the effect of rock composition, wetlands and vegetation. *Chem. Geol.* **2007**, *242*, 255–277. [CrossRef]
86. Raymond, P.A.; McClelland, J.W.; Holmes, R.M.; Zhulidov, A.V.; Mull, K.; Peterson, B.J.; Striegl, R.G.; Aiken, G.R.; Gurtovaya, T.Y. Flux and age of dissolved organic carbon exported to the Arctic Ocean: A carbon isotopic study of the five largest arctic rivers. *Glob. Biogeochem. Cycles* **2007**, 21. [CrossRef]
87. Anisimov, O.A.; Shiklomanov, N.I.; Nelson, F.E. Variability of seasonal thaw depth in permafrost regions: A stochastic modeling approach. *Ecol. Model.* **2002**, *153*, 217–227. [CrossRef]
88. Dankers, R.; Burke, E.J.; Price, J. Simulation of permafrost and seasonal thaw depth in the JULES land surface scheme. *Cryosphere* **2011**, *5*, 773–790. [CrossRef]
89. Stendel, M.; Christensen, J.H. Impact of global warming on permafrost conditions in a coupled GCM. *Geophys. Res. Lett.* **2002**, *29*, 1632. [CrossRef]
90. Babkina, E.A.; Leibman, M.O.; Dvornikov, Y.A.; Fakashchuk, N.Y.; Khairullin, R.R.; Khomutov, A.V. Activation of cryogenic processes in Central Yamal as a result of regional and local change in climate and thermal state of permafrost. *Russ. Meteorol. Hydrol.* **2019**, *44*, 283–290. [CrossRef]
91. Botch, M.S.; Kobak, K.I.; Vinson, T.S.; Kolchugina, T.P. Carbon pools and accumulation in peatlands of the former Soviet Union. *Glob. Biogeochem. Cycles* **1995**, *9*, 37–46. [CrossRef]
92. Liss, O.L.; Abramova, L.I.; Avetov, N.A.; Berezina, N.A.; Inisheva, L.I.; Kurnishnikova, T.V.; Sluka, Z.A.; Tolpysheva, T.Y.; Shvedchikova, N.K. *Wetland Systems of West Siberia and Their Importance for Nature Conservation*; Grifi K Publisher: Tula, Russia, 2001; 584p. (In Russian)
93. Gentsch, N.; Mikutta, R.; Alves, R.J.E.; Barta, J.; Capek, P.; Gitte, A.; Hugelius, G.; Kuhry, P.; Lashchinskiy, N.; Palmtag, J.; et al. Storage and transformation of organic matter fractions in cryoturbated permafrost soils across the Siberian Arctic. *Biogeosciences* **2015**, *12*, 4525–4542. [CrossRef]
94. Kaiser, C.; Meyer, H.; Biasi, C.; Rusalimova, O.; Barsukov, P.; Richter, A. Conservation of soil organic matter through cryoturbation in arctic soils in Siberia. *J. Geophys. Res.* **2007**, *112*, 9–17. [CrossRef]
95. Kawahigashi, M.; Kaiser, K.; Kalbitz, K.; Rodionov, A.; Guggenberger, G. Dissolved organic matter in small streams along a gradient from discontinuous to continuous permafrost. *Glob. Chang. Biol.* **2004**, *10*, 1576–1586. [CrossRef]
96. Mergelov, N.S.; Targulian, V.O. Accumulation of organic matter in the mineral layers of permafrost-affected soils of coastal lowlands in East Siberia. *Eurasian Soil Sci.* **2011**, *44*, 249–260. [CrossRef]
97. Oosterwoud, M.R.; Temminghoff, E.J.M.; van der Zee, S.E.A.T.M. Quantification of DOC concentrations in relation with soil properties of soils in tundra and taiga of Northern European Russia. *Biogeosci. Discuss.* **2010**, *7*, 3189–3226. [CrossRef]
98. Anisimov, O.; Kokorev, V.; Zhil'tsova, Y. Temporal and spatial patterns of modern climatic warming: Case study of Northern Eurasia. *Clim. Chang.* **2013**, *118*, 871–883. [CrossRef]
99. Romanovsky, V.E.; Drozdov, D.S.; Oberman, N.G.; Malkova, G.V.; Kholodov, A.L.; Marchenko, S.S.; Moskalenko, N.G.; Sergeev, D.O.; Ukraintseva, N.G.; Abramov, A.A.; et al. Thermal state of permafrost in Russia. *Permafr. Periglac. Proc.* **2010**, *21*, 136–155. [CrossRef]

100. Vasiliev, A.A.; Streletskaya, I.D.; Shirokov, R.S.; Oblogov, G.E. Evolution of cryolithozone of coastal zone of western Yamal during climate change. *Kriosf. Zemli* **2011**, *2*, 56–64.
101. Ernakovich, J.; Hopping, K.A.; Berdanier, A.B.; Simpson, R.T.; Kachergis, E.J.; Steltzer, H.; Wallenstein, M.D. Predicted responses of arctic and alpine ecosystems to altered seasonality under climate change. *Glob. Chang. Biol.* **2014**, *20*, 3256–3269. [CrossRef] [PubMed]
102. Karlsson, J.; Serikova, S.; Vorobyev, S.N.; Rocher-Ros, G.; Denfeld, B.; Pokrovsky, O.S. Carbon emission from Western Siberian inland waters. *Nature Comm.* (submitted).

water

MDPI

Review

The Main Features of Phosphorus Transport in World Rivers

Vitaly S. Savenko [1] and Alla V. Savenko [2,*]

[1] Faculty of Geography, Lomonosov Moscow State University, 119991 Moscow, Russia; Vitaly_Savenko@rambler.ru

[2] Faculty of Geology, Lomonosov Moscow State University, 119991 Moscow, Russia

* Correspondence: Alla_Savenko@rambler.ru

Abstract: Data on the geochemistry of phosphorus in the continental runoff of dissolved and solid substances were systematized and generalized, with a separate consideration of the processes of runoff transformation in river mouth areas. It has been established that atmospheric deposition, which many authors consider to be an important source of phosphorus in river runoff and not associated with mobilization processes in catchments, actually contains phosphorus from soil-plant recycling. This is confirmed by the fact that the input of phosphorus from the atmosphere into catchments exceeds its removal via water runoff. An analysis of the mass ratio of phosphorus in the adsorbed form and in the form of its own minerals was carried out. It was shown that the maximum mass of adsorbed phosphorus is limited by the solubility of its most stable minerals. The minimum concentrations of dissolved mineral and total phosphorus were observed in the rivers of the Arctic and subarctic belts; the maximum concentrations were confined to the most densely populated temperate zone and the zone of dry tropics and subtropics. In the waters of the primary hydrographic network, the phosphorus concentration exhibited direct relationships with the population density in the catchments and the mineralization of the river water and was closely correlated with the nitrogen content. This strongly suggests that economic activity is one of the main factors in the formation of river phosphorus runoff. The generalization of the authors' and the literature's data on the behavior of phosphorus at the river–sea mixing zone made it possible to draw a conclusion about the nonconservative distribution of phosphorus, in most cases associated with biological production and destruction processes. The conservative behavior of phosphorus was observed only in heavily polluted river mouths with abnormally high concentrations of this element.

Keywords: geochemistry of phosphorus; continental runoff; river mouth

Citation: Savenko, V.S.; Savenko, A.V. The Main Features of Phosphorus Transport in World Rivers. *Water* **2022**, *14*, 16. https://doi.org/10.3390/w14010016

Academic Editor: Liudmila S. Shirokova

Received: 27 November 2021
Accepted: 19 December 2021
Published: 22 December 2021

Publisher's Note: MDPI stays neutral with regard to jurisdictional claims in published maps and institutional affiliations.

1. Introduction

In the second half of the last century, the uncontrolled growth of economic activity led to a significant disruption in the natural migration of chemical elements, which can be eliminated or optimized only by controlling the fluxes of matter in the environment. In this regard, knowledge of the basic laws and physicochemical mechanisms of chemical element migration in the global hydrological cycle, which links the objects of the biosphere into an integrated dynamic system, is of paramount importance. Here, based on numerous former studies of phosphorus transport within the global hydrological cycle, we searched for the general patterns and physicochemical mechanisms of the aqueous migration of chemical elements in the global hydrological cycle. The objective of this work is to present the general features of the phosphorus biogeochemical cycle and describe the physicochemical mechanisms controlling phosphorus migration in the aqueous systems of the earth's surface, notably river runoff within the context of the global hydrological cycle.

2. Phosphorus Mobilization at the Stage of River Runoff Formation

The initial stage in the formation of the chemical composition of surface waters is often associated with atmospheric precipitation on the earth's surface and their subsequent interaction with soil and vegetation cover and rocks. However, due to the constant presence of terrigenous aerosols (mainly the products of the wind erosion of soils) in the surface air layers, this interaction begins already in the atmosphere immediately after the condensation of water vapor. Therefore, it is expedient to divide the mobilization of dissolved substances at the initial stage of river runoff formation into the mobilization in the atmosphere and in the catchments.

2.1. Phosphorus Mobilization in the Atmosphere

Chemical elements are delivered from the atmosphere to the catchments in the form of wet (rain, snow) and dry (aerosols) precipitation. The chemical composition of wet precipitation is due to leaching from the atmosphere and the partial dissolution of aerosols, which are represented by substances of terrigenous and marine genesis. The contribution of marine aerosols to the transport of phosphorus into the land is apparently small. This is indicated by an exponential decrease in the content of aerosol phosphorus in the lower atmosphere when moving from the coast to the central regions of the ocean and a significantly lower content of phosphorus in the rains over the ocean compared to land [1–3]. The estimates of phosphorus input into the atmosphere from various sources confirm this conclusion and show that the main role is played by the aeolian erosion of the soil cover and the combustion of terrestrial vegetation (Table 1). Another source associated with the products of plant metabolism (spores, pollen, volatile organic compounds, and small particles of plant residues) is currently not quantifiable, but observations unambiguously indicate the widespread occurrence of plant metabolism products present in atmospheric aerosols.

Table 1. Sources of phosphorus in the atmosphere [4].

Source	Mass of Aerosols Entering the Atmosphere, Gt/yr	Phosphorus Content in Aerosols, %	Phosphorus Input into the Atmosphere, Mt/yr
Aeolian soil erosion	4.6–8.3	0.08	3.7–6.6
Splashing of seawater	1.8–1.9	0.001	~0.02
Burning of vegetation	0.15–0.60	1.5 [1]	2.3–9.0
Volcanism	–	–	≤0.003
Combustion of solid fuels	0.03–0.06	0.1	0.03–0.06
Combustion of liquid fuels	0.003	0.01	0.0003

[1] By the composition of terrestrial vegetation at 10% ash content.

The total phosphorus concentration in aerosols varies from 600 to 4700 μg/g, averaging ~2000 μg/g [5], which is 2–5 times higher than the phosphorus content in the rocks of the earth's crust and soils, the main sources of terrigenous material in the atmosphere. The increased phosphorus concentrations in atmospheric aerosols are logically explained by the presence of the solid products of plant biomass combustion in the amount of 0.6–1.1% of the total aerosol mass [4]. A significant part of the phosphorus in aerosols is present in water-soluble form, which, as a rule, accounts for 20–50% of its total content [6]. Apparently, the soluble forms of phosphorus in aerosols are associated with the products of combustion of plant biomass and its destruction.

In atmospheric precipitation, the concentrations of mineral and total phosphorus (P_{min} and P_{total}) are distributed in accordance with the lognormal law. The average median concentrations of these forms are 15 and 33 μg/L and the values of their input with atmospheric precipitation to the earth's surface are 0.11 and 0.25 kg/ha yr (Table 2). The percentage of the soluble forms of the total phosphorus in atmospheric precipitation is in the range of 20–80%, with an average value of 55% [6], which is in good agreement with the percentage of soluble phosphorus in aerosols, the main source of dissolved substances.

Table 2. Concentration of phosphorus in atmospheric precipitation and its input on the earth's surface with wet precipitation [7].

Parameter	Concentration, µg/L		Input, kg/ha yr	
	P_{min}	P_{total}	P_{min}	P_{total}
Number of observation sites	130	97	104	77
Arithmetic mean	30	73	0.22	0.36
Median mean	13	34	0.093	0.23
Geometric mean	15	33	0.11	0.25

Atmospheric precipitation is considered by many authors as an important source of phosphorus in river runoff, which is not associated with the processes of its mobilization in the catchments. However, the balance of total phosphorus in the catchments shows that the input of this element with atmospheric precipitation usually exceeds the removal with water runoff [4]. The opposite situation, when the phosphorus runoff exceeds its input, is observed, as a rule, under the conditions of a strong anthropogenic load. The positive value of the difference between the phosphorus input from the atmosphere and its removal from the catchments is an artifact that is associated with the lack of reliable methods for quantifying the masses of substances remobilized from the earth's surface into the atmosphere and returned back as a part of atmospheric precipitation (Figure 1).

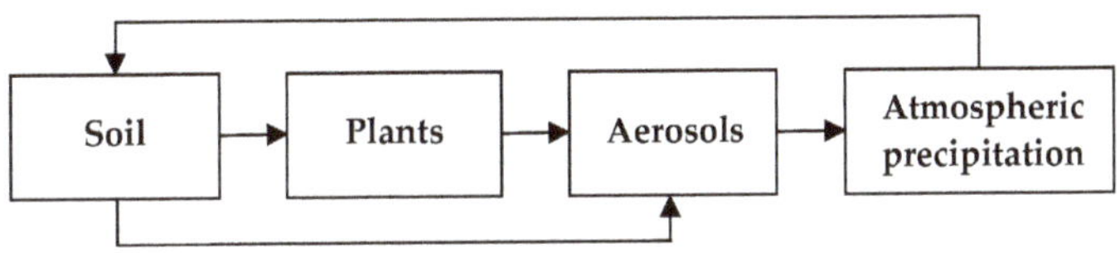

Figure 1. Scheme of phosphorus fluxes in the catchments.

2.2. Phosphorus Mobilization in the Catchment Areas

The primary sources of phosphorus are igneous, metamorphic, and sedimentary rocks, which differ significantly in the content of this element (Table 3). The maximum phosphorus concentrations are characteristic of basic and intermediate magmas; with an increase and decrease in acidity, the phosphorus content in igneous rocks decreases. In sedimentary rocks, the phosphorus concentration does not vary so much, and in general, for the sedimentary deposits it is slightly higher than in granites. In metamorphic processes, phosphorus behaves as an inert component, and its content is inherited from the parent rocks.

Table 3. Content of phosphorus in rocks [8,9].

Rock	[P], µg/g	Rock	[P], µg/g
Igneous rocks:		*Sedimentary rocks:*	
Ultrabasic rocks	280	Sands and sandstones	620
Basic rocks	1300	Clays and clay shales	790
Andesites, diorites	1350	Carbonate rocks	480
Granodiorites	980	Siliceous rocks	660
Granites	600	Evaporites	4
Syenites	800	Volcanic rocks	900
Volcanic rocks	900	Sedimentary deposits in general	710
		Surface of the continental lithosphere [1]	690

[1] Based on results of [9].

Apatite is the main phosphorus mineral in all types of igneous, metamorphic, and sedimentary rocks. The abundance of two other important phosphorus minerals, xenotime YPO_4 and monazite $CePO_4$, is 100–1000 times lower than that of apatite and can reach

10% only in acid rocks [10]. According to mineralogical analysis, in igneous rocks, apatite accounts for 1.7–5.7% of the total phosphorus, whereas in sedimentary carbonate, clayey, and sandy rocks, apatite contains 22.9, 0.5, and 7.1% of phosphorus, respectively [10,11]. In magmatic and metamorphic silicates, phosphorus can isomorphically replace silicon with charge compensation ($Na^+ + P^{5+} = Ca^{2+} + Si^{4+}$ and $Al^{3+} + P^{5+} = 2\,Si^{4+}$) or with the formation of cation vacancies. In the Critical Zone, the bulk of phosphorus is in the sorbed state, as well as in the form of apatite and various iron and aluminum phosphates. The composition of apatite is different for various types of rocks. Fluorapatite predominates in igneous and metamorphic rocks, with fluoro-carbonate-apatite pervading in sedimentary rocks, and bone phosphate is represented by hydroxyl-apatite and carbonate-hydroxyl-apatite.

Biological metabolites and the products of dead organisms' destruction are an important source of phosphorus in continental runoff. The phosphorus content in land plants (on average 1500–2000 μg/g dry weight) is almost an order of magnitude lower than the content in animals and bacteria [12–14]. Therefore, the destruction of animal and bacterial biomasses can lead to the emergence of high local concentrations of dissolved phosphorus.

The most obvious factor in phosphorus mobilization is the solubility of the phosphorus-containing mineral phases. According to calculations [15], the concentration of dissolved mineral phosphorus in the waters of the Rhine and Rhone rivers is controlled by the solubility of hydroxylapatite. However, under the conditions of the earth's surface, hydroxylapatite is unstable and transforms into a less soluble fluoro-carbonate-apatite. The dissolution of fluorapatite in fresh waters leads to a concentration of dissolved mineral phosphorus at the level of 14 ± 3 μg/L [16], which, as will be shown below, approximately corresponds to the average median value for the world rivers.

The acidity of the aquatic environment is apparently the main factor controlling the stability of the mineral forms of phosphorus. In a moderately alkaline medium, the stable phase is fluoro-carbonate-apatite; in a moderately acidic medium, iron (III) and aluminum phosphates are stable under oxidizing conditions and iron (II) and aluminum phosphates are stable under reducing conditions [6]. According to the experimental data [17], in waters with a reaction close to neutral, the monophosphates of iron (III) and aluminum transform into more stable iron-calcium and aluminum-calcium phosphates with the hypothetical chemical formulas $CaFe(OH)_3HPO_4$ and $CaAl(OH)_3HPO_4$. In the neutral medium, the dissolved iron (III) and aluminum are mainly in the form of the electroneutral hydroxocomplexes $Fe(OH)_3^0$ and $Al(OH)_3^0$, and the bulk of the phosphorus is represented by HPO_4^{2-}. Therefore, in accordance with the dissolution reactions

$$CaFe(OH)_3HPO_4 = Ca^{2+} + Fe(OH)_3^0 + HPO_4^{2-}, \quad (1)$$

$$CaAl(OH)_3HPO_4 = Ca^{2+} + Al(OH)_3^0 + HPO_4^{2-} \quad (2)$$

an inverse relationship between the logarithms of the concentrations of mineral phosphorus and calcium is observed (Figure 2).

In equilibrium with iron-calcium and aluminum-calcium phosphates, the concentration of dissolved mineral phosphorus is significantly higher than its content in river and ground waters. Therefore, it should be assumed that the presence of these solid phases is possible only where high local concentrations of dissolved phosphorus can be maintained for a long time. These can be bottom sediments with an extremely slow rate of water exchange or soils in which a high concentration of dissolved phosphorus is provided by the destruction of organic matter during the biological cycle. In all other cases, iron-calcium and aluminum-calcium phosphates must be replaced by hydroxides containing adsorbed phosphorus.

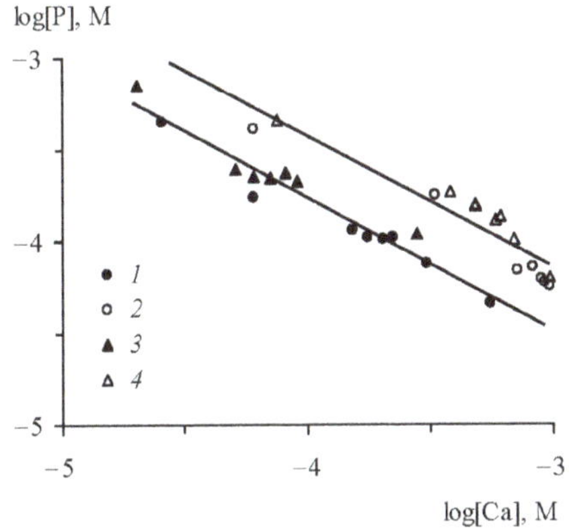

Figure 2. Relationship between logarithms of the concentrations of dissolved phosphates and calcium in the interaction of $FePO_4$ and $AlPO_4$ with fresh waters [17]. $FePO_4$: (*1*) water from the Moscow River, (*2*) water from the Don River mouth; $AlPO_4$: (3) water from the Moscow River, (*4*) water from the Don River mouth.

At sufficiently high concentrations of phosphates, arising, for example, during the destruction of animal or bacterial biomass, the silicate phosphatization reaction can occur, in which the silicon of the solid phase is replaced by phosphorus from the solution. This process was experimentally studied by us, using rock-forming minerals of different structural types (hornblende, orthoclase, labradorite, kaolinite, and montmorillonite) and background buffer solutions with variable concentrations of orthophosphates (0.25–6.0 mM), maintaining the pH at ~1.8, 3.7, 4.9, 6.8, 7.8, and 8.8 [18–20].

The results of the experiments demonstrated the following features. First, all the samples were characterized by approximately equivalent variations in the concentrations of phosphorus and silicon in the solution in the pH range of 3.7–8.8:

$$\Delta[\mathrm{Si}] \approx -\Delta[\mathrm{P}], \tag{3}$$

whereas at pH 1.8 the supply of the dissolved silicon was 1.3–2 times higher than the removal of the phosphates (Figure 3), which was likely explained by the change in the stoichiometry of the phosphatization reaction. Second, the amount of phosphorus absorbed by the silicates was linearly dependent on its final concentration in the solution,

$$-\Delta[\mathrm{P}] = k[\mathrm{P}]_{\mathrm{final}} \tag{4}$$

with almost the same values of the proportionality coefficient k for the different minerals, slightly decreasing with a decrease in the acidity of the medium (Table 4).

Table 4. Proportionality coefficient k in Equation (4) as a function of solution pH.

pH	1.8	3.7	4.9	6.8	7.8	8.8
k	0.57	0.57	0.54	0.51	0.46	0.46

According to the data in Table 5, the amount of silicon removed from the studied silicates and replaced by phosphates at pH 3.7–8.8 reached 6.5–11.0% of the initial silicon content in the minerals. Even more silicon (up to 9.4–19.9%) entered the solution at pH 1.8, when the process of phosphatization was accompanied by the acid leaching of silicates, which led to an additional release of silicon and violation of equivalence (3). Such large amounts of removed silicon and absorbed phosphorus, which were much higher than the limiting values of the sorption removal of phosphates, definitely indicated the occurrence of a chemical reaction which replaced the silicate with a phosphate mineral.

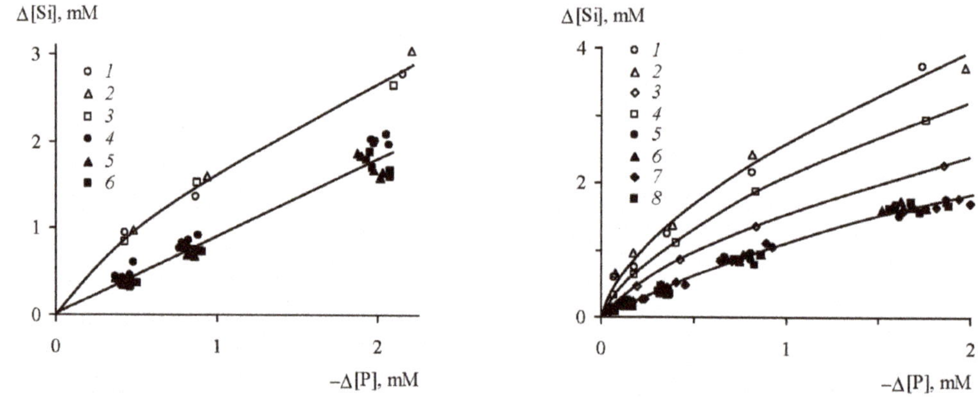

(a) (b)

Figure 3. Correlation between variations in the concentrations of phosphorus and silicon in the solution upon phosphotization of silicates [19]. (**a**) pH 1.8: (*1*) hornblende, (*2*) orthoclase, (*3*) labradorite; pH 3.7–8.8: (*4*) hornblende, (*5*) orthoclase, (*6*) labradorite. (**b**) pH 1.8: (*1*) kaolinite, Glukhovetsk, (*2*) the same, Podol'sk, (*3*) montmorillonite, Askania, (*4*) the same, near Askania; pH 3.7–8.8: (*5*) kaolinite, Glukhovetsk, (*6*) the same, Podol'sk; (*7*) montmorillonite, Askania, (*8*) the same, near Askania.

Table 5. Amount of silicon passed into solution in the experiments on phosphatization of silicates at the maximum initial concentration of phosphates [1], % of the initial concentration in the mineral [19].

Mineral	pH	
	1.8	**3.7–8.8**
Hornblende	19.4	11.0
Orthoclase	11.2	8.1
Labradorite	12.3	8.0
Kaolinite, Glukhovetsk	18.0	7.8
As above, Podol'sk	19.9	8.6
Montmorillonite, Askania	9.4	6.5
As above, near Askania	12.4	6.9

[1] 5 mM for hornblende, orthoclase, and labradorite; 6 mM for kaolinite and montmorillonite.

In previous studies [21–25], the negative correlation between the variations in the concentrations of dissolved phosphates and silicon was associated with the adsorption exchange of phosphate ions and silica on the surface of silicates. Since the duration of the experiments did not exceed several days, this time was sufficient to establish the adsorption equilibrium but was not long enough for noticeable progress in the phosphatization reaction of the bulk silicate phase. Our experiments proceeded for more than one year, so the amount of phosphorus absorbed from the solution and the silicon displaced from the solid phase indicated the participation in the process of not only the surface layer, but also the volume of the solid phase.

The same quantitative characteristics of the process of phosphatization for all the studied silicates, corresponding to different structural types and with different chemical compositions, were an unusual result. It can be assumed that the initial minerals were not subject to phosphatization, but that the secondary silicate phases formed during the interaction of the silicates with water and were stable in a certain pH range. The parameters of the phosphatization reaction at pH 1.8 varied due to the stability under these conditions of the surface silicate phase, which was different to that in the area of higher pH values.

A powerful factor of the phosphorus mobilization in the Critical Zone is the activity of living organisms. Primary producers annually synthesize about 140 Gt of dry organic matter on land, 98–99% of which is mineralized. With the average phosphorus content in plants equal to ~1500 μg/g dry matter, about 210 Mt of phosphorus participates in the biotic cycle, which forms soluble phosphates at the stage of mineralization and becomes a potential source of dissolved phosphorus in continental runoff. However, mineralized phosphorus is almost completely reincluded in the biotic cycle and used to create new organic matter. The highest degree of completeness of the biotic cycle is inherent in mature biogeocenoses (Table 6).

Table 6. Phosphorus input with litter and removal with subsurface runoff in forest biogeocenoses [26].

Process	Oak Forest	Aspen Forest
Input with litter, kg/ha yr	7.85	9.9
Removal with subsurface runoff, kg/ha yr	~0.001	~0.0015
Phosphorus removal, % of input with litter	0.013	0.015

If all mineralized phosphorus was a part of river runoff, the volume of which is 41,700 km^3/yr [27], its concentration due to this source alone would be 5 mg/L. Such high concentrations of dissolved phosphorus are extremely rare and usually associated with the reducing conditions of the environment or anthropogenic pollution. The average concentration of dissolved phosphates in unpolluted river waters is equal 30–50 μg P/L [28–30], which is 0.6–1.0% of the calculated value of 5 mg P/L. This means that continental runoff contains a very small portion of the labile phosphorus that is formed as a result of organic matter degradation.

It is known that when phosphorus fertilizers are applied to soils, the behavior of the phosphorus differs significantly depending on the properties of the soil and the fertilizers themselves. Poorly soluble phosphorite flour increases the content of biologically available phosphorus if the soil conditions are conducive to the transformation of apatite into more soluble forms. With the addition of highly soluble fertilizers, over time, phosphorus immobilization occurs due to chemosorption and the formation of poorly soluble compounds, including apatite phases. It is assumed that phosphates of iron, aluminum, and calcium make up ~90% of the immobilized phosphorus of fertilizers [31].

Whereas the final products of the transformation of fertilizers are represented by poorly soluble mineral phases, there is usually no direct relationship between the amount of applied phosphorus and its removal. The amount of removed phosphorus from fertilizers, as a rule, does not exceed 1–2% [32–35].

Formally, the mobility of chemical elements in the Critical Zone is characterized by the coefficients of water migration K_i, equal to the ratio of the concentrations of element i in the dry residue of water (a_i) and in drained rocks (m_i):

$$K_i = a_i/m_i. \tag{5}$$

Phosphorus belongs to the group of low-mobility elements with $0.01 < K_i < 0.1$ [36].

In (5), it is implicitly assumed that all the substances in the dry residue of water enter it as a result of the dissolution of drained rocks. However, there are two other powerful sources of dissolved matter: cyclic sea salts, transported from the ocean to land through the atmosphere, and anthropogenic substances. Taking into account the contribution of these sources leads to a significant change in the values of the coefficients of water migration, in particular, to an approximately tenfold increase of this coefficient for phosphorus (Table 7).

Table 7. Coefficients of water migration of chemical elements in the Critical Zone taking into account the contribution of cyclic sea salts and anthropogenic substances [37].

Cl	S	Na	F	C	Mg	Ca	K	P	Si	Mn	Fe	Ti	Al
9.6	9.2	4.0	3.3	3.3	2.4	2.4	0.93	0.81	0.27	0.22	0.02	0.01	0.01

3. Phosphorus in River Runoff

3.1. Phosphorus in the Waters of the Primary Hydrographic Network

The primary hydrographic network consists of small catchments, which are characterized by the significant spatial variability of the chemical composition of the waters, caused by the territorial heterogeneity of geomorphological, lithological, and biological soil conditions. The enlargement of rivers and pooling of small catchments leads to the "averaging" of the local conditions for runoff formation. Therefore, the larger-scale regularities associated with the implementation of the periodic law of geographic zonality are acquiring decisive importance.

The lithological characteristics of the catchments have a strong influence on the phosphorus concentration in the waters of the primary hydrographic network, because rocks are the main source of dissolved phosphorus. The highest concentrations of dissolved mineral phosphorus are found in catchments located on basalts, in which the phosphorus content is greater than in other types of rocks (Table 8). The runoff of dissolved phosphorus from drainage basins composed of sedimentary rocks is usually greater than for igneous rocks. In the small, almost completely forested catchments on the Canadian Crystalline Shield, the dissolved phosphorus runoff was 4.8 (2.5–7.7) mg/m^2 yr for igneous rocks and twice as large (10.7 (6.0–14.5) mg/m^2 yr) for sedimentary rocks [38].

Table 8. Relationship between the concentration of dissolved mineral phosphorus in the waters of the primary hydrographic network and phosphorus content in the catchment rocks.

Lithological Composition of Catchments	Phosphorus Concentration	
	in Water, µg/L [29]	in Rock, µg/g [8,9]
Sandstones	2	620
Granites	3	600
Limestones	2	480
Basalts	20	1300
Carbonaceous shales	3	–
Mica schists	4	550
Gypsum-bearing clays	1	–

Another important factor of phosphorus migration is the climate, which affects the rate of weathering and, consequently, the intensity of the phosphorus mobilization from rocks. For example, the runoff of dissolved phosphorus from the territory of Karelia (NW Russia, temperate climate) due to the pure weathering of crystalline rocks is 2 mg/m^2 yr [39], while the average intensity of dissolved phosphorus removal during the weathering of crystalline rocks for three catchments in Brazil (humid tropical climate) is 5 times higher: 10 (5–14) mg/m^2 yr [40].

The presence of areas with slow water exchange in catchments leads to a decrease in the phosphorus content in the waters of the primary hydrographic network. Indeed, Conley et al. [41] showed an exponential dependence of the concentration of total dissolved phosphorus ([P_{total}], µg/L) on the relative area of lakes (S, %) in catchments:

$$[P_{total}] = 2.18e^{-0.096S}. \quad (6)$$

Data on the content of dissolved phosphorus in the waters of the primary hydrographic network, to which catchments with an area ≤50 km^2 were assigned, were collected during observations that lasted for at least a year and were systematized in [42]. Based on the differences in the sources of phosphorus input, the conditions of runoff formation, and the processes in the catchments, all catchments were divided into four groups: (1) natural (forest) catchments; (2) mixed agricultural–forest catchments with land use <50%; (3) agricultural catchments with land use >50%; (4) urban catchments. For a number of catchments, the group could not be determined due to the lack of the necessary data.

Table 9 shows that the values of the arithmetic and median mean concentrations of mineral and total phosphorus in solution for all the accounted catchments differed several times, indicating the positive asymmetry of the probability distribution functions, which corresponds to the lognormal law. When the small catchments were combined into groups, the asymmetry of the probability distribution functions for the phosphorus concentrations remained. Therefore, the average median concentrations of mineral and total phosphorus, equal to 31 and 95 μg/L, can be considered as the global average concentrations of these forms of dissolved phosphorus in the waters of the primary hydrographic network under modern conditions.

Table 9. The average content and concentration range of dissolved phosphorus (μg/L) in the waters of the primary hydrographic network [42].

Component	Number of Catchments	Arithmetic Mean	Median Mean	Minimum	Maximum
		Forest catchments			
P_{min}	67	15	7	0	114
P_{total}	40	58	28	3	806
		Mixed agricultural–forest catchments			
P_{min}	23	88	48	6	435
P_{total}	24	142	90	17	589
		Agricultural catchments			
P_{min}	34	218	116	2	1145
P_{total}	26	535	250	7	3255
		Urban catchments			
P_{min}	6	708	700	101	1572
P_{total}	5	1605	1500	163	3300
		All accounted catchments, including catchments of unknown type			
P_{min}	137	115	31	0	1572
P_{total}	103	301	95	3	3300

The lowest concentrations of the dissolved forms of mineral and total phosphorus in the surface waters (7 and 28 μg/L) were observed in the forest landscapes with the least anthropogenic impact. As the economic activity intensified, the phosphorus content increased. For the mixed agricultural–forest catchments, the average concentrations of P_{min} and P_{total} were 48 and 90 μg/L, while for the agricultural catchments they increased to 116 and 250 μg/L. An even higher content of dissolved phosphorus was characteristic of the urban catchments, where the average concentrations of P_{min} and P_{total} reached 700 and 1500 μg/L. In general, there was a tendency towards an increase in the concentrations of P_{min} and P_{total} in the surface waters of the small catchments as the population density increased (Figure 4).

3.2. Phosphorus in River Waters

The rivers of the world carry into the ocean ~3 Gt/yr of dissolved matter and 15–20 Gt/yr of solid matter. The phosphorus runoff in the form of particulate suspended matter significantly exceeds its dissolved flux, which plays an extremely important role for biota and biogeochemical processes.

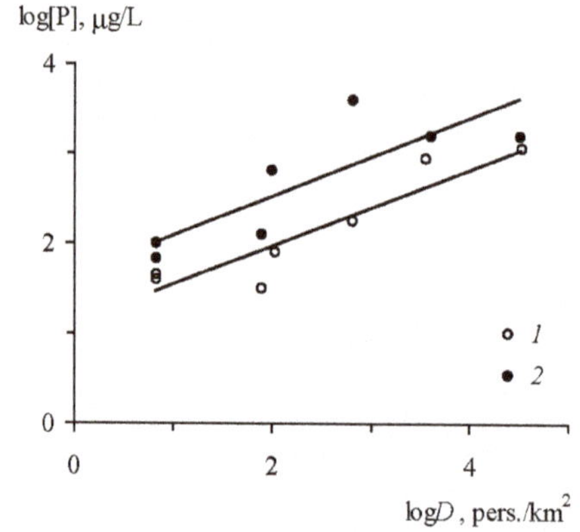

Figure 4. Relationship between the average annual concentrations of dissolved forms of mineral (*1*) and total (*2*) phosphorus in the waters of the primary hydrographic network and the population density *D* [6].

3.2.1. Phosphorus of Suspended Matter and Bed Load

The distribution function of the phosphorus content in suspensions of 77 large, medium, and small rivers of the world corresponds to a lognormal law; the arithmetic and geometric mean concentrations of phosphorus equal 1500 and 1000 μg/g [43], respectively, which is close to the estimate [44]: 1270 μg/g. About 3% of the phosphorus in river suspended matter is represented by bioavailable soluble/exchangeable forms that can be used by living organisms [45,46].

Phosphorus runoff in the form of suspended solids is affected by the ratio of fine and coarse fractions. The phosphorus content in the fine fractions of the suspended matter and bottom sediments of rivers is 2–10 times higher than that in the coarse fractions. The consequence of this is apparently a decrease in the phosphorus concentration in river suspensions, with an increase in the total content of suspended solids (turbidity), which is accompanied by an increase in the proportion of the coarse fractions (Figure 5). The highest phosphorus concentrations (~4000 μg/g) are observed at a turbidity <20 mg/L, while at a turbidity >100 mg/L, the phosphorus concentration begins to decline sharply, reaching 400 μg/g at a suspended matter content of 1000 mg/L. The same reason leads to an inverse relationship between the concentration of phosphorus in suspended matter and the water discharge or erosion rate. At small discharges during the low-water period, the relative contribution of fine suspensions increases and the phosphorus concentration reaches its maximum values, while in the high-water period, the bulk of suspended solids are represented by coarse suspensions with low phosphorus content.

The use of fertilizers is accompanied by an immobilization of the phosphorus in the upper soil horizons, which are the main supplier of suspended matter. As a result, the phosphorus content in suspensions denudated from cultivated lands is approximately 2 times higher than in the runoff of solids from forest catchments: 2500 and 1100 μg/g, respectively [47].

Forests prevent the erosion of the earth's surface and should reduce phosphorus runoff. This is confirmed by the data for seven small catchments in Southern Quebec [48], where the relationship between the concentration of suspended phosphorus ([P_{susp}], μg/L) and the degree of forest coverage of the territory (X, %) was established:

$$[P_{susp}] = 10.2 - 0.056X. \quad (7)$$

Deforestation should lead to an increase in suspended phosphorus runoff on a global scale, but it is still very difficult to quantify this effect.

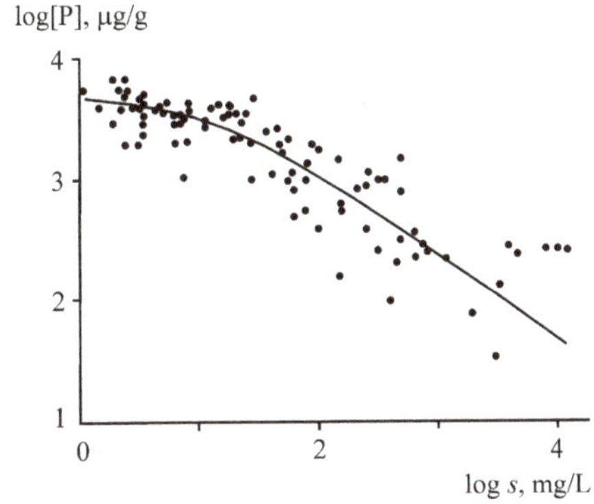

Figure 5. Relationship between the phosphorus content in suspended matter and the turbidity (s) of river waters [43].

It is estimated that 10 to 30% of the river runoff of solid matter is carried in the form of bed load, in which the phosphorus content is on average 800 μg/g [49]. This value is lower than the phosphorus content in river suspended matter (1000 μg/g), which corresponds to the larger hydraulic size of the bed load.

3.2.2. Dissolved Phosphorus

In [50,51], the average annual and long-term average annual data on the content of the dissolved forms of mineral and total phosphorus in 179 rivers of the world (>200 observation stations) are summarized. The arithmetic and median mean concentrations of dissolved mineral phosphorus are 113 and 28 μg/L, respectively, and those of total dissolved phosphorus are equal to 241 and 85 μg/L (Table 10). The distribution of the concentrations of dissolved phosphorus obeys the lognormal law; therefore, the median mean concentrations are preferred for obtaining average values.

Table 10. The average content of dissolved phosphorus (μg/L) in the river waters of different geographic zones [50,51].

Geographic Zone	Number of Stations	Arithmetic Mean	Median Mean
	P_{min}		
Arctic and subarctic zones	7	76	6
Temperate zone	123	132	32
Humid tropics and subtropics	19	39	16
Dry tropics and subtropics	33	93	31
Whole world	182	113	28
	P_{total}		
Arctic and subarctic zones	3	235	19
Temperate zone	68	247	106
Humid tropics and subtropics	12	226	91
Dry tropics and subtropics	14	228	175
Whole world	97	241	85

Most of the natural factors affecting the content of chemical elements in river water are closely related to the geographic zonality, which determines the features and intensity of the weathering processes, biological activity, etc. In this regard, to analyze the spatial distribution of phosphorus content, all rivers were divided into four groups according to their geographical zones: Arctic region and subarctic zone, temperate zone, humid tropics and subtropics, and dry tropics and subtropics.

The minimum median mean concentrations of the dissolved forms of mineral and total phosphorus were observed in the rivers of the Arctic and subarctic belts, where the biological cycle of elements is much slower and the anthropogenic impact on the aquatic environment is not pronounced, given that there are no extensive sources of phosphorus input associated with agricultural industries, fewer large cities, and, therefore, less industrial and domestic wastewater. The highest median mean concentrations of dissolved mineral phosphorus were characteristic of the rivers of the temperate zone and the zone of dry tropics and subtropics. This is explained by the powerful anthropogenic impact on the nature of these regions, as well as the favorable conditions for the involvement of phosphorus in the biological cycle and its rapid turnover therein. A similar situation is typical for total dissolved phosphorus.

The average annual concentrations of the dissolved forms of mineral and total phosphorus for the rivers of the world correlate with the mineralization of river water (r = 0.94 and 0.89, respectively) and with the concentration of total nitrogen (r = 0.81 and 0.79, respectively) [51]. The cycles of nitrogen and phosphorus are closely linked in the biological cycle of matter and liable to similar anthropogenic changes. Like phosphorus, nitrogen is used in mineral fertilizers and its concentration in wastewater also increases tens and hundreds of times. A rather close correlation between dissolved phosphorus and the mineralization of waters is interesting. It can be assumed that it arises due to an increase in the mineralization of river water in the north–south direction parallel to an increase in the population density, which is an indicator of the anthropogenic load and, in particular, of the intensity of anthropogenic phosphorus sources. Indeed, the average concentrations of mineral and total phosphorus in the river water regularly increase with an increase in the population density in the catchments (Table 11).

Table 11. The average content of dissolved phosphorus in the water of rivers with different population densities in their catchments [51].

Population Density *D*, pers./km^2	**Concentration, μg/L**	
	P_{min}	P_{total}
<1	21	81
1–10	28	76
10–50	34	157
50–100	39	139
100–200	193	120
200–700	556	598

According to [52], for large rivers there is only a weakly expressed tendency towards an increase in the runoff of dissolved mineral phosphorus with an increase in the population density in the catchments. However, if one takes into account the presence of a directly proportional dependence of phosphorus removal from catchments on the value of specific water discharge, a significant correlation (r = 0.78) is found between the runoff of dissolved mineral phosphorus and the population density in the catchments, normalized to the specific water discharge.

The intensification of economic activity is accompanied by an increase in the phosphorus content in river runoff. Systematic observations carried out in 1936–1980 on the territory of the USSR showed a noticeable increase in the concentration and runoff of dissolved mineral phosphorus over time (Table 12). The same was established for other large rivers of the world, including the coastal parts of the sea basins into which these rivers flow [53].

Table 12. Change in water runoff (Q, km^3/yr), concentration ([P_{min}], µg/L) and runoff (J_{Pmin}, thous. t/yr) of dissolved mineral phosphorus in the USSR in 1936–1980 [54].

Drainage Basin	1936–1970			1970–1980		
	Q	[P_{min}]	J_{Pmin}	Q	[P_{min}]	J_{Pmin}
Arctic Ocean	2746	6.2	16.9	2849	13.5	38.5
Pacific Ocean	866	12.8	11.1	726	27.2	19.8
Atlantic Ocean	261	24.6	6.4	235	38.2	9.0
Aral–Caspian	381	29.4	11.2	315	45.3	14.2
Former USSR territory	4250	10.7	45.6	4120	19.8	81.5

Environmental protection measures can not only stop the increase in dissolved phosphorus concentrations but also cause its significant decrease. In particular, due to a reduction in the volumes of municipal wastewater and the use of phosphorus-containing detergents, the total phosphorus runoff into Lake Erie decreased from 27.9 to 10.5 thous. t/yr during 1968–1981 [55]. The deepening of wastewater treatment and a decrease in its volume led to a decrease in the phosphorus runoff into the Rhine and Elbe rivers from 51.1 and 20.5 thous. t/yr, respectively, in 1983–1987 to 20.5 and 12.5 thous. t/yr in 1993–1997 [56].

3.3. Phosphorus in Groundwater in the Zone of Active Water Exchange

The surface waters of the primary hydrographic network, rivers and lakes, are in direct hydrodynamic connection with groundwater, which plays an important role in the formation of the chemical composition of the continental runoff of dissolved matter. The greatest influence is exerted by the groundwater of the zone of active water exchange, the discharge of which is the main source of river runoff during the low-water period. The phosphorus content in groundwater is of the same order of magnitude as in the waters of the primary hydrographic network.

The average content of dissolved mineral phosphorus in the groundwater of the Critical Zone varies within the same order of magnitude: from 18 to 191 µg/L (Table 13). The maximum concentrations (191 and 127 µg P/L) were found in the waters of bog landscapes and steppes (dry savannah). The lowest phosphorus content was observed in the waters of permafrost zones and mountainous areas, in which the fluorine mobilization from rocks is impeded by the low temperature and relatively high water velocity, respectively. The concentrations of dissolved mineral phosphorus in the groundwater in areas of leaching and continental salinization, despite the significant difference in their mineralization, are relatively equal, amounting to 56.9 and 62.6 µg/L, respectively.

Table 13. The content of dissolved phosphorus in groundwater of the Critical Zone [57].

Groundwater Type	[P_{min}], µg/L
Groundwater of the provinces of permafrost	
Northern bog landscapes	26.3
Tundra landscapes	19.1
Northern taiga landscapes	21.7
Groundwater of the provinces of temperate climate	
Bog landscapes	191
Mixed forest landscapes	59.5
Southern taiga landscapes	57.1
Forest-steppe and steppe landscapes	75.8
Groundwater of the provinces of tropical and subtropical climate	
Wet savannah landscapes	29.4
Rainforest landscapes	65.3
Subtropical forest landscapes	58.7
Landscapes of dry savannah and steppes	127

Table 13. *Cont.*

Groundwater Type	$[P_{min}]$, µg/L
Groundwater of the provinces of arid climate	
Landscapes of the temperate continental zone:	
soda waters	20.6
sulphate waters	63.3
chloride waters	21.7
Landscapes of the dry tropical zone	76.7
Groundwater of the mountainous areas	
High-mountain and mountain–meadow landscapes	18.0
Mountain–forest and mountain–taiga landscapes	40.1
Mountain–steppe landscapes	46.8
Average concentrations	
Groundwater of the leaching areas:	56.9
permafrost	22.6
temperate climate	98.2
tropical and subtropical climate	71.8
mountainous areas	34.9
Groundwater of the areas of continental salinization	62.6
Average for groundwater of the Critical Zone	58.0

3.4. *Integral Characteristic of the Phosphorus River Runoff*

3.4.1. Phosphorus Runoff in the Composition of Suspended Matter and Bed Load

The average phosphorus concentrations in the suspended matter and bed load of world rivers are 1000 and 800 µg/g, respectively [49]. The most detailed calculations of the global runoff of suspended matter give a value of 15.5 Gt/yr [58,59]. The mass of the bed load, according to various estimates, is from 10 to 30% of the mass of suspended matter, and 20% can be taken as an average value. Hence, the total continental runoff of suspended and drawn phosphorus is equal to 18.0 Mt/yr. This value is in close agreement with earlier estimates, 16.1 [44] and 20.4 [60] Mt P/yr; however, these did not take into account the runoff of bed load.

Phosphorus is also carried out from land via ice runoff and coastal abrasion. Here, phosphorus is mainly contained in the lithogenic material, while the contribution of its dissolved forms is negligible. The phosphorus content in the products of glacial erosion and coastal abrasion can be taken to equal that in the rocks of the land surface: 690 µg/g. A.P. Lisitsyn [61] estimates the removal of the solid products of ice runoff and coastal abrasion to be 1.5 and 0.5 Gt/yr, which corresponds to a phosphorus mass of 1.4 Mt/yr.

3.4.2. Dissolved Phosphorus in River Runoff

A detailed assessment of river phosphorus runoff was made in [50,51], where data for more than 100 medium and large rivers of the world were used and a correction for the value of the accounted water runoff for each continent was applied (Table 14). The total volume of continental water runoff in these works was taken to be equal to 38,500 km^3/yr. The more correct value is 41,700 km^3/yr [27], which would mean that the river runoff of dissolved mineral and total phosphorus increases to 1.6 and 4.5 Mt/yr, respectively.

3.4.3. Dissolved Phosphorus in Groundwater Runoff

Groundwater phosphorus runoff is difficult to estimate due to the limited amount of available information. According to calculations [62], the proportions of the dissolved forms of mineral and total phosphorus in the groundwater and river runoff into the seas of the Russian Arctic are approximately the same, at 11–13% (Table 15). A similar proportion of dissolved mineral phosphorus in groundwater and river runoff also follows from the global estimates [63]. With an average concentration of phosphorus in the groundwater of the Critical Zone of 58 µg/L [57] and a groundwater runoff value of 2200 km^3/yr, the phosphorus removal into the ocean is 0.13 Mt/yr, or ~8% of the river runoff of dissolved mineral phosphorus.

Table 14. River runoff of dissolved forms of mineral and total phosphorus [50,51] [1].

Continent	Water Runoff, km^3/yr	% of Accounted Water Runoff		Phosphorus Runoff, thous. t/yr			
				P_{min}		P_{total}	
		P_{min}	P_{total}	Accounted	Full	Accounted	Full
Europe	2365	80	52	92.3	116	160	310
Asia	10,152	53	–	484	913	–	1094
North America	7840	12	24	9.8	82	237	988
South America	11,700	84	22	124	148	286	1300
Australia and Oceania	2370	4	1	1.9	48	1.2	205
Africa	4110	49	34	58	120	85	249
Whole world	38,537 (41,700)	52	19	770	1481 (1603)	769	4154 (4495)

[1] Values in parentheses are calculated for water runoff of 41,700 km^3/yr.

Table 15. The proportion of dissolved phosphorus in groundwater and river runoff into the marginal seas of the Russian Arctic [62].

Receiving Water Body	Proportion of Groundwater Runoff from River Runoff, %		
	Water	P_{min}	P_{total}
White and Barents Seas	14.6	15.3	15.5
Kara Sea	10.3	12.3	–
Laptev Sea	7.6	6.9	5.1
East Siberian Sea	6.7	5.8	6.6
Average for the Arctic seas of Russia	10.0	12.8 [1]	10.7 [1]

[1] Weighted mean for water runoff.

4. Phosphorus in the Mixing Zone of River and Sea Waters

The final stage of the transformation of the river runoff of dissolved and suspended matter is carried-out in the mouth area of rivers, as a result of which the ratio of the dissolved and suspended forms of chemical elements entering the ocean changes.

The dissolved components with conservative behavior are involved in intrabasin chemical and biological processes to an insignificant extent, and their content linearly depends on the ratio of the proportions of the sea and river water masses in the mixing zone. The components with nonconservative behavior are also added into the solution or are removed from it as a result of their involvement in different processes occurring in the water column or at the water–air and water–bottom boundaries. In this case, the linearity of the relationship between the component concentration and the ratio of the water mass proportions is violated. The best indicator of the ratio of the sea and river water mass proportions is the isotopic composition of water; however, the concentration of chemically inert chlorides, which is more accessible for measurements, is used more often.

The conservative behavior of the dissolved component i in the mixing zone of river and sea waters is described by the linear relationship between its concentration $[i]_{mix}$ and chloride content $[\mathrm{Cl}]_{mix}$:

$$[i]_{mix} = a + b[\mathrm{Cl}]_{mix}, \tag{8}$$

where $a = \frac{[i]_{rw}[\mathrm{Cl}]_{sw} - [i]_{sw}[\mathrm{Cl}]_{rw}}{[\mathrm{Cl}]_{sw} - [\mathrm{Cl}]_{rw}} \approx [i]_{rw}$ is a constant parameter; $b = \frac{[i]_{sw} - [i]_{rw}}{[\mathrm{Cl}]_{sw} - [\mathrm{Cl}]_{rw}}$ is the slope ratio taking positive or negative values at a higher or lower concentration of the component i in seawater in comparison with river runoff; and the subscripts "*rw*" and "*sw*" denote the concentrations in river and sea water, respectively. If the component i is removed from the solution or, on the contrary, its internal source is present, the line showing the actual distribution of the concentrations of the relevant component is located, respectively, below or above the calculated line of conservative behavior (Figure 6). Equation (8) is widely used to determine the type of behavior of chemical components in the mixing zones of river and sea waters.

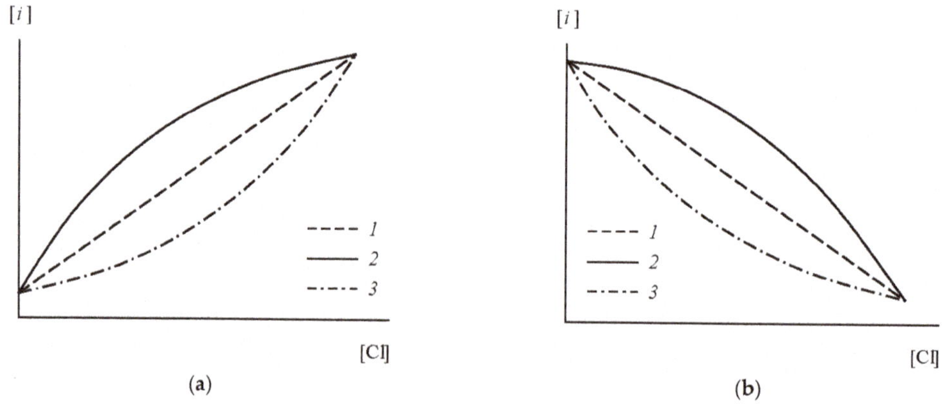

Figure 6. Relationships between the concentration of dissolved component *i* and chloride content by the conservative behavior (*1*) and availability of processes of additional input (*2*) or removal (*3*) of this component in the mixing zone of river and sea waters: (**a**,**b**) are cases when the concentration of component *i* in the river water is accordingly below or above that in the seawater.

Active participation in biological processes brings about the nonconservative behavior of phosphorus, which is observed in most river mouths of the world. The consumption of dissolved mineral phosphorus by phytoplankton leads to a decrease in its content in the water down to "analytical zero". The mineralization of the precipitated organic detritus causes the input of phosphorus into a solution in the lower layers of the water column and at the water–bottom boundary. Approximately half of the phosphorus entering the bottom as part of the organic detritus, after the mineralization of organic matter, can return back to the water with circulating currents or during the stirring up of the bottom sediments [64]. In addition to the production–destruction processes, an important role in the transformation of dissolved phosphorus runoff in the mixing zone of river and sea waters is played by the transformation processes of phosphorus-containing solid phases, sorption–desorption, and coprecipitation, as well as the diagenetic processes that control the phosphorus fluxes at the water–bottom boundary [6].

Table 16 summarizes the data on the distribution of the concentrations of the dissolved forms of mineral, organic, and total phosphorus in the mouth areas of large and small rivers of the world. The behavior of phosphorus in river mouths can be both nonconservative and conservative. In some cases, a complex type of phosphorus distribution was established exhibiting different behavior in various parts of the mixing zone.

4.1. Nonconservative Behavior of Phosphorus

Biological processes are the main reason for the nonconservative behavior of dissolved phosphorus in the mixing zones of river and sea waters. This is shown in the interrelated change in the phosphorus content and the various characteristics of the biological processes: the concentrations of nutrients, oxygen, chlorophyll, and organic detritus, and the pH value.

Table 16. The behavior of dissolved phosphorus in the mouth areas of world rivers.

River	Receiving Sea Area	Phosphorus Form	Observation Period	Salinity, ‰		[P], μg/L		Phosphorus Runoff Transformation, %	Supposed Cause of Nonconservative Behavior	Reference
				Observation Zone	Transformation Zone	River Boundary	Sea Boundary			
Severn	Bristol Channel of the Atlantic Ocean	P_{min}	June 1975	0–28	16–26	610	80	+10	Wastewater inflow	[65]
		"	February 1976	0–32	5–28	310	100	+20	As above	"
Clyde	Irish Sea	P_{min}	April 1973	0–32	1–8	242	15	−68	Sorption on suspended matter	[66]
		"	March 1974	0–32	–	40	242	~0	–	"
Scheldt	North Sea	P_{min}	January 1978	0–35	–	280	62	~0	–	[67]
Rhine–Meuse	As above	P_{min}	"	0–35	–	465	62	~0	–	"
Ems	"	P_{min}	October 1981	14–30	–	–	40	~0	–	[68]
Weser	"	P_{min}	"	0–32	–	520	62	~0	–	"
Elbe	"	P_{min}	"	0–31	–	200	68	~0	–	"
Neva	Baltic Sea	P_{min}	August 1990	0–5	0–3	13	31	−82	Biological consumption	[69]
		P_{total}	"	0–5	0–3	36	17	−49	As above	"
Keret	White Sea	P_{total}	July 1992	0–23	0–18	5.0	14	−62	"	[70]
Pulonga	As above	P_{total}	"	0–25	0–5	7.0	14	−48	"	"
Knyazhaya	"	P_{min}	June 2000	0–26	–	5.3	3.3	~0	–	[71]
Niva	"	P_{total}	July 1992	0–20	–	9.0	11	~0	–	[70]
		P_{min}	"	0–20	0–10	3.8	1.3	−33	Biological consumption	"
		"	June 2000	0–27	0–10	9.7	3.5	−46	As above	[71]
		"	June 2016	0–19	0–13	5.5	12	+35	Destruction of organic matter	A.S. [1]
Kolvitsa	"	P_{min}	June 2000	0–28	–	5.7	3.3	~0	–	[71]
		"	June 2016	0–21	0–13	5.4	12	+35	Destruction of organic matter	A.S.

Table 16. *Cont.*

River	Receiving Sea Area	Phosphorus Form	Observation Period	Salinity, ‰		[P], μg/L		Phosphorus Runoff Transformation, %	Supposed Cause of Nonconservative Behavior	Reference
				Observation Zone	Transformation Zone	River Boundary	Sea Boundary			
Stream in the Por'ya Inlet	"	P_{min}	July 2008	0–23	0–12	4.4	2.1	−21	Biological consumption	[72]
Kuzreka	"	P_{min}	February 2020	0–23	0–2	2.5	6.8	+70	Destruction of organic matter	A.S.
Indera	"	P_{min}	September 2008	0–24	0–12	5.6	2.1	−25	Biological consumption	[72]
Chavanga	"	P_{min}	"	0–24	0–12	3.0	2.1	−46	As above	"
Strelna	"	P_{min}	February 2010	0–10	0–10	4.3	2.5	−18	"	A.S.
Onega	"	P_{min}	August 2004	0–21	0–12	11	4.7	−20	"	"
		"	June 2011	0–18	0–2	1.4	0.5	−43	"	[73]
		"	January 2017	0–6	–	6.9	12	~0	–	A.S.
		"	August 2017	0–18	0–10	3.0	5.0	+56	Destruction of organic matter	"
		P_{org}	June 2011	0–18	0–10	3.9	5.6	+51	As above	[73]
		"	January 2017	0–6	0–4	43	7.0	+100	"	A.S.
		"	"	0–6	4–6	43	7.0	−30	Mineralization	"
		"	August 2017	0–18	0–6	16	6.0	−21	As above	"
		P_{total}	June 2011	0–18	0–10	5.3	6.1	+26	Biological consumption and destruction of organic matter	[73]
		"	January 2017	0–6	0–4	50	19	+44	Destruction of organic matter	A.S.
		"	"	0–6	4–6	50	19	−40	Mineralization	"
		"	August 2017	0–18	0–4	19	11	−10	As above	"
Kyanda	"	P_{min}	August 2016	0–21	0–18	16	5.3	−8	Biological consumption	"
		P_{org}	"	0–21	0–4	25	6.5	−15	Mineralization	"
		P_{total}	"	0–21	0–14	41	12	−18	Biological consumption and mineralization	"

Table 16. *Cont.*

River	Receiving Sea Area	Phosphorus Form	Observation Period	Salinity, ‰		[P], μg/L		Phosphorus Runoff Transformation, %	Supposed Cause of Nonconservative Behavior	Reference
				Observation Zone	Transformation Zone	River Boundary	Sea Boundary			
Northern Dvina	"	P_{min}	April 2003	0–25	0–18	32	9.0	−17	Biological consumption	[74]
		"	June 2003	0–23	0–6	18	7.3	−22	As above	"
		"	July 2016	0–14	0–2	3.1	2.3	−22	"	A.S.
		P_{org}	"	0–14	0–7	11	5.3	+110	Destruction of organic matter	"
		P_{total}	"	0–14	0–7	14	7.6	+93	As above	"
Mezen	"	P_{min}	July 2009	0–21	0–1	27	10	+88	Input from bottom sediments	[73]
		"	August 2015	0–21	0–1	25	6.0	+93	As above	[75]
		P_{org}	July 2009	0–21	–	2.5	0.5	~0	–	[73]
		P_{total}	"	0–21	0–1	30	11	+82	Input from bottom sediments	"
Semzha	"	P_{min}	August 2018	0–17	–	8.7	6.6	~0	–	A.S.
Indiga	Barents Sea	P_{min}	April 1981	0–30	–	2.0	18	~0	–	[76]
		"	September 1981	0–30	–	6.0	4.0	~0	–	"
Pechora	Pechora Sea	P_{min}	July 1984	0–30	0–7	16	20	−75	Biological consumption	"
Ob	Kara Sea	P_{min}	September 1993	0–34	0–20	43	50	−28	As above	[77]
		"	"	0–34	20–34	43	50	+120	Destruction of organic matter	"
		"	August 1999	0–33	0–15	45	90	−33	Biological consumption	"
		"	"	0–33	15–33	45	90	+150	Destruction of organic matter	"

Table 16. *Cont.*

River	Receiving Sea Area	Phosphorus Form	Observation Period	Salinity, ‰		[P], μg/L		Phosphorus Runoff Transformation, %	Supposed Cause of Nonconservative Behavior	Reference
				Observation Zone	Transformation Zone	River Boundary	Sea Boundary			
Yenisei	As above	P_{min}	September 1993	0–34	0–20	5.0	50	~0	–	"
		"	"	0–34	20–34	5.0	50	+50	Destruction of organic matter	"
		"	September 2009	0–30	0–12	20	33	−30	Biological consumption	[78]
		"	"	0–30	12–30	20	33	+100	Destruction of organic matter	"
		"	September 2010	0–17	0–2	7.1	1.5	−57	Biological consumption	"
		"	April 2016	0–26	0–10	19	22	−6	As above	"
		"	"	0–26	10–26	19	22	+47	Destruction of organic matter	"
Rhone	Mediterra-nean Sea	P_{min}	1983–1984	1–37	–	170	6	~0	–	[79]
Tiber	Tyrrhenian Sea	P_{min}	July 1984	0–37	–	93	280	~0	–	[80]
		"	November 1984	0–37	–	93	22	~0	–	"
		P_{total}	July 1984	0–37	–	93	310	~0	–	"
		"	November 1984	0–37	–	155	22	~0	–	"
Salgir	Azov Sea	P_{min}	February 2016	0–15	0–8	540	116	−29	Biological consumption	A.S.
		P_{org}	"	0–15	0–4	70	90	+54	Destruction of organic matter	"
		P_{total}	"	0–15	0–4	610	206	−19	Biological consumption and destruction of organic matter	"
Chernaya (Crimea)	Black Sea	P_{min}	February 2004	0–17	0–2	7.1	1.8	−65	Biological consumption	[81]
Anapka	As above	P_{min}	May 2014	0–16	0–3	18	13	+700	Input from bottom sediments	[82]
		P_{org}	"	0–16	0–7	8.8	0.7	−58	Mineralization	"
		P_{total}	"	0–16	0–3	27	14	+450	Input from bottom sediments	"

Table 16. *Cont.*

River	Receiving Sea Area	Phosphorus Form	Observation Period	Salinity, ‰		[P], µg/L		Phosphorus Runoff Transformation, %	Supposed Cause of Nonconservative Behavior	Reference
				Observation Zone	Transformation Zone	River Boundary	Sea Boundary			
Ashamba	"	P_{min}	July 2010	0–15	0–11	14	11	+35	Destruction of organic matter	"
		"	August 2010	2–13	2–9	82	25	−11	Biological consumption	"
		"	January 2011	0–10	0–3	2.2	14	+950	Input from bottom sediments	"
		P_{org}	July 2010	0–15	0–11	1.0	5.0	−320	Mineralization	"
		"	August 2010	2–13	2–9	21	6.4	−8	As above	"
		"	January 2011	0–10	0–1	1.7	1.4	−80	"	"
		P_{total}	July 2010	0–15	0–11	15	16	~0	–	"
		"	August 2010	2–13	2–9	103	31	−8	Biological consumption	"
		"	January 2011	0–10	0–3	3.9	15	+430	Input from bottom sediments	"
Mezyb	"	P_{min}	September 2010	0–15	2–9	5.5	7.7	+180	Desorption from suspended matter	"
		P_{org}	"	0–15	0–11	4.0	5.3	−37	Mineralization	"
		P_{total}	"	0–15	2–5	9.5	13	+100	Desorption from suspended matter	"
Hotetsai	"	P_{min}	September 2010	0–12	2–9	4.7	7.1	+170	As above	"
		P_{org}	"	0–12	–	0.3	0.3	~0	–	"
		P_{total}	"	0–12	2–9	5.0	7.4	+170	Desorption from suspended matter	"
Vulan	"	P_{min}	July 2006	0–16	0–2	8.0	1.9	−32	Biological consumption	"
Volga	Caspian Sea	P_{min}	August 2003	0–11	0–1	37	10	−60	As above	[83]
		"	August 2004	0–10	0–1	12	4.0	−63	"	"
		"	August 2005	0–5	0–1	41	2.6	−91	"	"
		"	September 2006	0–9	0–1	14	3.4	−64	"	"
Ural	As above	P_{min}	April 2016	0–8	0–5	16	29	+150	Input from bottom sediments	[84]
		"	April 2017	0–5	0–3	1.6	1.1	+300	As above	"

Table 16. *Cont.*

River	Receiving Sea Area	Phosphorus Form	Observation Period	Salinity, ‰		[P], μg/L		Phosphorus Runoff Transformation, %	Supposed Cause of Nonconservative Behavior	Reference
				Observation Zone	Transformation Zone	River Boundary	Sea Boundary			
Mandovi	Arabian Sea	P_{min}	February 1981	0–36	0–12	3.0	6.0	−100	Sorption on suspended matter	[85]
Cauvery	Bay of Bengal	P_{min}	July 1986	0–15	0–2	267	108	−98	Unspecified	[86]
Chilka Lake	As above	P_{min}	November 1988	0–21	0–6	22	14	−24	Sorption on suspended matter	[87]
Mekong	South China Sea	P_{min}	November 1987	0–26	0–2	25	9.0	−84	As above	[88]
		"	"	0–26	2–15	25	9.0	+300	Desorption from suspended matter	"
Yangtze	East China Sea	P_{min}	June 1980	0–27	18–22	15	<3	−100	Biological consumption	[89]
		"	November 1981	0–33	0–24	7.0	11	+63	Desorption from suspended matter	"
		"	August 1981	0–32	25–28	23	<1	−44	Biological consumption	[90]
Razdolnaya	Sea of Japan	P_{min}	July 2001	0–32	0–25	60	<5	−27	Biological consumption	[91]
		"	"	0–32	0–9	80	<3	−70	As above	"
		"	August 2005	0–32	0–24	25	<3	−52	"	[92]
		"	February 2008	0–34	0–7	75	15	−28	"	[93]
Serebryan-ka	As above	P_{min}	July 2009	1–18	1–15	8.4	8.4	+13	Desorption from suspended matter	[94]
		P_{org}	"	1–18	–	0.3	5.2	~0	–	"
		P_{total}	"	1–18	1–15	8.7	14	+12	Desorption from suspended matter	"
Usalgin	Sea of Okhotsk	P_{min}	July 2016	0–25	–	54	9.0	~0	–	[95]
		P_{total}	"	0–25	–	70	25	~0	–	"
Uda	As above	P_{min}	"	0–27	0–24	9.0	29	+100	Destruction of organic matter	"
		P_{total}	"	0–27	0–24	10	30	+70	As above	"
Sacramento	Pacific Ocean	P_{min}	May 1976	0–29	0–20	84	43	+74	Unspecified	[96]
		"	July 1976	0–30	0–30	77	43	+56	As above	"
		"	September 1976	0–30	0–28	81	15	+108	"	"
		"	November 1976, March 1977	0–29	0–18	62	62	+125	"	"

Table 16. *Cont.*

River	Receiving Sea Area	Phosphorus Form	Observation Period	Salinity, ‰		[P], μg/L		Phosphorus Runoff Transformation, %	Supposed Cause of Nonconservative Behavior	Reference
				Observation Zone	Transformation Zone	River Boundary	Sea Boundary			
Mississippi	Gulf of Mexico	P_{min}	March 1971	10–33	–	71	15	~0	–	[97]
		"	January 1973	6–33	–	81	5.0	~0	–	"
		"	May 1973	0–20	8–20	90	5.0	−68	Biological consumption	"
		"	October 1983	0–34	0–3	248	3.0	+24	Dissolution of calcium phosphates	[98]
Peace	As above	P_{min}	October 1981	0–30	0–25	2790	496	−80	Biological consumption	[99]
Raritan	Atlantic Ocean	P_{min}	June 1982	0–22	0–3	12	50	−100	Sorption on suspended matter	[100]
		"	October 1982	0–27	0–9	600	31	−50	As above	"
Hudson	As above	P_{min}	March 1974	0–29	0–4	25	46	+156	Wastewater inflow	[101]
		"	August 1974	0–29	0–15	50	56	+250	As above	"
		"	August 1988	0–29	0–12	62	68	+90	"	"
Old Mill Creek	"	P_{min}	December 1978	0–27	0–9	6.0	12	+195	Desorption from suspended matter	[102]
Orinoco	"	P_{min}	November 1979	0–35	0–6	7.0	<3	+42	As above	[89]
		"	"	0–35	6–20	7.0	<3	−100	Biological consumption	"
Amazon	"	P_{min}	May 1976	0–35	8–15	15	3.0	−80	Biological consumption	[103]
		"	December 1982	0–36	0–2	20	6.0	−15	Sorption on suspended matter	[104]
		"	"	0–36	2–20	20	6.0	+118	Desorption from suspended matter	"
		"	May 1983	0–32	–	28	12	~0	–	"
Zaire (Congo)	"	P_{min}	November 1976	0–35	0–20	28	6.0	+100	Desorption from suspended matter	[105]
		"	May 1978	0–36	0–25	28	6.0	+100	As above	"

[1] Data of A.V. Savenko.

The leading role of production–destruction processes in the transformation of dissolved phosphorus runoff has been established for the mouths of many rivers (Table 16): Neva, the small rivers of the Kandalaksha Bay of the White Sea, Onega, Kyanda, Northern Dvina, Mezen, Pechora, Ob, Yenisei, most of the small rivers in the Black and the Azov Sea catchments, Volga, Ural, Yangtze, the Far Eastern rivers Razdolnaya and Uda, and Peace (USA). Due to the seasonal and interannual dynamics of phytoplankton development, the distribution of nutrients in the mixing zones of river and sea waters is also subject to seasonal and interannual variability.

The consumption of dissolved mineral phosphorus by biota occurs in the surface layer or in the vertically mixing water column throughout the entire salinity range and varies from 6 to 100% of its content in river runoff. The most intensive consumption of mineral phosphorus by aquatic organisms is observed during the vegetative season and is most often accompanied by the extraction from the solution of another biogenic element, silicon [71–74,82,83,91,92]. Along with this, in the mouths of some rivers (Strelna of the Kandalaksha Bay of the White Sea, Chernaya of the Sevastopol Bay of the Black Sea, and Razdolnaya of the Amur Bay of the Sea of Japan), the removal of dissolved mineral phosphorus was found even in winter, with relatively low biological activity.

Under stratification conditions, the plots of river and sea water mixing show the influence of two processes that regulate the concentration of dissolved mineral phosphorus at different depths. Either the behavior of phosphorus in surface waters is conservative, or it is removed as a result of biological assimilation, while in the deep layers, phosphorus, on the contrary, enters the solution due to the destruction of the organic matter deposited on the bottom. As a result, the relationships between the concentration of dissolved mineral phosphorus and the chlorinity (salinity) acquire orderliness only when the points are grouped along the horizons [77,78,103,106]. This distribution is typical for the estuaries of the Ob and Yenisei Rivers, in which aquatic organisms consume 6–57% of the mineral phosphorus supplied via river runoff, and the destruction of organic matter provides an increase in its concentration by 50–150%, relative to the content in the river water. A distinctive feature of the release of mineral phosphorus into the solution during the destruction of organic matter is the simultaneous entry into the water column of the mineral forms of nitrogen [77,96].

The predominance of destruction processes over production processes is also often found throughout the entire water column, periodically occurring in the mouths of the Onega River, some rivers of the Kandalaksha Bay of the White Sea, the Ashamba River of the Black Sea coast of Russia, and the Uda River of the Sea of Okhotsk, and leading to an increase in the flux of dissolved mineral phosphorus by 35–100%. An even greater transformation of river runoff under the influence of destruction processes occurs when pore solutions from stirred up sediments enter the mixing zone. This situation is typical for the mouths of the Mezen and the Ural Rivers and in some periods is observed in the mouths of the Black Sea rivers Anapka and Ashamba, where the additional input of mineral phosphorus into the solution exceeds its removal via river runoff by 90%, 150–300%, 7.0 times, and 9.5 times, respectively. Firstly, the organic matter in the bottom sediments is remineralized with the release of phosphates, and secondly, it can reduce iron (III) phosphates to iron (II) phosphates, which causes the input of part of the phosphorus into a dissolved state:

$$12FePO_4 + 3CH_2O + 38H_2O = 4Fe_3\left(PO_4\right)_2 \cdot 8H_2O + 4HPO_4^{2-} + 3HCO_3^- + 11H^+. \quad (9)$$

The behavior of dissolved organic phosphorus characterizes the process of the destruction of organic matter in the mixing zone of river and sea waters. Its additional intake indicates the excess of the rate of the release of dissolved organic forms over the rate of complete phosphorus recycling and is observed in the mouths of the Northern Dvina and Salgir Rivers. A decrease in the concentration of dissolved organic phosphorus relative to the line of the conservative mixing of water masses occurs at a higher rate of its mineralization than its release into the solution during destruction and was noted in the mouths of the Kyanda River and rivers of the Black Sea coast of Russia (Anapka, Ashamba, Mezyb, and Hotetsai). Both losses and excesses of dissolved organic phosphorus were recorded at the mouth of the Onega River in different periods and at different salinity intervals. The distribution of dissolved organic phosphorus is close to conservative at the Mezen River estuary and the Serebryanka River mouth (Sikhote Alin Reserve).

A significant part of the phosphorus in suspended matter in rivers is represented by reactive mineral and organic compounds, which can be sources of dissolved phosphorus in the mixing zones of river and sea waters. From the experimental data [107], it follows that the amount of phosphorus passing from river suspensions into the dissolved state increases with an increasing salinity.

Some curves of the distribution of dissolved phosphorus concentration have a deflection in the salinity range from 0 to 9‰ [103]. In this case, the similar form of such relationships for dissolved iron and aluminum in the absence of clear signs of the biological consumption of phosphorus indicates the physicochemical removal of the latter from the solution as a result of coprecipitation with iron and aluminum hydroxides.

It can be expected that the terrigenous hydroxophosphates of iron and aluminum in the marine environment will be unstable due, firstly, to an increase in the pH value, accompanied by the displacement of phosphates by hydroxyl ions, and, secondly, to an increase in the concentration of dissolved calcium, which promotes the formation of apatite phases typical for ocean sediments. This assumption is confirmed by the results of observations [108], which showed that iron–calcium hydroxophosphates in oceanic pelagic sediments decompose during diagenesis into more stable iron oxyhydroxides and apatite. The replacement of the terrigenous hydroxophosphates of iron and aluminum with apatite should begin already in the river mouth areas and cause an increase in the concentration of dissolved phosphorus, according to the reaction

$$5\mathrm{CaMe}_n(\mathrm{OH})_{3n}\mathrm{HPO}_4 + \mathrm{F}^- = \mathrm{Ca}_5(\mathrm{PO}_4)_3\mathrm{F} + 5n\mathrm{Me(OH)}_3 + 2\mathrm{HPO}_4^{2-} + 5\mathrm{H}^+, \quad (10)$$

where Me = Fe(III), Al. Therefore, the input of iron and aluminum phosphates into the salinized portion of the river mouth area as a component of the suspended matter and bed load can lead to a partial release of dissolved phosphorus and, in some cases, be the cause of its nonconservative behavior. From these positions, one can explain the decrease in phosphorus content with the increase in salinity in the bottom sediments of the mouths of the Pamlico and Potomac Rivers [109,110].

Phosphorus removal as a result of sorption and coprecipitation processes occurs, as a rule, at the initial stage of river and sea water mixing during the period of low biological activity and is observed in the mouths of the rivers Clyde (Great Britain), Mandovi and Chilka Lake (India), Mekong (Vietnam), Raritan (USA), and Amazon, accounting for 15–100% of the removal of dissolved mineral phosphorus via river runoff (Table 16). The input of dissolved phosphorus due to desorption from river suspensions penetrating into the marine environment is also a common phenomenon established for the mouths of the Black Sea rivers Mezyb and Hotetsai, Mekong in the area of medium salinity, Yangtze, Serebryanka (Sikhote Alin Reserve), Old Mill Creek (USA), Orinoco (Venezuela), Amazon at medium salinity, and Zaire (Table 16). The maximum desorption values (13–300% of the content in river water, or 1–75 μg P/L) are reached at a salinity of 7–15‰, and the mixing curves have a convex shape.

The spatial separation of the processes of phosphorus sorption and desorption (predominance of sorption in the freshwater part of the mixing zone, and desorption in the area of intermediate salinity) confirms the distribution of phosphorus and iron in the bottom sediments of the river mouth areas. Thus, at the Pamlico River mouth, the concentrations of phosphorus and iron in the bottom sediments of the riverine part of the mixing zone closely correlate with each other (r = 0.98), whereas when approaching the sea boundary of the mouth area, this relationship becomes less pronounced (r = 0.86–0.77). The same concentrations of iron correspond to lower phosphorus concentrations in the marine part of the mouth, which indicates the release of the latter from the bottom sediments [111]. This combination of sorption and desorption in the estuaries is called the phosphate buffer mechanism [60,112–114]. Many authors have tried to determine the concentration of dissolved mineral phosphorus at which an equilibrium between the water and bottom sediments is established. According to experiments [99,112,113], the equilibrium concentrations of dissolved phosphates are in the range of 22–46 μg P/L. Convincing results of field observations proving the existence of such a concentration limit have not yet been obtained, although in the studied estuaries, with the exception of the Mekong River mouth, the concentrations of desorbed phosphorus do not really exceed the values recorded in the experiments and amount to 1–28 μg/L [82,89,94,102,104,105].

4.2. Conservative Behavior of Phosphorus

Despite the active participation of phosphorus in the intrabasin biological and chemical processes, cases of its conservative behavior were established (Table 16), which were observed either under conditions of the severe pollution of the aquatic environment (the mouths of the European rivers Clyde, Scheldt, Rhine–Meuse, Ems, Weser, Elbe, Rhone, and Tiber), or during periods of low biological activity (the mouths of the Mississippi River; Amazon River; the rivers of the Arctic; and the Far Eastern rivers Knyazhaya, Niva, Kolvitsa, Onega, Semzha, Indiga, Pechora, and Usalgin).

The concentration of dissolved mineral phosphorus in the waters of polluted rivers (93–520 μg/L) is an order of magnitude higher than its average content in the rivers of the world. Increased concentrations of dissolved mineral phosphorus were also found at the sea boundary of the mouth areas of these rivers: up to 242 and 280 μg/L on the near-shore zone of the Clyde and Tiber River mouths. The conservative behavior of phosphorus in the mouths of these European rivers can be explained by the fact that the absolute values of the fluctuations in its concentration in the mixing zones at such a high content in river or sea waters are comparable to the amount of phosphorus involved in the intrabasin processes. In addition, most of the observations in which the conservative behavior of phosphorus was recorded were carried out in the autumn–winter period, when the intensity of production processes in the temperate zone decreases with the intensification of the biological processes, and the conservative behavior of phosphorus can turn into nonconservative behavior within several weeks, which was noted, for example, for the Rhone River delta [115]. The conservative distribution of dissolved mineral phosphorus in the mouths of the Onega, Mississippi, and Amazon Rivers that appeared in the winter–spring period is also, apparently, caused by the low activity of aquatic organisms.

Separately, we should consider the conservative behavior of dissolved phosphorus during the vegetative season in the mouths of the small Arctic and Far Eastern rivers that are not subject to strong anthropogenic impact. The concentration of suspended matter in the mouths of these rivers in spring and early summer is small due to the slow thawing of soils in the catchments and the low water temperature preventing phytoplankton development, which limits the participation of phosphorus in physicochemical and biological processes.

Thus, the conservative behavior of dissolved phosphorus in the mixing zone of river and sea waters is an atypical phenomenon and occurs in special conditions when the biological and chemical processes at river mouths are suppressed as a result of an unfavorable combination of natural and anthropogenic factors.

4.3. Phosphorus Balance in the Mixing Zones of River and Sea Waters

An analysis of the mixing curves indicates the complex nature of the dynamics of dissolved phosphorus fluxes in river mouth areas, with the combination of conservative and nonconservative distribution and the spatiotemporal variability of the latter, including a change in the direction of transformation. Therefore, the calculations of the values of the removal or input of phosphorus in the mixing zones of river and sea waters based on data for relatively short time intervals turn out to be insufficiently representative for balance estimates.

To obtain more reliable estimates, Savenko and Zakharova [116] summarized the results of balance studies carried out in river mouths and bays for a year or more (Table 17). As follows from the data presented, on average, a significant part of the phosphorus is removed per year. The maximum removal (80–94%) is characteristic of the total phosphorus, including the suspended fraction (P_{dissol} + P_{susp}), and only a third of this value is associated with physical sedimentation, while the rest of the phosphorus is removed as a result of biosedimentation [117,118]. Biological processes also play a major role in the extraction of mineral phosphorus from the solution (40–80%) due to its transfer to the composition of suspended organic matter, which is subsequently deposited at the bottom. Total dissolved phosphorus (P_{total}) is retained in river mouths in much smaller amounts (7–38%). This is corresponds to the observational data presented in Table 16, according to which the losses of the dissolved forms of mineral and total phosphorus during biological consumption in the mixing zones of river and sea waters are on average 46 and 25%.

Table 17. Balance estimates of phosphorus losses in the river mouth areas and bays.

Object	Phosphorus Form	Observation Period	[P], μg/L		Phosphorus Balance, %	Supposed Cause of Transformation	Reference
			River Boundary	Sea Boundary			
Gulf of Riga, Baltic Sea	P_{min}	1989	–	–	−57	Biological consumption	[119]
	P_{total}	"	–	–	−7	As above	"
Vigo Bay, Spain	P_{min}	1986	394	12	−40	Biological consumption and sedimentation	[64]
	P_{total}	"	–	–	−38	As above	"
Mikawa River mouth, Japan	P_{dissol} + P_{susp}	July 1983	930	–	−80	"	[118]
Delaware River mouth, USA	P_{min}	April 1986–July 1988	–	–	−61	Biological consumption	[120]
	"	As above, winter	–	–	−75	As above	"
	"	As above, spring	–	–	−135	"	"
	"	As above, summer	–	–	−70	"	"
	"	As above, autumn	–	–	+33	Destruction of organic matter	"
	P_{total}	April 1986–July 1988	–	–	−23	Biological consumption	"
	"	As above, winter	–	–	−53	As above	"
	"	As above, spring	–	–	−56	"	"
	"	As above, summer	–	–	−37	"	"
	"	As above, autumn	–	–	+52	Destruction of organic matter	"
Potomac River mouth, USA	P_{min}	1979–1981	–	–	−81	Biological consumption	[121]
Huizache-Caimanero Lagoon, Mexico	P_{dissol} + P_{susp}	1969–1981	93	170	−94	Biological consumption and sedimentation	[117]

5. Conclusions

The formation of the chemical composition of surface waters begins already in the atmosphere during the interaction of aerosols with the condensates of water vapor: cloudy water and the water of atmospheric precipitation. The average median concentrations of mineral and total phosphorus in atmospheric precipitation are 15 and 33 μg/L, respectively; the values of the input of these forms into the earth's surface are equal to 0.11 and 0.25 kg/ha yr. The content of the soluble forms of total phosphorus in atmospheric precipitation is in the range of 20–80%, with an average value of 55%.

The average median concentrations of dissolved mineral and total phosphorus in the waters of the primary hydrographic network are 31 and 95 μg/L, respectively. The concentrations of both forms increase with an increasing anthropogenic load: natural (forest) catchments < agricultural–forest catchments with land use less than 50% < agricultural catchments with land use over 50% < urban catchments. The concentration of dissolved mineral phosphorus, all other conditions being equal, increases with an increase of the phosphorus content in the catchment rocks.

The average median concentrations of dissolved mineral and total phosphorus in world rivers are 28 and 85 μg/L, respectively. The minimum values are observed in the rivers of the Arctic and subarctic zone; the maximum values are found in the most densely populated temperate zone and the zone of dry tropics and subtropics. The anthropogenic load is a dominant factor for riverine export, which is confirmed by the presence of a direct relationship between the concentrations of mineral and total phosphorus, on the one hand, and the population density, on the other hand.

The distribution of dissolved mineral and total phosphorus in the mixing zones of river and sea waters in the overwhelming majority of cases corresponds to nonconservative behavior. The conservative type of distribution is rarely observed and is found in the mouths of polluted rivers with high phosphorus concentrations, which significantly exceed the possible changes that occur as a result of biological and chemical processes. The decreases in the fluxes of dissolved mineral and total dissolved phosphorus at the river–sea geochemical barrier are 40–80% and 7–38%, respectively.

Author Contributions: V.S.S. conceived the study; V.S.S. and A.V.S. jointly carried out the research and wrote the manuscript. All authors have read and agreed to the published version of the manuscript.

Funding: The study of the phosphorus geochemistry in the continental runoff of dissolved and solid matter was supported by the RFBR, project 18-05-60219; the study of the transformation of phosphorus fluxes in river mouths was carried out with the support of the RFBR, project 20-05-00802.

Data Availability Statement: Data supporting reported results can be found in the literature cited in the manuscript.

Conflicts of Interest: The authors declare no conflict of interest.

References

1. Williams, P. Sea surface chemistry: Organic carbon and organic and inorganic nitrogen and phosphorus in surface films and subsurface waters. *Deep. Sea Res. Oceanogr. Abstr.* **1967**, *14*, 791–800. [CrossRef]
2. Graham, W.F.; Duce, R.A. Atmospheric inputs of phosphorus to remote tropical islands. *Pac. Sci.* **1981**, *35*, 241–255.
3. Graham, W.F.; Duce, R.A. The atmospheric transport of phosphorus to the western North Atlantic. *Atmos. Environ. (1967)* **1982**, *16*, 1089–1097. [CrossRef]
4. Savenko, V.S. *Atmospheric Input of Phosphorus and Its Runoff from Catchment Areas*; Transactions Doklady-Russian Academy of Sciences Earth Science Sections; Scripta Technica, Inc., A Wiley Company: New York, NY, USA, 1997; Volume 5, pp. 783–785.
5. Savenko, V.S. Atmospheric aerosols as a source of phosphorus in aquatic ecosystems. *Water Resour.* **1995**, *22*, 170–179.
6. Savenko, V.S.; Savenko, A.V. *Geochemistry of Phosphorus in the Global Hydrological Cycle*; GEOS: Moscow, Russia, 2007; pp. 1–248. (In Russian)
7. Savenko, V.S. Phosphorus in atmospheric precipitation. *Water Resour.* **1996**, *23*, 169–179.
8. Ovchinnikov, L.N. *Applied Geochemistry*; Nedra: Moscow, Russia, 1990; pp. 1–248. (In Russian)
9. Ronov, A.B.; Yaroshevsky, A.A.; Migdisov, A.A. *Chemical Constitution of the Earth's Crust and Geochemical Balance of the Main Elements*; Nauka: Moscow, Russia, 1990; pp. 1–182. (In Russian)
10. Grigor'ev, N.A. The average mineral composition of the sedimentary layer of the continental crust's. *Lithosphere* **2003**, *3*, 3–14. (In Russian)
11. Grigor'ev, N.A. Average concentrations of chemical elements in rocks of the upper continental crust. *Geochem. Int.* **2003**, *41*, 711–718.
12. Bowen, H.J.M. *Environmental Chemistry of the Elements*; Acad. Press: London, UK, 1979; pp. 1–333.
13. Romankevich, E.A. Living matter of the Earth (biogeochemical aspects of the problem). *Geochemistry* **1988**, *2*, 292–306. (In Russian)
14. Frossard, E.; Brossard, M.; Hedley, M.J.; Metherell, A. Reactions controlling the cycling of P in soils. In *Phosphorus in the Global Environment*; Tiessen, H., Ed.; J. Willey: Chichester, UK, 1995; pp. 107–137.

15. Golterman, H.L.; Meyer, M.L. The geochemistry of two hard water rivers, the Rhine and the Rhone, Part 4: The determination of the solubility product of hydroxyl-apatite. *Hydrobiologia* **1985**, *126*, 25–29. [CrossRef]
16. Savenko, A.V.; Savenko, V.S. Stability of accessory fluorine minerals in the process of formation of dissolved matter continental runoff. *Process. GeoMedia* **2019**, *1*, 82–86. (In Russian)
17. Savenko, A.V. Experimental study of the transformations of iron and aluminum phosphates at the river–sea geochemical barrier. *Geochem. Int.* **2005**, *43*, 414–419.
18. Savenko, A.V. The possibility of phosphatization of silicates in the supergene zone. *Geochem. Int.* **2014**, *53*, 87–94. [CrossRef]
19. Savenko, A.V. Regularities in low-temperature phosphatization of silicates. *Dokl. Earth Sci.* **2018**, *478*, 67–69. [CrossRef]
20. Savenko, A.V. Experimental study of silicate phosphatization under supergene zone conditions: Hornblende, orthoclase, and labradorite. *Geochem. Int.* **2019**, *57*, 722–727. [CrossRef]
21. Reifenberg, A.; Buckwold, S.J. The release of silica from soils by the orthophosphate anion. *Eur. J. Soil Sci.* **1954**, *5*, 106–115. [CrossRef]
22. Gorbunov, N.I.; Shchurina, G.N. Significance of the chemical composition, dispersion degree, and mineral structure for phosphate absorption. *Pochvovedenie* **1970**, *12*, 142–153. (In Russian)
23. Rajan, S.S.S. Phosphate adsorption and the displacement of structural silicon in an allophane clay. *J. Soil Sci.* **1975**, *26*, 250–256. [CrossRef]
24. Rājan, S.S.S.; Fox, R.L. Phosphate adsorption by soils: II. Reactions in tropical acid soils. *Soil Sci. Soc. Am. J.* **1975**, *39*, 846–851. [CrossRef]
25. Rajan, S.S.S.; Perrott, K.W. Phosphate adsorption by synthetic amorphous aluminosilicates. *J. Soil Sci.* **1975**, *26*, 257–266. [CrossRef]
26. Remezov, N.P. On the relationship between biological accumulation and the eluvial process under the forest canopy (based on research in the Voronezh State Reserve). *Pochvovedenie* **1958**, *6*, 3–12. (In Russian)
27. *World Water Balance and Water Resources of the Earth*; Gidrometeoizdat: Leningrad, Russia, 1974; pp. 1–525. (In Russian)
28. Meybeck, M. Carbon, nitrogen, and phosphorus transport by world rivers. *Am. J. Sci.* **1982**, *282*, 401–450. [CrossRef]
29. Meybeck, M. C, N, P and S in rivers: From sources to global inputs. In *Interactions of C, N, P and S Biogeochemical Cycles and Global Change*; Springer: Berlin, Germany, 1993; pp. 163–193.
30. Gordeev, V.V. *Geochemistry of the River—Sea System*; IP I.I. Matushkina: Moscow, Russia, 2012; pp. 1–452. (In Russian)
31. Hanley, P.; Murphy, M.D. Soil and fertilizer phosphorus in the Irish ecosystem. *Water Res.* **1973**, *7*, 197–210. [CrossRef]
32. Shikula, N.K.; Lomakin, M.M. Loss of nutrients from gray podzolized soils with surface runoff. *Pochvovedenie* **1978**, *4*, 113–121. (In Russian)
33. Nazarov, G.V. Runoff of biogenic substances from arable land. *Water Resour.* **1991**, *6*, 60–72. (In Russian)
34. Nazarov, G.V.; Kuznetsov, V.K. Zonal features of the nutrients removal from natural and agricultural lands. In Proceedings of the 5th All-Union Hydrological Congress, Leningrad, Russia, 20–24 October 1986; Gidrometeoizdat: Leningrad, Russia, 1991; Volume 5, pp. 184–189. (In Russian).
35. Sharpley, A.N.; Hedley, M.J.; Sibbesen, E.; Hillbricht-Ilkowska, A.; House, W.A.; Ryszkowski, L. Phosphorus transfers from terrestrial to aquatic ecosystems. In *Phosphorus in the Global Environment*; J. Willey: Chichester, UK, 1995; pp. 171–199.
36. Perel'man, A.I.; Kasimov, N.S. *Geochemistry of Landscape*; Astreya-2000: Moscow, Russia, 1999; pp. 1–768. (In Russian)
37. Savenko, V.S. Geochemistry of continental link of the global hydrological cycle. In *Global Changes in the Natural Environment—2001*; Publ. House of the Siberian Branch of RAS: Novosibirsk, Russia, 2001; pp. 274–287. (In Russian)
38. Dillon, P.; Kirchner, W. The effects of geology and land use on the export of phosphorus from watersheds. *Water Res.* **1975**, *9*, 135–148. [CrossRef]
39. Lozovik, P.A.; Basov, M.I.; Litvinenko, A.V. Assessment of chemical input into the drainage network from Karelian river basins. *Water Resour.* **2005**, *32*, 532–536. [CrossRef]
40. Gardner, L. The role of rock weathering in the phosphorus budget of terrestrial watersheds. *Biogeochemistry* **1990**, *11*, 97–110. [CrossRef]
41. Conley, D.J.; Stalnacke, P.; Pitkänen, H.; Wilander, A. The transport and retention of dissolved silicate by rivers in Sweden and Finland. *Limnol. Oceanogr.* **2000**, *45*, 1850–1853. [CrossRef]
42. Savenko, V.S.; Zakharova, E.A. Phosphorus in the water of the primary hydrographic network. *Water Resour.* **1997**, *24*, 266–273.
43. Savenko, V.S. Phosphorus discharge with suspended load. *Water Resour.* **1999**, *26*, 41–47.
44. Martin, J.M.; Meybeck, M. The content of major elements in the dissolved and particulate load of river. In *Biogeochemistry of Estuarine Sediments*; Forstner, U., Muller, G., Stoffers, P., Eds.; UNESCO Press: Paris, France, 1978; pp. 95–110.
45. Chase, E.; Sayles, F. Phosphorus in suspended sediments of the Amazon River. *Estuar. Coast. Mar. Sci.* **1980**, *11*, 383–391. [CrossRef]
46. van Eck, G.T.M. Forms of phosphorus in particulate matter from the Hollands Diep/Haringvilet, The Netherlands. In *Sediment/Freshwater Interaction. Developments in Hydrobiology*; Dumont, H.J., Ed.; Junk Publ.: Hague, The Netherlands, 1982; Volume 9, pp. 665–681.
47. Kohonen, T. Influence of sampling frequency on the estimates of runoff water quality. *Publ. Water Res. Inst.* **1982**, *27*, 1–30.
48. Prairie, Y.T.; Kalff, J. Particulate phosphorus dynamics in headwater streams. *Can. J. Fish. Aquat. Sci.* **1988**, *45*, 210–215. [CrossRef]
49. Savenko, V.S. *Chemical Composition of World River's Suspended Matter*; GEOS: Moscow, Russia, 2006; pp. 1–175. (In Russian)
50. Savenko, V.S.; Zakharova, E.A. Phosphorus in river runoff. *Doklady Akademii Nauk.* **1995**, *345*, 682–685. (In Russian)

51. Savenko, V.S.; Zakharova, E.A. Principal features of phosphorus behavior in river flow. *Water Resour.* **1997**, *24*, 139–147.
52. Caraco, N.F. Influence of human populations on P transport to aquatic systems: A regional scale study using large rivers. In *Phosphorus in the Global Environment*; Tiessen, H., Ed.; J. Willey: Chichester, UK, 1995; pp. 235–244.
53. Justić, D.; Rabalais, N.N.; Turner, R. Stoichiometric nutrient balance and origin of coastal eutrophication. *Mar. Pollut. Bull.* **1995**, *30*, 41–46. [CrossRef]
54. Tarasov, M.N.; Smirnov, M.P.; Kryuchkov, I.A.; Laki, G.I. River runoff of nutrients from the territory of the USSR and its change in time (1936–1980). *Hydrochem. Mater.* **1988**, *103*, 49–66. (In Russian)
55. Fraser, A.S. Tributary and point source total phosphorus loading to Lake Erie. *J. Great Lakes Res.* **1987**, *13*, 659–666. [CrossRef]
56. Behrendt, H. Estimation of the nutrient inputs into medium and large river basins—A case study for German rivers. *LOICZ Newsl.* **1999**, *12*, 1–4.
57. Shvartsev, S.L. *Hydrogeochemistry of the Hypergenesis Zone*; Nedra: Moscow, Russia, 1998; pp. 1–366. (In Russian)
58. Dedkov, A.P.; Mozzherin, V.I. *Erosion and Sediment Yield on the Earth*; Kazan University Publ. House: Kazan, Russia, 1984; pp. 1–264. (In Russian)
59. Dedkov, A.P.; Mozzherin, V.I.; Safina, G.R.; Gusarov, A.V. Global sediment yield and its variability. In *Universities of Russia: Fundamental Research. Geography*; Moscow University Publ. House: Moscow, Russia, 2000; p. 70. (In Russian)
60. Gordeev, V.V. *River Runoff into the Ocean and Specifics of Its Geochemistry*; Nauka: Moscow, Russia, 1983; pp. 1–160. (In Russian)
61. Lisitzin, A.P. *Ocean Sedimentation*; Nauka: Moscow, Russia, 1974; pp. 1–438. (In Russian)
62. Gordeev, V.V.; Dzhamalov, R.G.; Zektser, I.S.; Zhulidov, A.V.; Bryzgalo, V.A. Assessment of nutrient discharge with river and groundwater flow into marginal seas of the Russian Arctic regions. *Water Resour.* **1999**, *26*, 206–211.
63. Savenko, V.S. Global hydrological cycle and geochemical balance of phosphorus in the ocean. *Oceanology* **2001**, *41*, 360–366.
64. Prego, R. Biogeochemical pathways of phosphate in a Galician ria (north-western Iberian Peninsula). *Estuar. Coast. Shelf Sci.* **1993**, *37*, 437–451. [CrossRef]
65. Morris, A. The chemistry of the Severn Estuary and the Bristol Channel. *Mar. Pollut. Bull.* **1984**, *15*, 57–61. [CrossRef]
66. Mackay, D.W.; Leatherland, T.M. Chemical processes in an estuary receiving major inputs of industrial and domestic wastes. In *Estuarine Chemistry*; Burton, J.D., Liss, P.S., Eds.; Acad. Press: London, UK, 1976; pp. 185–218.
67. van Bennekom, A.J.; Wetsteijn, F.J. The winter distribution of nutrients in the Southern Bight of the North Sea (1961–1978) and in the estuaries of the Scheldt and the Rhine/Meus. *Netherl. J. Sea Res.* **1990**, *25*, 75–87. [CrossRef]
68. Kronberg, B.I.; Lammers, U.; Schutt, M.; Jonge, V.N. Valdivia Cruise October 1981: Mineral nutrients in the Elbe, Weser and Ems Rivers and German Bight. *Mitt. Aus Dem Geol.-Paläontologisch Inst. Univ. Hambg.* **1982**, *52*, 655–666.
69. Pitkänen, H.; Tamminen, T.; Kangas, P.; Huttula, T.; Kivi, K.; Kuosa, H.; Sarkkula, J.; Eloheimo, K.; Kauppila, P.; Skakalsky, B. Late summer trophic conditions in the north-east Gulf of Finland and the River Neva Estuary, Baltic Sea. *Estuar. Coast. Shelf Sci.* **1993**, *37*, 453–474. [CrossRef]
70. Zakharova, E.A.; Savenko, V.S. Phosphorus in the waters of the river mouths of the Kandalaksha Bay of the White Sea. *Water Resour.* **1997**, *24*, 503–507. (In Russian)
71. Savenko, A.V. Hydrochemical structure of mouth areas of the small rivers discharging into Kandalaksha Bay of the White Sea. *Oceanology* **2001**, *41*, 799–807.
72. Savenko, A.V.; Pokrovsky, O.; Kozhin, M.N. Transformation of the dissolved components runoff in the mouth areas of small watersheds of the southern coast of the Kola Peninsula. *Oceanology* **2011**, *51*, 785–795. [CrossRef]
73. Savenko, A.V.; Demidenko, N.A.; Pokrovskii, O.S. Chemical transformation of the runoff of dissolved matters in the mouth areas of the Onega and Mezen' rivers. *Geochem. Int.* **2016**, *54*, 439–448. [CrossRef]
74. Savenko, A.V.; Shevchenko, V.P. Seasonal variability of the distribution of dissolved forms of biogenic elements and alkalinity in the Northern Dvina mouth. *Water Resour.* **2005**, *32*, 417–421. [CrossRef]
75. Savenko, A.V.; Demidenko, N.A.; Pokrovsky, O.S. Spatial and temporal variability of the transformation of dissolved matter runoff in the Mezen River estuary. *Oceanology* **2019**, *59*, 199–207. [CrossRef]
76. *Hydrometeorology and Hydrochemistry of the Seas of the USSR. Vol. II. Barents Sea. Issue II. Hydrochemical Conditions and Oceanological Foundations of Bioproductivity Formation*; Gidrometeoizdat: Leningrad, Russia, 1992; pp. 1–265. (In Russian)
77. Rachold, V.; Beeskow, B.; Gordeev, V.V. Geochemistry of the Ob and Yenisey estuaries: A comparative study. *Ber. Polar-Meeresforsch.* **2007**, *565*, 1–235. [CrossRef]
78. Savenko, A.V.; Pokrovsky, O.S. Distribution of dissolved matter in the Yenisei estuary and adjacent Kara Sea areas and its inter-annual variability. *Geochem. Int.* **2019**, *57*, 1201–1212. [CrossRef]
79. Arfi, R. Hydrologie et charge nutritive à l'embouchure du Rhône (Saison froide). *Bull. Ecol.* **1987**, *18*, 123–130.
80. Pagnotta, R.; Blundo, C.M.; La Noce, T.; Pettine, M.; Puddu, A. Nutrient remobilisation processes at the Tiber River mouth (Italy). *Hydrobiologia* **1989**, *176*, 297–306. [CrossRef]
81. Badyukov, D.D.; Korneeva, G.A.; Savenko, A.V. Transformation of structural and functional characteristics of the mainland runoff of the river Chernaya and the sea waters of the Sevastopol Bay in winter. *Probl. Reg. Ecol.* **2014**, *3*, 69–75. (In Russian)
82. Savenko, A.V.; Pokrovsky, O.S. Transformation of major and trace element composition of the dissolved matter runoff in the mouths of medium and small rivers of the Black Sea coast of Russia. *Oceanology*, **2022**, *62*, *in press*.
83. Savenko, A.V.; Brekhovskikh, V.F.; Pokrovskii, O.S. Migration of dissolved trace elements in the mixing zone between Volga River water and Caspian seawater: Results of observations over many years. *Geochem. Int.* **2014**, *52*, 533–547. [CrossRef]

84. Savenko, A.V.; Pokrovsky, O.S. Transformation of dissolved matter runoff in the Ural River mouth. *Geochem. Int.* **2020**, *58*, 947–958. [CrossRef]
85. De Sousa, S. Studies on the behaviour of nutrients in the Mandovi Estuary during premonsoon. *Estuar. Coast. Shelf Sci.* **1983**, *16*, 299–308. [CrossRef]
86. Ramanathan, A.L.; Subramanian, V.; Vaithiyanathan, P. Chemical and sediment characteristics of the upper reaches of Cauvery Estuary, east coast of India. *Ind. J. Marine Sci.* **1988**, *17*, 114–120.
87. Panda, D.; Tripathy, S.K.; Patnaik, D.K.; Choudhury, S.B.; Gouda, R.; Panigraphy, R.C. Distribution of nutrients in Chilka lake, east coast of India. *Ind. J. Marine Sci.* **1989**, *18*, 286–288.
88. Shulkin, V.M.; Khristoforova, N.K.; Chernova, E.N. *Variability of Chemical Parameters in the Mekong River Estuary*; VINITI Deposit: Moscow, Russia, 1989; Volume 730, pp. 1–24. (In Russian)
89. Edmond, J.M.; Spivack, A.J.; Grant, B.; Minghui, H.; Zexiam, C.; Sung, C.; Xiushau, Z. Chemical dynamics of the Changjiang estuary. *Cont. Shelf Res.* **1985**, *4*, 17–36. [CrossRef]
90. Tian, R.; Hu, F.; Martin, J. Summer nutrient fronts in the Changjiang (Yantze River) Estuary. *Estuar. Coast. Shelf Sci.* **1993**, *37*, 27–41. [CrossRef]
91. Zvalinsky, V.I.; Nedashkovsky, A.P.; Sagalayev, S.G.; Tishchenko, P.J.; Shvetsova, M.G. Nutrients and primary production in the estuary of the Razdol'naya River (Amur Bay, Sea of Japan). *Russ. J. Mar. Biol.* **2005**, *31*, 91–101. [CrossRef]
92. Zvalinsky, V.I.; Tishchenko, P.P.; Tishchenko, P.Y.; Lobanov, V.B.; Sagalaev, S.G.; Shvetsova, M.G.; Volkova, T.I.; Sergeev, A.F.; Propp, L.N. Hydrochemical and production parameters of Amursky Bay during of high water of Razdol'naya River in August 2005. In *Current Environmental Condition and Tendencies of Its Change in the Peter the Great Bay, Sea of Japan*; GEOS: Moscow, Russia, 2008; pp. 199–229. (In Russian)
93. Maryash, A.A.; Zvalinsky, V.I. Production–destruction processes in the estuary of the Razdol'naya River during the freeze-up period. In *Organic Matter and Nutrients in Inland Water Bodies and Sea Waters*; Publ. House of the Karelian Scientific Center of RAS: Petrozavodsk, Russia, 2012; pp. 346–350. (In Russian)
94. Savenko, A.V.; Pokrovsky, O.S. Migration of dissolved matter at Serebryanka R. Mouth, the basin of the Sea of Japan (Sikhote Alin Reserve). *Water Resour.* **2014**, *41*, 671–676. [CrossRef]
95. Semkin, P.Y.; Tishchenko, P.P.; Tishchenko, P.Y.; Pavlova, G.Y.; Sagalaev, S.G.; Khodorenko, N.D.; Shkirnikova, E.M.; Shvetsova, M.G. Characterization of production/destruction processes in the Uda and Usalgin estuaries of the Sea of Okhotsk during the summer flood. *Vestnik Dal'nevostochnogo Otdeleniya Rossiyskoy Akademii Nauk* **2020**, *2*, 88–96. [CrossRef]
96. Peterson, D.H.; Smith, R.E.; Hager, S.W.; Harmon, D.D.; Herndon, R.E.; Schemel, L.E. Interannual variability in dissolved inorganic nutrients in northern San Francisco Bay estuary. In *Temporal Dynamics of an Estuary: San Francisco Bay*; Springer: Singapore, 1985; Volume 30, pp. 37–58.
97. Ho, C.L.; Barrett, B.B. Distribution of nutrients in Louisiana's coastal waters influenced by the Mississippi River. *Estuar. Coast. Mar. Sci.* **1977**, *5*, 173–195. [CrossRef]
98. Fox, L.E.; Sager, S.L.; Wofsy, S.C. Factors controlling the concentrations of soluble phosphorus in the Mississippi estuary. *Limnol. Oceanogr.* **1985**, *30*, 826–832. [CrossRef]
99. Froelich, P.; Kaul, L.; Byrd, J.; Andreae, M.; Roe, K. Arsenic, barium, germanium, tin, dimethylsulfide and nutrient biogeochemistry in Charlotte Harbor, Florida, a phosphorus-enriched estuary. *Estuar. Coast. Shelf Sci.* **1985**, *20*, 239–264. [CrossRef]
100. Maest, A.S.; Crerar, D.A.; Stallard, R.F.; Ryan, J.N. Metal and nutrient behavior in the Raritan estuary, New Jersey, USA: The effect of multiple freshwater and industrial waste inputs. *Chem. Geol.* **1990**, *81*, 133–149. [CrossRef]
101. Clark, J.F.; Simpson, H.J.; Bopp, R.F.; Deck, B. Geochemistry and loading history of phosphate and silicate in the Hudson estuary. *Estuar. Coast. Shelf Sci.* **1992**, *34*, 213–233. [CrossRef]
102. Eastman, K.W.; Church, T.M. Behaviour of iron, manganese, phosphate and humic acid during mixing in a Delaware salt marsh creek. *Estuar. Coast. Shelf Sci.* **1984**, *18*, 447–458. [CrossRef]
103. Edmond, J.; Boyle, E.; Grant, B.; Stallard, R. The chemical mass balance in the Amazon plume I: The nutrients. *Deep. Sea Res. Part A Oceanogr. Res. Pap.* **1981**, *28*, 1339–1374. [CrossRef]
104. Fox, L.E.; Sager, S.L.; Wofsy, S.C. The chemical control of soluble phosphorus in the Amazon estuary. *Geochim. Cosmochim. Acta* **1986**, *50*, 783–794. [CrossRef]
105. van Bennekom, A.J.; Berger, G.W.; Helder, W.; Vries, R.T.P. Nutrient distribution in the Zaire estuary and river plume. *Netherl. J. Sea Res.* **1979**, *12*, 296–323. [CrossRef]
106. Kudo, I.; Matsunaga, K. Behavior of nutrients and heavy metals in a high productive estuary. *Bull. Jpn. Soc. Sci. Fish.* **1989**, *55*, 957–962. [CrossRef]
107. Chambers, R.M.; Fourqurean, J.W.; Hollibaugh, J.T.; Vink, S.M. Importance of terrestrially-derived, particulate phosphorus to phosphorus dynamics in a west coast estuary. *Estuaries* **1995**, *18*, 518–526. [CrossRef]
108. Dubinin, A.V. Geochemistry of iron–calcium hydroxophosphates in pelagic sediments: Origin and compositional evolution in the course of diagenesis. *Geochem. Int.* **2001**, *39*, 585–596.
109. Upchurch, J.B.; Edzwald, J.K.; O'Melia, C.R. Phosphates in sediments of Pamlico Estuary. *Environ. Sci. Technol.* **1974**, *8*, 56–58. [CrossRef]
110. Glenn, J.L. Bottom sediments and nutrients in the tidal Potomac system, Maryland and Virginia. *Bottom Sediments Nutr. Tidal Potom. Syst. Md. Va.* **1988**, *2234F*, 1–74. [CrossRef]

111. Williams, P.J. Primary productivity and heterotrophic activity in estuaries. In *River Input to Ocean System*; Martin, J.M., Burton, J.D., Eisma, D., Eds.; UNEP and UNESCO: New York, NY, USA, 1981; pp. 243–258.
112. Pomero, L.R.; Smith, E.E.; Grant, C.M. The exchange of phosphate between estuarine water and sediments. *Limnol. Oceanogr.* **1965**, *10*, 167–172. [CrossRef]
113. Butler, E.I.; Tibbitts, S. Chemical survey of the Tamar estuary I. Properties of the waters. *J. Mar. Biol. Assoc. UK* **1972**, *52*, 681–699. [CrossRef]
114. Liss, P.S. Conservative and non-conservative behavior of dissolved constituents during estuarine mixing. In *Estuarine Chemistry*; Acad. Press: London, UK, 1976; pp. 93–130.
115. Denant, V.; Saliot, A. Seasonal variations of nutrients (NO_3, NO_2, NH_4, PO_4 and $Si(OH)_4$) and suspended matter in the Rhone delta, France. *Oceanol. Acta.* **1990**, *13*, 47–52.
116. Savenko, V.S.; Zakharova, E.A. Phosphorus within the mixing zone of sea and river waters. *Water Resour.* **1998**, *25*, 293–301.
117. Arenas, V.F.; de la Lanza, G.E. Annual phosphorus budget of a coastal lagoon in the northwest of Mexico. *Environ. Biogeochem. Ecol. Bull.* **1983**, *35*, 341–440.
118. Matsukawa, Y.; Sasaki, K. Budgets of nitrogen, phosphorus and suspended solid in an intertidal flat. *Bull. Japan. Soc. Sci. Fish.* **1986**, *52*, 1791–1797. [CrossRef]
119. Yurkovskis, A.; Wulff, F.; Rahm, L.; Andruzaitis, A.; Rodriguez-Medina, M. A nutrient budget of the Gulf of Riga; Baltic Sea. *Estuar. Coast. Shelf Sci.* **1993**, *37*, 113–127. [CrossRef]
120. Lebo, M.E.; Sharp, J.H. Modeling phosphorus cycling in a well-mixed coastal plain estuary. *Estuar. Coast. Shelf Sci.* **1992**, *35*, 235–252. [CrossRef]
121. Bennett, J.P. Nutrient and sediment budgets for the tidal Potomac River and Estuary. *IAHS Publ.* **1983**, *141*, 217–227.

MDPI
St. Alban-Anlage 66
4052 Basel
Switzerland
Tel. +41 61 683 77 34
Fax +41 61 302 89 18
www.mdpi.com

Water Editorial Office
E-mail: water@mdpi.com
www.mdpi.com/journal/water

www.ingramcontent.com/pod-product-compliance
Lightning Source LLC
Chambersburg PA
CBHW040826250926
51850CB00007BA/1268
9783036567440